T0342466

_BEAUTIFUL DATA

Beautiful Data__ A HISTORY OF

 Experimental Futures. TECHNOLOGICAL LIVES, SCIENTIFIC ARTS, ANTHROPOLOGICAL VOICES. *A series edited by Michael M. J. Fischer and Joseph Dumit*

VISION *and* REASON *since* 1945   **ORIT HALPERN**

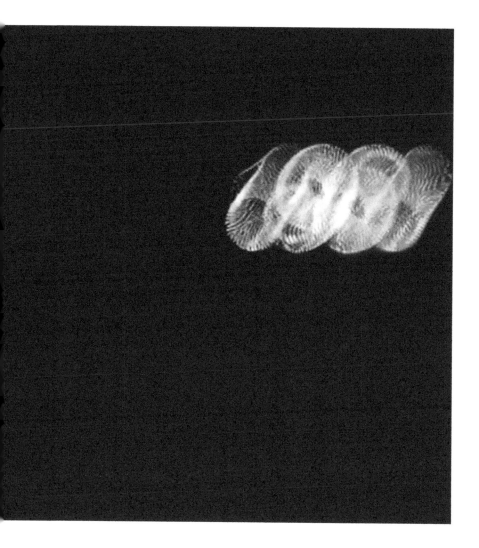

Duke University Press__Durham and London__2014

© 2014 **Duke University Press.** All rights reserved.

Printed and bound by CPI Group (UK) Ltd, Croydon, CR0 4YY
Typeset in Helvetica Neue and Minion by Tseng Information Systems, Inc.

Library of Congress Cataloging-in-Publication Data
Halpern, Orit, 1972–
Beautiful data : a history of vision and reason since 1945 / Orit Halpern.
pages cm — (Experimental futures)
Includes bibliographical references and index.
ISBN 978-0-8223-5730-8 (cloth : alk. paper)
ISBN 978-0-8223-5744-5 (pbk. : alk. paper)
1. Cybernetics. 2. Big data. 3. Perception. 4. Cognition.
I. Title. II. Series: Experimental futures.
Q310.H35 2014
003'.54 — dc23
2014013032

*Duke University Press gratefully acknowledges the support
of the Warner Publication Fund at Columbia University, which
provided funds toward the publication of this book.*

Cover art: Images from Kepes, *New Landscape*: patterns emerging
from charged particles; aerial survey of Chicago; oscilloscope
patterns of an Analogue Computer (1956). Background image:
George Maciunas, Expanded Arts Diagram, (1966), detail.
© All Rights Reserved, George Maciunas Foundation, Inc., 2013.

Contents

Acknowledgments

This book resulted from a complex network of people, dispersed both geographically and temporally. I give thanks to all of them.

This work began when I was a graduate student at Harvard's Department of the History of Science, and I am particularly grateful to Peter Galison, who first encouraged me to follow my interest in digital media and histories of knowledge and vision. He has been steadfast in encouraging new forms of scholarship and experimentation throughout my career. I also want to give special thanks to David Rodowick, who was critical in introducing me to new ways to think about media and the productive and imaginative potentials of engaging with critical and poststructural theory. Allan Brandt and Charis Thompson were ever-supportive readers, and provided intellectual guidance throughout. My fellow graduate students also offered support and intellectual inspiration, both during my time at Harvard and, perhaps more significantly, afterward. I am particularly grateful to Sharrona Pearl, Colin Milburn, Carla Nappi, Jimena Canales, Debbie Weinstein and Hannah Shell.

I was fortunate enough to have a postdoctoral fellowship at Duke University from 2006–2007 at the John Hope Franklin Center, dedicated to digital media and its study. I profited enormously from the prolonged engagement and support of a wonderful working group headed by Timothy Lenoir and Priscilla Wald, both of whom provided an incredible environment for encouraging new forms of scholarship and who have continued to be immensely supportive of my scholarship. I also benefited greatly from input and exchange with numerous members of the group, particularly Robert Mitchell, Cathy Davidson, David Liu, Harry Halprin, and Mitali Routh.

At the New School for Social Research (NSSR), Eugene Lang College, and Parsons School of Design I have enjoyed being in a truly emergent environment, where change and intellectual experimentation has been regularly encouraged by my colleagues and friends. I want to thank the members of the Visual Culture Working Group, who have provided an intellectual bedrock since my arrival. Victoria Hattam, Deborah Levitt, David Brody, Janet Kraynak, Margot Bauman, Pooja Rangan, Jeffrey Lieber, and Julie Napolin have all provided an amazing and fostering space to develop my thought and writing. I must also thank my many colleagues who have so thoughtfully engaged my work and supported my research: Ann Stoler, Janet Roitman, Ken Wark, Hugh Raffles, Shannon Mattern, Jeremy Varon, Inessa Medzihbovskya, Julia Ott, and Banu Bargu.

In the course of the past few years, I have also had the great fortune of receiving both financial and intellectual support from a number of organizations and foundations. Throughout the years 2010–2012, I was a member of Columbia University's Seminar on the Theory and History of Media. I give sincere thanks to the members of the seminar, particularly Felicity Scott, Rheinhold Martin, Brian Larkin, Stefan Andriopoulus, and Alexander Galloway. I was also given subvention assistance as a result of the seminar from the Warner Publication Fund at Columbia, for which I am very grateful.

In 2010–2013 I was a beneficiary of a unique program—The Poiesis Project—of the Institute for Public Knowledge at New York University, sponsored by the BMW Stiftung/Herbert Quandt and Gerda Hankel Stiftung. I give thanks to these foundations and to IPK for their support. I worked throughout the period with a wonderful group; I am particularly thankful to Ash Amin, Saskia Sassen, Jesse LeCavalier and Nerea Calvillo for sharing their thoughts and practices with me.

Lorraine Daston and the Max Planck Institute for the History of Science in Berlin also deserve special credit for supporting this project by providing space and time (always at a limit) for writing and research. Raine also provided an incredible environment for fostering discussion and thought. I have never been in a better space for encouraging scholarship than in her abteilung at the Planck. I had the good fortune of being in residence there twice, the first time during the year 2001 as a graduate student, the second time in 2009 for my sabbatical. Other members of the Planck were also invaluable as colleagues and readers—Josh Berson, Rebecca Lemov, Henning Schmidgen, Susanne Bauer, and John Carson were generous in commenting upon and supporting my work.

I would also especially like to thank the people I have been closest to over the years and who have continually mentored and supported my work as friends and colleagues: Joseph Dumit, Natasha Myers, Patricia Clough, Jonathan Bellar, Michelle Murphy, Hannah Landaecker, Jim Bono, Wendy Hui Kyong Chun, Luciana Parisi, Nina Wakeford, Celia Lury, Thomas LaMarre, Michael Fisch, Martha Poon, and Dooeun Choi.

I also want to thank the members of the Instrument group, of which I am a member, and which is part of the Aggregate Architecture and Theory collective: Zeynep Çelik Alexander, John May, Michael Ousman, John Harwood, Matthew Hunter, and Lucia Allais have given input on parts of this project.

A number of archivists have also helped me. David Hertsgaard and Genevieve Fong at the Eames Office have been very generous. Heidi Coleman at the Noguchi Archive and Museum assisted me greatly with the collections there. I also want to thank the Smithsonian Archive of American Art, the Library of Congress, the American Philosophical Society, MIT Archives and Special Collections, Harvard University Archives, the IBM Corporate Archives, Imre Kepes and Juliet K.Stone, and the Archigram Archives.

At Duke University Press, I want to thank Ken Wissoker, whose guidance, support, and assistance have been invaluable to realizing this project, and Elizabeth Ault and Sara Leone, who have both done so much to help ensure that this book comes to fruition. Finally, I want to thank my family, Mordechai, Atara, Galia, Iris, and Tal—without them I would never have started, much less completed, this book.

PROLOGUE_**Speculating on Sense**

This book is about the historical construction of vision and cognition in the second half of the twentieth century. It posits that our forms of attention, observation, and truth are situated, contingent, and contested and that the ways we are trained, and train ourselves, to observe, document, record, and analyze the world are deeply historical in character. The narrative traces the impact of cybernetics and the communication sciences after World War II on the social and human sciences, design, arts, and urban planning. It documents a radical shift in attitudes to recording and displaying information that produced new forms of observation, rationality, and economy based on the management and analysis of data; what I label a "communicative objectivity." Furthermore, the book argues that historical changes in how we manage and train perception and define reason and intelligence are also transformations in governmentality. My intent is to denaturalize and historically situate assumptions about the value of data, our regular obsession with "visualization," and our almost overwhelming belief that we are in the midst of a digital-media-driven "crisis" of attention that can only be responded to through recourse to intensifying media consumption.

To begin to interrogate this past and its attendant stakes, I would like to offer an example in the present. I want to open with the largest private real estate development on earth.[1] One hour's drive southwest from Seoul, the new city of Songdo is being built from scratch on land reclaimed from the ocean (fig. P.1).[2] It is a masterpiece of engineering, literally emerging from a previously nonexistent territory. Beneath this newly grafted land lies a massive infrastructure of conduits containing fiber optic cables. Three feet wide, these tunnels are far larger than in most western European and American cities. They are largely empty spaces waiting, in theory, to provide some of the highest bandwidth on earth. To the eye of a New Yorker this is a strange landscape of inhuman proportions. Nowhere in the United States are there construction sites even approximating this size.

Part of the newly established Incheon Free Economic Zone (IFEZ), Songdo is one of three developments—the other two go by the labels "logistics" and

FIG. P.1__Frontier architecture. Songdo, Incheon Free Economic Zone, South Korea.
Image: author, July 4, 2012.

"finance/leisure"—to be rolled out as the latest testing grounds for the future of human habitation.[3] It is perhaps telling that this free trade zone is built on an extension of the same beaches that marked the successful American invasion of Korea during the war in 1950; where one invasion occurred in the name of containment, now airports and free trade zones rise in the name of global integration. The Incheon Free Economic Zone and its commodity cities are interfaces and conduits into networks linked to other territories.[4] Conceived as a zone integrating finance, airport and logistics, high technology, and lifestyle by the South Korean government in the midst of the Asian currency crisis, the area is being developed in collaboration with Gale, a Boston-based real estate development company, and Cisco Systems, a major network infrastructure provider based in San Jose, California, now seeking to enter management consulting and telepresence service provision.[5] These cities made to hold hundreds of thousands, even millions, of people are sold for export by engineers and consultants. Marketed as machines for the perfect management and logistical organization of populations, services, and resources with little regard for the specific locale, these products are the latest obsession in urban planning.[6] They are massive commodities.

Songdo is a special class of such spatial products. The city's major distinguishing feature is that it is designed to provide ubiquitous physical computing infrastructure. Marketed as a "smart" city, it is sold as the next frontier in computing—an entire territory whose sole mandate is to produce interactive data fields that, like the natural resources of another era, will be mined for wealth and produce the infrastructure for a new way of life. Cisco's strategic planners envision the world as interface, an entire sensory environment where human actions and reactions, from eye movements to body movements, can be traced, tracked, and responded to in the name of consumer satisfaction and work efficiency (whatever these terms may denote, and they are always ill defined and malleable, as are, perhaps not incidentally, "intelligence" or "smartness").[7] Every wall, room, and space is a conduit to a meeting, a building, a lab, or a hospital in another place. The developers thus envision an interface-filled life, where the currency of the realm is human attention at its very nervous, maybe even molecular, level. (Engineers speak candidly of transforming the laws of South Korea to allow the construction of medical grade networks to allow genetic and other data to flow from labs in the home to medical sites in order to facilitate the proliferation of home-health care services.) Accompanying the provision of computing infrastructure, the South Korean government also offers tax incentives to global high-tech and biotechnology companies to build research and development facilities that leverage the data structures and bandwidth of the location. Samsung's biotech division has already relocated, along with POSCO, a major steel refining conglomerate, IBM/KYOBO e-book storage and web sales, Cisco's urban management division, and numerous other companies.[8]

As some of the city's more enthusiastic proponents write, "as far as playing God. . . . New Songdo is the most ambitious instant city since Brasília 50 years ago. . . . It has been hailed since conception as the experimental prototype city of tomorrow. A green city, it was LEED-certified from the get-go, designed to emit a third of the greenhouse gases of a typical metropolis its size. . . . And it's supposed to be a 'smart city' studded with chips talking to one another." The article goes on to address the role of Cisco in the project and their plans to "wire every square inch with synapses."[9] The developers, financiers, and media boosters of this city argue for a speculative space ahead of its time that operates at the synaptic level of its inhabitants, linking the management of life at a global and ecological level to the very modulation of nerves.

FIG. P.2__**Bandwidth = Life.** Image of control room in Songdo, monitoring environmental data, traffic movement, security cameras, and emergency response systems. Image: author, September 1, 2013. As the marketers explain: "life in the Incheon Free Economic Zone is peaceful and abundant with parks and broad fields of green covering more than 30 percent of the city. There is a new city waste incinerating facility, a treated sewage recycling system and other systems, which work beyond eyeshot." Incheon Free Economic Zone marketing materials, July 2012.

The government and the corporations developing this space hope to create value around this systemic (human, machine, and even environmental) attentive capacity. They speak of "monetizing" bandwidth, implying that terms like "information" and "communication" can be seamlessly translated into rates of bits transmitted[10] and into the amount of attentive, even synaptic, time consumers dedicate to unspecified applications in business, medicine, and education.[11] This is a landscape where bandwidth and sustainability are fantasized as organizing life through a proliferation of interfaces to the point of ubiquity (fig. P.2). What constitutes "intelligence" and "smartness" is now linked to the sensorial capacity for feedback between the users and the environment: bandwidth and life inextricably correlated for both profit and survival.

Songdo arguably demonstrates a historical change in how we apply ideas of cognition, intelligence, feedback, and communication into our built environments, economies, and politics. It is a city that is fantasized as being about reorganizing bodies, down to the synaptic level, and reorienting them into global data clouds or populations with other similarly reorganized nervous systems globally.[12] These populations are not directly linked back to individual bodies but are agglomerations of nervous stimulation; compartmentalized units of an individual's attentive, even nervous, energy and credit.[13] Furthermore, it is imagined as a self-regulating organism, using crowdsourcing and sensory data to administer the city and limit (in theory) the necessity for human, or governmental, intervention. Songdo's speculators who are banking on the big data sets to be collated from such spaces no longer deal with consumers as individual subjects but rather as recombinable units of attention, behavior, and credit. This form of political economy is often labeled "biopolitics" for making life its object and subject of concern, and it produces a range of new forms of administration, management, and productivity.[14]

The fantasy of managing life itself by bandwidth, and the often unquestioned assumption that data presents stability, wealth, and sensorial pleasure is not solely the privy of real estate speculators. Today "big data" is regularly touted as the solution to economic, social, political, and ecological problems; a new resource to extract in a world increasingly understood as resource constrained.[15]

This ubiquitous data that is so valuable, even without a set referent or value, is also often explicitly labeled "beautiful." In the pamphlets of technology corporations touting the virtues of a "smart" planet and in prominent textbooks in computer science and blogs by computer research groups, stories abound about "elegant data solutions." These narratives come with labels such as "Beautiful Data" and "Beautiful Evidence." Opening with the premise that the web today is above all about the collection of personal data, many data visualization sites and textbooks urge the designers, engineers, and programmers of our future to address the important aesthetic component of making this data useful, which is to say, "beautiful." But data is not always beautiful. It must be crafted and mined to make it valuable and beautiful.[16] Despite the seeming naturalness of data and its virtues, therefore, there is nothing automatic, obvious, or predetermined about our embrace of data as wealth. There is, in fact, an aesthetic crafting to this knowledge, a performance necessary to produce value.

These discourses of data, beauty, and "smartness" should, therefore, present us with numerous critical historical questions of gravity, such as: how did space become sentient and smart? How did knowledge come to be about data analysis, perhaps even in real time, not discovery? How did data become "beautiful"? How did sustainability and environment come to replace structure, class, and politics in the discourses of urban planning, corporate marketing, and governmental policy? To summarize, how did perception, understood as a capacity to consume bandwidth, come to reorganize life itself?

There is much at stake in these questions. In tying the management of the future of life so tightly to computation and digital media, Songdo is a particular instantiation of how emerging infrastructures of knowledge and per-

FIG. P.3__**Visible**: demonstration control room, Tomorrow City, Songdo. Image: author, July 4, 2012. **Ubiquitous**: "smart" ubiquitous home prototype; the table and the walls are all projection-responsive interfaces, along with sensors for environmental control and telemedicine, at SK Telcom "U" (for ubiquitous) products showroom, Seoul. Image: author, July 3, 2012. **Smart**: "smart" pole, with sensors installed for movement detection, Internet wi-fi hotspot, surveillance cameras and sensors linked to police, fire, and hospital for emergencies, and "smart" LED screens. The poles play music to passersby, provide direct-to-consumer advertising, and enhance, according to

ception are involved in the reformulation of population and in the transformation, if not disappearance, of space and territory. But these cities are also massive prototypes, not-yet-realized instantiations of futures that may or may not come to pass. Part of rethinking these futures is renegotiating their past.

The philosopher and cultural critic Walter Benjamin was among the most prominent thinkers to realize that a history of perception can transform the future. "Architecture," he once wrote, in his essay on art in the age of mechanical reproduction, "has always represented the prototype of a work of art the reception of which is consummated by a collectivity in a state of distraction. The laws of its reception are most instructive."[17] For Benjamin architecture was the spatial key to a temporal problem—how to denaturalize the present and thus reimagine the future? The laws of reception stipulated by Benjamin, however, can no longer be received, as they hide inside protocols, storage banks, and algorithms. The terms "attention" and "distraction" are inadequate to describe a sensorium now understood as infinitely extendable.

I have opened, therefore, with this example that is seemingly distant from any history of cybernetics, visuality, or reason because it demonstrates the complexity and urgency of interrogating this present and its biopolitical rationalities. But Songdo is a disposable architecture, whose material manifestations are banal and constantly mutating. The city is not a space full of top

the designers, "Emotional Happiness." Image: Nerea Calvillo, July 2, 2012, Digital Media City, South Korea. **Cute**: bunnies in the petting zoo in the "central park." Songdo possesses some curious, almost farcical, features. There is, for example, a small zoo with large rabbits for children in the middle of a park that planners argue is based on "Central Park" in New York. This curious set of elements, somewhat touching, almost cute, also idiosyncratic and darkly humorous, are the interfaces to our present. Image: author, July 4, 2012.

architectural names and monumental features. What it is full of is screens and interfaces. Apartments come replete with surfaces that allow users to engage with building management systems and import telemedical and other data. The urban landscape is full of LED screens, and vast control rooms monitor the cities' activities, even though human intervention is rarely necessary (fig. P.3). Big data and visualization are key concerns to planners and engineers attempting to use the data generated from these systems for better planning and for sale. As Keller Easterling notes, digital capitalism is sneaky, contagious, and often costumed in its material manifestations (see fig. P.3).[18] To begin contemplating what it even means to see or to think in such a space, where every interface is only a conduit into ongoing interactions, demands placing a history of design, planning, and aesthetics alongside a history of knowledge, communication, and science. This book will do so by tracing the historical relationship between cybernetics, vision, knowledge, and power, culminating in contemporary concerns with biopolitics. It will draw a map beginning with early cybernetic ideas developed at MIT in the late 1940s in the work of mathematician Norbert Wiener concerning vision, perception, and representation. I will trace the influence of these ideas on American designers and urban planners who reformulated design education and practice in the 1950s. The book then turns to the cybernetic impact on social and human sciences, particularly psychology, political science, and organizational management. The narrative vacillates between on the one hand examining attitudes to visualization, measurement, and cognition in the communication and human sciences and on the other hand examining attitudes to vision and attention in design practice. A central focus of this narrative is to demonstrate how ideas about human sense perception are intimately linked to a transformation in the definition of intelligence and rationality; and that it is precisely this merger between vision and the reformulation of reason that underpins contemporary biopolitics. My interest is in giving equal weight to both the histories of art and design and the histories of science and technology, in order to examine how each coproduces the other, and to offer an account of how aesthetic *and* epistemological discourses combine to reformulate power and population simultaneously. This is a history of our contemporary infrastructures of sense and knowledge.

INTRODUCTION_**Dreams for Our Perceptual Present**

There is a long history linking utopian ideals of technology and calculation with governance, of which Songdo and its sister "smart" cities are but the latest additions. For example, *New Atlantis*, written in 1624 by the English philosopher and statesman Francis Bacon, posited an ideal space governed by education, inductive reason, and empirical experimentation as a scientific practice. This utopia was invented to address the transformations in religion, knowledge, and power in the England of his day and to encourage his ideals of natural philosophy and governance.[1] In the late eighteenth century, the British social reformer and philosopher Jeremy Bentham presented an ideal architecture—the panopticon (fig. 1.1)—to demonstrate his ideal of a link between visuality, the rational and calculated management of space, and democratic government. Bentham posited a perfectly organized space where power could be wielded without force as part of a utopian reconceptualization of politics.[2]

Modern utopias have also often reflected the media, technology, and scientific methods of their time. The famous French architect Le Corbusier, for example, imagined cities of tomorrow in 1923 (fig. 1.1) that would be perfectly statistically managed, showcase the latest technologies, and eliminate disorganization and could be built and replicated through systemic, machine-like principles and the application of careful statistical social science. Le Corbusier invented a method of proportions that allowed his designs to be implemented at different scales—from individual homes to entire cities.[3] His plans went on to shape the future of cities like Brasília and Chandigarh and to define the future of public housing globally in the postwar years. According to the architectural historian Robert Fishman, Le Corbusier imagined that the industrialist and engineer had built the perfectly rationalized mode of production, and therefore architecture and planning had to provide a city that refracted and advanced modern technology and capital in the early to mid-twentieth century.[4]

If modernity had "a machine for living," to quote Le Corbusier's definition of his home design, by the 1970s architecture itself was being envisioned as

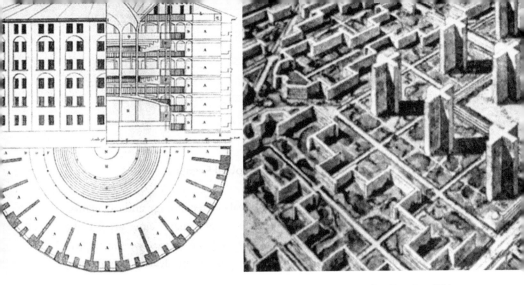

FIG. I.1__Jeremy Bentham's Panopticon penitentiary, drawn by Willey Reveley, 1791. Wikipedia, http://en.wikipedia.org/wiki/File:Panopticon.jpg; City of Tomorrow (1923). From Le Corbusier, *City of To-morrow*, 173; Songdo, satellite imagery, with

a "machine," but a new one: a computational and artificially intelligent network composed of intimate feedback loops between designers, users, and computers. One of the key sites forwarding this vision of computational environments was the Massachusetts Institute of Technology (MIT) and the MIT Media Lab. In fact, MIT was initially supposed to be involved in the Songdo project, but there were problems in the collaboration that may have emanated from either capital constraints or ideological differences or both. The reason remains obscure, or at least my informants refused to elaborate. Many of the chief architects of the smart-city initiatives in South Korea claim MIT as inspiration and model, and much of our contemporary thinking about ubiquitous computing and smart cities in urban planning emanated from Nicholas Negroponte's Architecture Machine Group, which was started in 1967 at MIT with funding from major corporations and the Cybernetics Technology Division of the Advanced Research Projects Agency (ARPA, after 1972 DARPA), of the U.S. Department of Defense for the purposes of integrating computers into architecture and urban planning. Negroponte's ideas were popularized through the labs and a number of books introducing the idea of an "architecture machine" and later "soft architecture machines" in the early 1970s.[5]

Negroponte opened his text on "the architecture machine" with two premises. The first was that "computer-aided design cannot occur without machine intelligence" and that this intelligence must be "behavioral" and "must have a sophisticated set of sensors, effectors, and processors."[6] The fundamental re-

projected space to be reclaimed from the sea, and the outline of the projected topography of the official Incheon Free Economic Zone visible in white. Image: author, September 2, 2013.

organization of planning and architecture around computing did not, therefore, begin with any set of concepts usually linked to architecture. Instead, these manifestos opened with discussions of two elements: sensory capacity and intelligence. For Negroponte a true "architecture machine" would not be a modern machine serving human needs but an integrated system that was based on a new type of environmental intelligence related to the regular capacity to sense and respond to sensory inputs. His articles and books distilled a constellation of theories about intelligence and complexity to argue that design had to become process, a "conversation," in his words, between two intelligent species—human and machine—and not a linear cause-effect interaction.[7] "We are talking about a symbiosis that is a cohabitation of two intelligent species," he wrote.[8] He was not interested in computerizing design so much as rethinking the design process itself. This symbiosis was necessary to address both a human inability to deal with "large-scale problems"—beyond the protocols of architecture and planning, which were incapable of dealing with systemic problems, emergence, or changing contexts—and simultaneously architects' and planners' inability to handle large amounts of specific and local data.[9] Architecture as a machine was about design as a process that could mine data, find patterns, and produce new forms of emergent growth through feedback.

It is, therefore, not even to architecture that these original formulations of smart and sentient design and urban planning paid debt but rather to cybernetics, and to ideas of systems, behavioralism, and cognition that had emerged in the previous two decades out of work in the cognitive sciences and neural nets.[10] At the heart of Negroponte's manifestos for computer-aided design lay the work of cyberneticians, particularly the MIT mathematician Norbert Wiener, the neural net pioneer Warren McCulloch, and the British cybernetician Gordon Pask, along with influences from other pioneers in computer-aided design, such as Christopher Alexander.[11]

To begin, then, I want to start not with architecture but with cybernetics. In 1953, the MIT mathematician and cybernetician Norbert Wiener, in his memoir *Ex-Prodigy*, made a statement about diagrams that also imagined a new future into being, and that bears on our contemporary concerns with ubiquitous computing, data, and visualization. "I longed," he wrote, "to be a naturalist as other boys longed to be policemen and locomotive engineers. I was only dimly aware of the way in which the age of the naturalist and explorer was running out, leaving the mere tasks of gleaning to the next generation."[12] Developing this theme, he would later write, "even in zoology and botany, it was diagrams of complicated structures and the problems of growth and organization which excited my interest fully as much as tales of adventure and discovery."[13] In a series of popular books and technical manifestos, Wiener would go on to interrogate this "problem" that complexity poses. Written in a reflective moment after World War II, Wiener's comments sought to mark the passing of one age to another—the end of "exploration" and the emergence of another type of "organization."

This was no small claim. When situated in the context of his other works about communications theory and computing, this seemingly minute comment about personal memory gestured to a fervent hope: that an epistemic transformation involving the relations between temporality, representation, and perception was in process. Wiener indicated a desire to see an older archival order, adjoined to modern interests in taxonomy and ontology, rendered obsolete by another mode of thought invested in prediction, self-referentiality, and communication. Wiener's words anticipate the emergence in the coming decades of a machine design that might indeed surpass the hand or eye of the architect; he imagined a new form of visualization and knowledge.

Wiener dreamed of a world where there is no "unknown" left to discover, only an accumulation of records that must be recombined, analyzed, and processed. Wiener argued that in observing too closely and documenting too "meticulously," one is unable to deduce patterns, to produce in his words a "flow of ideas." He wrote that "if he [a student] decides to take notes at all, he has already destroyed much of his ability to grasp the argument in flight, and at the end of the course has nothing but a mass of illegible scribble. . . . It is far better to give up the idea of taking notes and to organize in his mind the material as it comes to him from the speaker."[14] *Ex-Prodigy*'s obsessive implication was this gap between thought and action, and not, as the autobiographical genre might lead us to expect, the need to document or account for past experiences. This subtle shift of emphasis away from concerns with documentary and personal experience opens a site to excavate the historical reformulation of relations between vision, cognition, and communication.[15]

Today, seated behind our personal computer monitors, constantly logged in to data networks through our personal devices, we stare at interfaces with multiple screens and no longer aspire to go out and explore the world. From the vast cityscapes of Songdo to our everyday use of numerous mobile devices, the environment is assumed to be an interface to elsewhere that will bring information to users. There is no "unknown" left to discover. We have come to assume that the world is always already fully recorded and archived; accessible at a moment's notice through the logics of computational searches. Wiener's words seemingly have been technologically realized, our relationship to historical time, documentation, and knowledge apparently reconfigured through the terms of communication and control. In the realms of neuroscience and the many attention deficit disorders we now cultivate as pathologies, this situation is ordained genetic. The speculation to build an architecture to harness this attention is at a frenzy. Humanity, it seems, always sought to communicate through screens, always wanted to garner ever more data from more locations, more immediately. It is the purpose of this work to denaturalize such assumptions.

Wiener's autobiography thus bridges late nineteenth- and early twentieth-century ideals of taxonomy, ontology, and archiving and post-mid-twentieth-century concepts of organization, method, and storage. He articulated a desire to see previous traditions in natural history and scientific representation replaced by a discourse of active diagrams, processes, and complexity. And Wiener did not dream alone. His memories found concrete expression in such diverse places as the new multimedia architectures of spectacular geopolitics and the minute neural nets of the mind.

In the postwar era, throughout the social sciences, neurosciences and cognitive sciences, computer sciences, arts and design, endless flow charts emerged producing images not of an outside world but of the patterns linking thought to action.[16] The social and human sciences turned to performance and visualization as a route to innovation.

Prominent designers, such as Gyorgy Kepes of MIT, for example, would exclaim that "the essential vision of reality presents us not with fugitive appearances but with felt patterns of order."[17] Arguing for a reality that is not "fugitive" and a beauty produced out of patterns rather than essence and forms, designers, engineers, and scientists propagated a discourse of a "new" vision emerging from informational abundance. This vision cannot be understood as solely concerning optics and eyes but rather, in Kepes's language, a "landscape of sense" produced through technologies like "radar and electronic computers"[18] that would organize perception, and practice, in many fields. This "felt order" would be the source of beauty, and would transform data from being set out in "terms of measured quantities" to being "set out in terms of recreated sensed forms . . . exhibiting properties of harmony, rhythm, and proportion.[19]

Kepes's compatriot and interlocutor, the designer and inventor Buckminster Fuller, the Cold War evangelist of a unity between art, design, and science through cybernetics in the 1960s and 1970s, vociferously propagated the concept of a renaissance "design scientist." In his effort to unify the varied fields of physical, social, biological, and design practices, he labeled the very process of inquiry a thing of "beauty" in and of itself. "It is one of our most exciting discoveries," he wrote, "that local discovery leads to a complex of further discoveries. Corollary to this we find that we no sooner get a problem solved than we are overwhelmed with a multiplicity of additional problems in a most beautiful payoff of heretofore unknown, previously unrecognized, and as-yet unsolved problems."[20] Complexity and problems, rather than solutions, became valuable. Implicitly, like Wiener, Fuller is calling for a valorization of process and method as material artifacts and objects, in the way that previously designers conceived of architecting a building or a chair. In his standard hypertextual fashion Fuller (known to give eight-hour lectures full of slides in a test of attention and repetition whose only goal was the inundation of data in a mimetic reperformance of this aesthetics of informational overload) argued that such practices fostered an "awareness of the processes leading to new degrees of comprehension." This search for awareness, he continued, "spontaneously motivates the writer to describe over and over again what—to the careless listener or reader—might seem to be tiresome repetition, but to the *successful explorer* is known to be essential mustering of *operational strategies* from which alone new thrusts of comprehension can be successfully accomplished."[21] Process, Fuller implies, *is* the site of exploration; generating in turn "strategies" that are also "beautiful." His argument for an optic of process and the beauty of method are the marks of a midcentury shift in the aesthetics and practices of information visualization. Fuller's pronouncements mark the rise of a new aesthetic and practice of truth; a valorization of analysis and pattern seeking that I label "communicative objectivity."

If the stereotype of the nineteenth century is that of a naturalist or industrialist extracting value from natural resources (including alienating labor from human bodies), these citations from the preeminent designers and pedagogues of the mid-twentieth century gesture to an aspiration and desire for data as the site of value to emerge from the seeming informational abundance once assumed to be the province of nature. Data, Kepes and Fuller implied, appeals to our senses and can be seen, felt, and touched with seemingly no relationship to its content. Behind this materialization of data as an object to be marveled at, however, lay an aesthetic infrastructure of sensorial training and a new imaginary of vision as a channel and a capacity that was autonomous, networked, and circulative.

Such cybernetically inflected attitudes also emerged in the social and human sciences. "All that is offered here," a prominent textbook in political science from the early 1960s argued, "is a point of view. Men have long and often concerned themselves with the power of governments, much as some observers try to assess the muscle power of a horse or an athlete. Others have described the laws and institutions of states, much as anatomists describe the skeleton or organs of a body. . . . [Political science must] concern itself less with the bones or muscles of the body politic than with its nerves—its channels of communication and decision."[22] Written by one of the preeminent political scientists of the time, the Yale professor Karl Deutsch, the book implied that the study of government would be a study of perception, a training in "a point of view," to be guided by nervous diagrams. The entire book calls on visual metaphors and presents flow charts of decision-making trees that emulate those of computer programming, also emerging at the time (fig. I.2). Like Wiener's drawings reconciling the slowness of the hand with the speed of thought, so in the study of organizations would the careful mapping of process synchronize the time of bureaucracy and the flow of information.

Rather than observe closely as an anatomist, Deutsch insisted on another type of vision. He wrote that "today we are learning in television [and other communications technologies] to translate any outline of a static or slow-changing thing, such as the edge of a mountain, or the edge of a human skull, or the lines of a human face, into a sequence of rapidly-changing little dots. . . . We learn, through scanning . . . how to put together these events, which move slowly but are strung out along a period of time, and to see them all at once."[23] Moving beyond a dialectic opposing close observation to theoretical abstractions, Deutsch's image world was simultaneously empirical and abstract. Be-

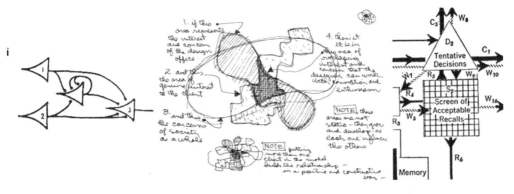

FIG. I.2__Neural net diagram by Warren McCulloch and Walter Pitts (1943), from McCulloch, *Embodiments of Mind*, 36; Design process flow chart, by Charles Eames, made for the exhibition "What Is Design," Musée des Arts Décoratifs, Paris (1969). The Work of Charles and Ray Eames, Manuscript Division, box 173, folder 9, Library of Congress. © 2013 Eames Office, LLC; Diagram of foreign policy decision-making process, by Karl Deutsch, from Deutsch, *Nerves of Government*, appendix.

tween the nearsightedness of the bureaucrat trapped in the trees for the forest and the abstracted metaphysics of the political theorists, the visual tactics of "scanning" and pattern seeking might create a bridge. These diagrams produced, in his words, a new "scale" of observation that turned discrete and nonsensical data points into coherent patterned flows.[24]

Deutsch's diagrams, which were ubiquitous during three decades of elite pedagogy in political science, linked local knowledge and global trends through a methodology of "scale" and "scanning" that mimed the television and communication technologies whose aesthetics they invoked. The purpose of these instructional images was to teach a cadre of elite future policy-makers, analysts, and legal thinkers how to see and scan for a new object of study— decision-making processes and managerial actions—to be able to reflexively use data to make the world visible and knowable.

Data visualization became a democratic virtue and moral good; reason was now understood as algorithmic, rule-bound, definitive, and fast. The reconfiguration of the eye of the technocrat came with the reformulation of the mind of the decision-maker and of the organization itself. The rise of cognition as a model for human thinking and organizational behavior and the rise of visualization as a virtue came hand in hand. Ideas of territory, population, and design were rethought in tandem with transformed ideas about knowledge, representation, and measurement in the social and human sciences. This book unites these two strains of history that are so closely merged in contemporary digital environments.

This book traces the trajectory laid out above to link design, architecture, and artistic practices with cybernetics and the human and social sciences. Excavating Wiener's initial concerns about the relationship between nineteenth-century science and archiving, and his own efforts in pioneering the science of communication and control that he labeled "cybernetics," I chart the relationship between contemporary obsessions with storage, visualization, and interactivity in digital systems to previous modernist concerns with archiving, representation, and memory. Postwar design and communication sciences, believing the world to be inundated with data, produced new tactics of management for which observers had to be trained and the mind reconceived. The result of this reformulation of vision and reason was the production of a range of new tactics, and imaginaries, for the management and orchestration of life.

In my description of contemporary ubiquitous computing environments and data-driven sciences, I have therefore specifically drawn attention to three elements that emerge prominently in cybernetic accounts: the way contemporary discourses on data revise epistemology, create temporalities, and produce aesthetics. The book genealogically traces these three aspects of our present that are so critical to this reformulation of observation and knowledge: first, the reconceptualization of the archive and the document in cybernetics and the human sciences; second, the reformulation of perception and the emergence of data visualization and the interface as central design concerns; and third, the redefinition of consciousness as cognition in the human, cognitive, and social sciences. These three loci—the reformulation of temporality and truth, the reformulation of attention and distraction into interactivity, and the reconfiguration of reason into rationality—structure the book. My argument is that the reconceptualization of evidence, vision, and cognition are the foundations for producing new techniques of calculation, measurement, and administration. I seek to account for this condition that finds itself most graphically demonstrated in such extreme prototypes as Songdo, but can be found more ubiquitously in our armory of electronic and graphical interfaces.

Histories of the Present

How, then, would one begin to comprehend this transformation in the treatment of the senses as commodities, technologies, and infrastructures? This book started over a decade ago, when the concepts of interactivity and histories of observation were still novel ideas. Works like Lev Manovich's *Language of New Media*, Alexander Galloway's *Protocol*, and Katherine Hayles's *How We Became Post-Human* were read alongside an emergent concern in the history of science and art in histories of the body, rationality, taste, and emotion. The rise of a history of observation, perception, and objectivity, written by figures like Jonathan Crary, Wolfgang Schievelbusch, and Lorraine Daston and Peter Galison, interested me as fully as the discussions about the nature of new media.[25]

In the intervening time, complexes like Songdo have moved from being at the margins of literary critics' imaginations to being built and circulated spatial products. As a historian grappling with the media subjects, new questions began to emerge: What makes such a space feasible and even seemingly natural? Are our models of software and hardware based on certain architectures of computing the most useful to account for these emergent formations? Do discourses of embodiment, or even materiality, account for a world where major corporations are also invested in object-oriented thinking? The world is alive with datafiable objects that are also commodities and bodies for the engineers at Cisco. What types of temporal narratives, then, would be able to produce a history of the senses and of observation and knowledge that might challenge the stability of the present without recourse to an imagined ahistoricity of objects and matter?

As Eugene Thacker and Alexander Galloway have put it in their most recent book, *The Exploit*, "again and again, poetic, philosophical, and biological studies ask the same question: how does this 'intelligent,' global organization emerge from a myriad of local, 'dumb' interactions?"[26] Swarms, clouds, black boxes—the question is not unimportant, and in tracing the specific tactics and forms that reason took in the last half a century, I seek to situate such questions. I am not answering why one would pose such a question but rather asking under what conditions would it be thinkable, and even virtuous, to pose such questions about stupidity, either from the perspective of media theory or engineering?

In the late 1940s and early 1950s, for example, the same phenomena—swarms, clouds, masses—were treated far differently. There was no effort to be morally neutral when it came to media protocols. For political theorists such as the German-Jewish émigré and prominent political theorist Hannah Arendt, such phenomena posed mortal threats to the future of life. In *The Human Condition*, published in 1958, Arendt is concerned with a world in which "*speech has lost its power*" to a "language of mathematical symbols which . . . in no way can be translated back into speech." This loss of a critical place for human activities to enter the realms of representation and subjectivity she aligns with "automation," particularly the automation of computational machines. These losses, of labor and of language, are for her fundamentally about losing connection and ability to act politically as individuals, not as masses. Her conception of freedom is fundamentally liberal, linked to the ability to represent and know the world and to act as a connected but independent agent. This is a vision absolutely antagonistic to the types of affective and networked theory often being supported in our present.[27] Her colleague, interlocutor, and fellow émigré the prominent Frankfurt School theorist Theodor Adorno, would write of the "culture industries" in a similar light. "The masses," Adorno wrote, "are not the measure but the ideology of the cultural industry." For Adorno, the culture industries turn concerns of class criticism or artistic imagination into a routinized format through the "dehumanizing" protocols of stardom. Ironically, these protocols appeal to aura and individuality while crafting unthinking masses. Thus, the effect of such media is "anti-enlightenment." In his words, "it impedes the development of autonomous, independent individuals who judge and decide consciously for themselves."[28] The culture industries are guilty of producing the masses while denying the right of the same populations to gain, for Adorno, freedom and agency. Similar discourses about hoards, swarms, and unthinking masses characterized Communism as well at the time. From the science fiction fantasies of body snatching and zombies to the characterization of brainwashing and the Chinese Army during the Korean War, the dominant view of Communism was as a force against agency, rendering individuals subservient to some common intelligence.[29]

I call attention to these debates, however, not to account for the reality of media but to ask about history. These commentators demonstrate that in 1948, ideas of feedback and information inundation were negatively associated with totalitarian regimes, and hardly considered virtuous. The social sciences had only begun to even contemplate the idea that communication was a social virtue, and many political theorists, cultural analysts, and popular psychologists would have found our contemporary valuation of collectivity, social networking, and analytics terrifying. These changes in the moral value and aesthetics assigned to different ideas are the markers of historical revisions in attitudes toward, and imaginaries of, the place of media technologies within societies and the constitution of knowledge and truth. I am particularly interested in the years between 1945 and the early 1970s because this is a period when one can still witness debates between modern and prewar conceptions of truth, certainty, and subjectivity and emerging ideas of communication and cybernetics.

I follow Deleuze, who asked in his cinema books not what is cinema but what is philosophy after cinema? My question is a derivative. I ask what is it to tell history under the conditions of digital media? The status of historicism is under duress, and the organization of temporality in this text is one of feedback and density, not orderly linear time. I examine how reason, cognition, and sentience were redefined in a manner that makes it logical and even valuable to pose such philosophical questions, and, on a more pragmatic note, to begin building territories, for example, based on ideals of distributed intelligence and a belief that space can be sentient, and smart, through (literally) so much "stupidity." These are pathways that produce an albeit limited, but at least speculative, history of concepts such as "interactivity," "beautiful data," and the interface.

To comprehend this transformation in the treatment of the senses as commodities, technologies, and infrastructures and the concomitant transformation in population and governance without recourse to technical determinism, I focus, therefore, not on realized technologies but on post–World War II ideas, pedagogies, and practices of observation in cybernetics, communication sciences, and their affiliated social and design sciences. Such an attitude follows the lead of figures such as the art historian Jonathan Crary[30] in his history of observation and media genealogists, such as Jonathan Sterne, who are not as interested in the realized technology of recording the senses as in the "possibilities" for producing media technologies.[31]

As Crary notes, "observer" means "'to conform one's action, to comply with,' as in observing rules, codes, regulations, and practices." He continues that observation is more than representational practice, rather "[the observer] is only an *effect* of an irreducibly heterogeneous system of discursive, social, technological, and institutional relations. There is no observing subject prior to this continually shifting field."[32] Crary wrote about the nineteenth century, but what is increasingly evident is that contemporary forms of observation and perception may not even be linked back to single bodies or unified subjects. Life here has to be considered as a set of mechanically calibrated movements and gestures operating at various embodied and even molecular levels. The sensory networks of train systems and smart cities are operating at multiple scales. But what likens these networks to practices of an "observer" is that they operate within certain conditions of possibility, "embedded in systems of conventions and limitations,"[33] and, I might add, affordances and capacities, that are historically situated. To produce this account, therefore, I insist on linking the transformation in attitudes to perception *with* the reformulation of ideas of reason and cognition, because this alignment between how we know and how we sense is critical to understanding and contesting contemporary attitudes to intelligence, data banking, and interactivity.

Vision, Visualization, Visuality, *Visibilités*

One of the curious elements of invoking terms like "vision" and "observation" in our present is that it complicates the very idea of sense perception itself. Moreover, in that no computer actually sees the way a human being does, one needs to ask what it is that is being invoked with the language of vision? As should by now be obvious from the opening of the book, "vision" in this text operates as a holding term for multiple functions: as a physical sense, a set of practices and discourses, and a metaphor that translates between different mediums and different communication systems. Vision is thus a discourse that multiplies and divides from within.

To offer a cartography of this complicated terrain, I want to start with one of the more popular words applied to contemporary data display— "visualization." According to the OED, the term is not ancient but rather a modern convention, only appearing in 1883 to depict the formation of mental images of things "not actually present in sight."[34] Throughout the next few decades, this term expanded to encompass any "action or processes of making visible."[35] Visualization slowly mutated from the description of human psychological processes to the larger terrain of rendering practices by machines, scientific instrumentation, and numeric measures. Most important, visualization came to define bringing that which is not already present into sight. Visualizations, according to current definition, make new relationships appear and produce new objects and spaces for action and speculation.

While the language of vision perseveres, it is important not to assume a direct correlation between vision as a sense and visualization as an object and practice. Married initially to psychology, and now digital computation and algorithmic logic, the substrate and content of this practice has often had little to do with human sense perception or the optic system. Moreover, with the rise of emphasis on haptic interactions and interactivity, visualizations also often take multisensorial modes. Vision cannot be taken, therefore, as an isolated form of perception, but rather must be understood as inseparable from other senses.

In the present, visualization is most often understood not only as a process but also an object, a subject and discipline, a vocation, a market, and an epistemology. For example, SAS, one of the major contemporary makers of data visualization software and enterprise solutions, on their website states that visualization is the practice of making complex data (also not defined in this case) "dynamic," "universal," and "valuable." The website enjoins future clients to believe that visualization software allows previously "invisible" relationships in market or other data to become "visible" and operable.[36] The prime figure behind the "smart" city and now "smart planet" mandate, IBM, repeats this definition in discussing analytics and visualization: "organizations are overwhelmed with data. On a smarter planet, the most successful organizations can turn this data into valuable insights about customers, operations, even pricing . . . for business optimization by enabling rapid, informed and confident decisions and actions." Visualization, IBM then insists, is part of making data actionable through representation while also facilitating the ongoing analysis of data.[37] Repeating the assumption of an "overwhelming" data landscape, visualization is understood to offer a map for action. At the same time visualization and analytics, comprehending and analyzing, are viewed as an integrated process.

Visualization, both marketing manuals and studies of digital images suggest, is the language for the act of translation between a complex world and a human observer.[38] Visualizations are about making the inhuman, that which is beyond or outside sensory recognition, relatable to the human being. One might understand "visualization" in this context as the formulation of an interaction between different scales and agents—human, network, global, nonhuman.

Visualization is also about temporal scales. For example, IBM and SAS assume that data only becomes valuable, or a site of action, once it is crafted into the realm of appearances. However, the realm of the image and the space of data are not in the same time. As in the nineteenth-century definitions, when visualization was solely about mental images and thus not synchronous with the world, in our present a visualization is understood as being out of time and space, nonsynchronic with the event it is depicting, translating, comprehending, and guiding.

This nonsynchronicity preoccupies our imaginings of "real-time" interactivity and data visualization, driving a constant redefinition of the temporal lags between collecting, analyzing, displaying, and using interfaces. Underpinning the contemporary frenzy to visualize is an implicit supposition that cognition, and value, lags behind the workings of networks and markets. The work of visualization is thus temporal—to modulate and manage this time lapse.

As the preeminent language for negotiating our data-filled world, "visualization" invokes a specific technical and temporal condition and encourages particular practices of measurement, design, and experimentation. Visualization, like the term "data," looms, therefore, as a never fully defined verb/noun that straddles the actual practices of depicting and modeling the world, the images that are used, and the forms of attention by which users are trained to use interfaces and engage with screens.

If the language of visualization organizes our present relationship to the interface, the term's relatively recent emergence at the end of the nineteenth century also poses historical questions. For historians of science and art, "visuality" is the language for asking about such historically specific formulations of sense. It is the language for inquiry into the historical, technical, social, physical, and environmental conditions that shape the experience of "seeing." The filmmaker Harun Farocki, for example, offers a clear-cut example of how the most physiological act of seeing is permeated with history. In a famous film, *Images of the World and the Inscription of War* (1988), Farocki offers a tableau. He asks how is it that American and British analysts looking at aerial surveillance photographs fail to see Auschwitz, and only identify Buna manufacturing plants in 1944, but two CIA analysts can later find the camp in the same images in 1978 after seeing melodramatic TV series about the Holocaust? This difference, as the film theorist Kaja Silverman has made explicit, is about how physiological capacities are conditioned and vary under historical conditions.[39] The same example could be repeated with attention, identity, or any number of other examples that demonstrate that while the actual sense of vision may traverse history, its organization and arrangement is historical, culturally, and technically specific. The emergence of panoramas, impressionist painting, abstraction in art, and so forth are all the trace markers of a history of visuality, where how we see comes into contact with the ideas, structures, and technologies of society.[40] Things that appear strange, ugly, or invisible in one era are not so at other times, and our forms of attention, distraction, beauty, disgust, and empathy are all physically and psychically real and simultaneously historically modulated.

When I speak of vision, then, it often encompasses the actual sensory-motor-cognitive apparatus of seeing, the eye, the brain, nerves, even if not always human. But vision, particularly within a Western tradition, also operates metaphorically as a term organizing how we know about and represent the world; a metaphor for knowledge, and for the command over a world beyond or outside subjective experience.[41] To be seen by another, to see, to be objective, to survey, all these definitions apply in etymology and philosophy to the Latin root: *videre*.[42] *Videre* is also at the root of the word "evidence," and I maintain the language of vision precisely because its etymology provides a space to begin asking how truth and knowledge are being reconstituted in different historical moments.

The history of evidence is also, for Deleuze, reading Foucault, one of vision. For Deleuze visuality is closely linked to *visibilités*, or what in English I will label "visibilities." Deleuze defines this term as "visualness,"[43] implying that vision cannot only be understood in a physiological sense but must also be understood as a quality or operation. For Deleuze visibilities are sites of production constituting an assemblage of relationships, enunciations, epistemologies, and properties that render agents into objects of intervention for power. Visibilities are historically stipulated apparatuses for producing evidence about bodies, subjects, and now, perhaps, new modalities of population.

The philosopher John Rajchman offers an example to illustrate this difficult idea. He reminds us of the two instances at the start of Foucault's *Discipline and Punish* and *Birth of the Clinic*: the careful description of the torture of the regicide and the close detailing of the bathing cure of a hysteric. "In both cases," Rajchman writes, "we have pictures not simply of what things looked like, but how things were made visible, how things were given to be seen, how things were 'shown' to knowledge or to power—two ways in which things became seeable. In the case of the prison, it is a question of two ways crime was made visible in the body, through 'spectacle' or through 'surveillance.' In the case of the clinic, it is a question of two ways of organizing 'the space in which bodies and eyes meet.'"[44] As this example demonstrates, visibilities are married to visuality as the historically situated conditioning infrastructure for how subjects come to be known to power. Visibilities are accumulations of a density of multiple strategies, discourses, and bodies in particular assemblages at specific moments. Therefore, visibilities are not merely "visual." Visibilities can be constituted through a range of tactics from the organization of space—both haptic and aural—to the use of statistics.

"Vision" is thus a term that multiplies—visualization, visuality, visibilities. These multiple permutations of the term "vision" demonstrate that vision cannot therefore be merely about the isolated sense of vision but must also be about what, following Walter Benjamin, I would label a technical condition— and what, following Foucault, makes the organization of the senses critical to understanding the tactics of governance and power at any historical moment. The "task of the history of aesthetic forms," the film theorist David Rodowick argues, "is to understand the specific set of formal possibilities—modes of envisioning and representing, of seeing and saying—historically available to different cultures in different times."[45] This study is ultimately dedicated to comprehending just such historical transformations of sense and the specific conditioning of attention under particular technical conditions.

In focusing on the relationship between epistemology and sense, I follow the lead of historians of science, art, and media that focus on histories of observation, knowledge, and aesthetics. The history of science has long been concerned with how instrumentation, standards, and measurement techniques are co-produced with new ideas of perception, observation, cognition, and life. From pharmaceutical trials and statistical instruments, to the complex photographic and cinematic apparatus necessary to capture, assess, and study the world, our idea of ourselves, and of others, is never separate from our practices of observation, documentation, and truth.[46] As Jimena Canales notes in her history of a tenth of a second, a history taken from the perspective of measurement begins to collapse clear-cut distinctions between the modern and nonmodern and makes visible the contests and heterogeneities that produce knowledge. Discourses concerning truth, facts, and representation demonstrate continuities and fissures in history. More important, problems of measurement allow us to focus on epistemic uncertainty and desire; on sites where cultural and social interest is invested before and outside of technical realization.[47]

In linking histories of the senses to those of visibility and measure, I can also begin to account for transformations in governance. In fact, the very etymology of the word "cybernetics" already suggests a relationship to histories of governance. Cybernetics is, in Wiener's words, an "emergent term" derived from the Greek *kubernetes*, or "steersman," the same Greek word from which we eventually also derived the word "governor."[48] Cybernetics is thus a science of control or prediction of future events and actions. From the start, despite disavowals by many prominent practitioners, the ideas of communication and control were applied to theorizing and reenvisioning systems, both sociological and biological. A history of cybernetics must therefore also extend to account for a history of governmentality, and to how governmentality links to ideals of knowledge and sense.

In his final lectures, Foucault defined "governmentality" as "the genesis of a political knowledge [savoir] that was to place at the center of its concerns the notion of population and the mechanisms capable of ensuring its regulation."[49] For Foucault, the particular form of political reason that emerges throughout the second half of the twentieth century comes under the rubric of biopolitics and is intimately tied to data, calculation, and economy, particularly neoliberal economics. He defines biopolitical governance as related to a "new type of calculation that consists in saying and telling government: I accept, wish, plan, and calculate that all this should be left alone."[50] In our present, this calculative rationality is certainly evident in the new smart cities, where ubiquitous computing is imagined as necessary to supplant, and displace, the role of democratic government. More critically, these technical systems serve a discourse of security and defense—of life, futurity, and value. It is a very thin line between the autonomous robotic systems of networked trains and smart sensor cameras monitoring traffic flow and consumer consumption to the more militarized drones or smart border fences that make up the landscapes of contemporary war and security.

But there is a longer history to security and politics that links itself to the cybernetic ideas of information and prediction. In the 1920s the economist Frank Knight isolated the term "uncertainty." Uncertainty, unlike risk, according to Knight, has no clearly defined endpoints or values.[51] Songdo is one potent example of this management of uncertainty. The city serves as a vacillating network awaiting purposes not yet assigned and preparing for disasters of environment and ecology that have not yet been assessed or definitively calculated and whose temporal horizons are eternally deferred. Interviewees I spoke with from government and Cisco repeated the same discourse—bandwidth is valuable even if its function, and monetization, has not yet been determined.[52] Contemporary technical networks reformulate governmentality through the production and manipulation of temporalities. Preemption through the management of uncertainty supports the increased penetration of computational interventions in the name of sustainability, and central to this capacity, as Wiener suggested from the start, is an ability to reenvision, visualize, and manage data in specific ways.

One of the central themes in this book is to trace just how the historical reorganization of vision and reason (or "intelligence") that began in the mid-twentieth century reformulated population and territory in ways that support (and sometimes contest) contemporary forms of biopolitical governance and economy.[53] While the manipulation and direct monetization and materialization of time as a commodity appears central to contemporary financial and technical systems, this book will demonstrate that these contemporary phenomena are intimately linked to transformations in knowledge, observation, and archiving that began already in the mid-twentieth century. In the work of individuals like Herbert Simon in business and finance, the designs of urban planners like Kevin Lynch at MIT, and the rising discourses of systems and networks, very quickly concerns about total war and risk were eclipsed into those of economy, consumption, and ecology, making life, as the sociologists Patricia Clough and Craig Willse frame it, in its emergent mode the very target of technical automation.[54]

Organization

This book traces this cybernetic trajectory and the reformulation of vision I have just mapped in order to situate our contemporary forms of perception and cognition in relationship to historical factors. It is also a narrative deeply concerned with the relationship between these historical forms of attention and thought as related to governmentality—particularly biopolitical rationality.

The chapters group themselves in clouds around particular themes I have identified—storage and archiving, the interface and the training of the observer, the transformation in attitudes to cognition and knowledge, and the assemblage of these components into a new structure for the attentive reorganization of territory and population. Structuring these territories is my effort as a historian to trace particular practices and concepts as they move and mutate between different locations. This is, of course, a partial endeavor; the breadth and impact of the communication sciences is too great to be fully accounted for. Rather, I have selected figures and practitioners who focused on topics of visuality, storage, cognition, and design.

The first chapter serves as an interface to the book and maps the work of Norbert Wiener and his colleagues, particularly in neuroscience and cognitive science at MIT, in relationship to nineteenth-century concepts of recording, memory, sense, and time. The chapter centers on a theme critical to both theories of governmentality and history—the archive. The chapter traces how cybernetic ideas of storage, time, and process reformulated older nineteenth-century concepts of documentation, knowledge, and perception. I make a case for a contested history of time in digital media and probe the emergent potentials of a tension between the archive and the interface that underpins contemporary desires for interaction, data storage, and data visualization.

Returning to Wiener's discourse of diagrams, I investigate a nascent series of debates about time, storage, and memory that can also be read as the traces of what Lorraine Daston and Peter Galison, on another register, label "epistemic anxiety or instability"; those moments where the value and virtue of what constitutes evidence is contested.[55] These discussions demonstrated new sites of inquiry and interest for the scientists involved. In cybernetics these debates increasingly were no longer framed in terms of reality or metaphysical truth, or even about objectivity as defined earlier in the century. The object of this book is to analyze and trace this transformation of ideas of mechanical objectivity, or even expert authority and trained judgment, into another form of methodological truth, that is, a truth about the strength and density of networks and the capacity to circulate information and action.[56]

If the archive organizes the first chapter, it is a history of the interface that organizes the second. This chapter traces how cybernetic concepts transformed aesthetic practice, urban planning, and engineering, business, and design education. Moving through a range of spaces from classrooms to urban redevelopment projects, I make a case for the reformulation of perception into interactivity. I trace the rise of a new epistemological ideal—"communicative objectivity"—emerging from the integration of design, cybernetics, and pedagogy in engineering and the arts.

The chapter maps the work of two designers and an urban planner—the aforementioned designer and artist Gyorgy Kepes, the urban planner Kevin Lynch, and the designer Charles Eames. These three figures were central to American modernism, postwar design, engineering education, and urban planning, and all of them engaged with cybernetics and the communication and cognitive sciences. Their work is landmark in creating infrastructures for postwar American life (and perhaps empire)—both attentive and physical.

In their respective projects, we can trace the reimagining of the observer as isolated but ecologically networked. This observer was linked to a new aesthetics of visualization and management. Interactivity as a personal mode of attention became associated with environment as a discourse for managing systems in fields ranging from marketing to urban planning. The chapter culminates with an examination of one site where practices in design, marketing, and management recombined, in the 1964–1965 New York World's Fair, with the innovative launch of the IBM installation "The Information Machine," which advertised the new information economy. The installation propagated an aesthetic of information inundation as a virtue at the same time that New York was undergoing massive transformations in transportation, suburbanization, economy, and race relations. I trace how environment and psychology[57] came to take the place of previous sociological discussions of systems and society, while new strategies of attention emerged as both the solutions and engines for a growing physical infrastructure of racial segregation and an emerging postindustrial economy.

These new forms of political, perhaps biopolitical governance, were not merely reductive and disciplinary. I also trace some of the new forms of producing and imagining urban space and human interrelationality that emerged (for example community gardening) by viewing landscapes as ecologies of psychic and informational interaction. These new strategies for social intervention emerged even as ongoing historical problems of race, class, and gender could now be repressed and reformulated through consumption and interactivity.

The third chapter explores the doppelgänger of perception in cybernetics — cognition. If designers and planners used cybernetic paradigms to rethink vision and environment, human and social scientists used the same ideas to transform techniques of measurement, assessment, and calculation. Read together, these two chapters demonstrate how aesthetics and perception were linked in new assemblages to revise how, to quote IBM, we "think," and how, to repeat the concepts of the engineers at Cisco, space becomes "smart" through new models of sense, measure, and calculation.

The chapter mirrors the first by diachronically mapping how nineteenth- and early twentieth-century ideas of consciousness in psychoanalysis and reason and computability in mathematics and logic were transformed into cognition and rationality. Starting with the conception of neural nets of the psychiatrist and cybernetician Warren McCulloch and the logician Walter Pitts, I examine how these new ideas about mind and communication entered fields ranging from government to economics to computing. I trace the networks of interchange between cybernetic ideas of mind and the work of political scientists, such as the aforementioned Harvard and Yale professor Karl Deutsch, the organizational management, finance, and artificial intelligence pioneer Herbert Simon, and a number of other human and social scientists. In turning to the reformulation of cognition, I also expand the discussion of vision to the territory of new methods for making data and populations visible as objects for study, surveillance, and management.

These nervous networks, while labeled rational, were also, in McCulloch's psychiatrically informed language, "psychotic." In the cybernetically informed human and social sciences, computational rationality was no longer Enlightenment reason. What could be algorithmically defined and computed must by logical definition be antagonistic to intuition, genius, or liberal agency. In a curious turn, however, policy-makers and social scientists, having turned to a nonreasonable rationality and logic to redefine the behavior of subjects and systems, repressed their discovery by valorizing data visualization as a technique to command and control what was increasingly understood to be a world of unknowns, chance, and unreasonable behavior. The chapter explores this mutual interaction between the reformulation of reason in terms of cognition and rationality and the rise of new models of visualizing data and society.

Visualization, here, is a set of techniques by which to manage, calculate, and act on a world of incomplete information.

While much has been written about psychosis and schizophrenia as symptoms of contemporary information economies and endemic to the nature of capital, my analysis is not an explicit theory of psychosis or capital.[58] Rather, I take the language that cognitive scientists, neuroscientists, and social scientists invoked quite literally. This chapter examines what work the discourses of psychosis did in the computational and social sciences to allow new types of knowledge to emerge, and to produce new methods for experiment, calculation, and measurement. The remaining question is why it has been forgotten that rationality was defined in terms of psychosis, not reason, throughout the 1950s? A massive number of media theorists continue to insist on the enduring legacy of enlightened and liberal reason in the present; these assumptions demand interrogation.[59] We must ask: what is at stake in our contemporary amnesia? While contemporary culture looks ever more frequently to neuroscience, behaviorism, and data mining to predict human behavior, economists, policy-makers and even the public also continue to insist on older nineteenth- and earlier twentieth-century definitions of consciousness and choice. Politics happens in this interstice between the memory of liberal reason and the embrace of psychotic logics. This interaction between historical forms of reason and contemporary beliefs in cognition and rationality drives the desire to produce computational approaches to intelligence, economy, and governance. The political question is, however, what defines computation and rationality? These questions, black-boxed in our present, were hotly debated in the 1950s and early 1960s in a range of social and human sciences.

The fourth chapter completes the book in a feedback loop by linking the transformations in cognition and perception with governance and rationality to ask how politics and aesthetics are linked through the valorization of beautiful data. Examining cybernetic work on vision and cognition done by McCulloch, the MIT neuroscientist Jerome Lettvin, and the psychologist George Miller in connection to the design practices of the prominent designers George Nelson and Charles and Ray Eames and the pioneer computer animator John Whitney, Sr., I make a case for the radical reformulation of the very tactics by which bodies, territories, and networks are governed through measurement and attention. The chapter centers on changing attitudes to perception and cognition in the late 1950s as applied to U.S. Information Agency (USIA) propaganda and to the staging of Cold War politics. The chapter ruminates on the past to speculate on the inevitability and organization of contemporary forms of war and terror.

The book ends in interrogating the ethical and political implications of making data beautiful and affective. In the epilogue, we find ourselves simultaneously inside the gardens of IBM's corporate headquarters in suburban New York and standing on hilltops in Jerusalem at the Israel Museum's sculpture garden, contemplating the work of another prominent midcentury artist, Isamu Noguchi, and considering the implications of a new information aesthetics that links the inside of corporations to the reformulation of territories. Like the small rabbits and the performative control rooms of Songdo (fig. P.3), these different landscapes pressure the present and create different possibilities for the future.

Taking seriously the aesthetics and methods of cybernetics, each chapter is an effort to find patterns between fields. Each chapter holds together a series of objects related by way of discourse and method in the interest of unearthing their commonalities while insisting on the irreducible differences and simultaneous heterogeneities between them.

Why Tell History Anyway?

Technology always presents historians with confusing spectacles of obsolescence and novelty. To my eyes, trained in urban planning and public health, Songdo appears part science fiction, part twentieth-century utopianism. The nostalgic forms of past urban developments—a seeming grotesque parody of modernist grids and skyscrapers—is merged with the speculative landscape of server buildings and amorphous blocks of high-tech and biotech corporate installations. Cisco's managers reminded me that they were well acquainted with Le Corbusier. Songdo, they argued, adopted the best of modern architecture without its utopian and failed elements.[60]

In fact, for all its shiny newness, Songdo proclaims its historical, perhaps even already obsolete, nature as a matter of economic logic. The city plan is full of direct reenactments of archival forms. There is a "central park" based on the one in New York with a petting zoo of large bunnies for children (fig. 1.3). The park is lined with communal kiosks containing books for sharing (the old paper ones, not the electronic readers). Particularly uncanny are the large control rooms dressed in bizarre trappings of Cold War science fiction awaiting the infusion of data from every system in the city—water, electricity, medical, traffic, environmental. The reality is that the humans who watch these screens are often passive observers (fig. 1.3). For the most part, these systems run themselves. This intelligence is not always (in fact usually not) humanly controlled. And often it is stupid. Many little sensors operating in local net-

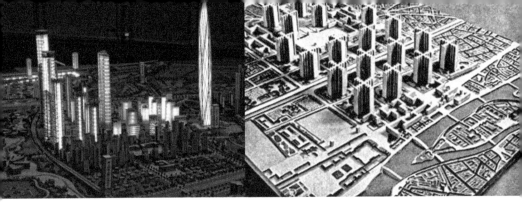

works making minute decisions about traffic lights, water flow, and subway exchanges are what constitute this city's "smartness."[61] One might even ask why, under such conditions, build so many interfaces and visualize at all?

This relationship between the archive and the interface, and between historical forms of attention and ideals of intelligence, is one of the central themes of the book, and key to substantiating contemporary fantasies of visualization, logistics, and control.[62] It is precisely the older memory of surveillance and knowledge that drives an unremitting desire to increase the penetration of sensors, recording instrumentation, and analytic techniques in these territories of ubiquitous computing.

Digital infrastructures, therefore, like the colonial archives depicted by the anthropologist Ann Stoler, are produced through "grids of intelligibility . . . fashioned from uncertain knowledge." These are spaces full of "disquiet and anxieties."[63] Songdo appears stunningly legible as a commodity. Its grids appear to clearly replicate the ideal cities of the earlier twentieth century envisioned by collectives like the Congrès Internationaux d'Architecture Moderne and regularly put into play in the urban redevelopment projects of the United States and Europe in the 1960s.

This is a deceptive legibility.[64] Form does not follow function in Songdo. The perfectly reasoned surface area ratios underpinning the modern towers are failing to produce value. The development has only lost money for the real estate developers. Engineers openly confess to never speaking to developers or urban planners, and admit that the city could take any form desired (circles, spires, anything really the surface does not reflect the infrastructure). At the same time, the developers are being forced to admit that their standard strategies are self-destructive. Banking on real estate while selling bandwidth, it's unclear what is actually more valuable or what is actually being purchased in such developments.

Ironically, the function and action of the territory may actually be one

FIG. I.3__**Prototype?** model for New Songdo City. Image: author, 2012. **Ideal?** Ville Radieuse–Le Corbusier (1924) from the article "AD Classics: Ville Radieuse/ Le Corbusier at http://www.archdaily .com/411878/ad-classics-ville-radieuse-le -corbusier/; **Real?** "Blow-out Village," Peter Cook. Archigram Archive © Archigram 1966.

that was envisioned in the fantastical projections of such countercultural and avant-garde urban designers as those of the group Archigram in the 1960s. The irony, of course, is that it is not the vision of the Congrès Internationaux d'Architecture Moderne (CIAM), so regularly linked in architectural history to rationalization, abstraction, colonialism, and the decontextualization of space and time, that is ascendant in this situation. Rather, it is the vision of the breakaway architectural movements, these avant-garde artists, that is realized in the function of this space.[65] Songdo is less playful than the fantasized cities of the London-based avant-garde group Archigram: walking cities that would roam the earth transporting their workers, or the "Blow-Out Village" (fig. I.3), described by its designers as an entire temporary city that can be inflated from a hovercraft and can rove the earth. However, in many ways, Songdo is just that: an elastic and plastic territory, infinitely mobile, networked to the information economy. But this independent 1960s group, with their embrace of technology, consumerism, and futurism, imagined a humorous but attractive mode of being. Capital has many guises, and reason, or rationalization, as we most regularly imagine it, is rarely the one being deployed in the second half of the twentieth century.[66]

There is an excess in these developments, somewhat like the imagined camouflage of the insects of the surrealist Roger Caillois, whose mimetic capacities are so potent, their ability to look like the environment so perfect, that others of their own species cannibalize them.[67] In this case the resemblance of the new development to modernist fantasies and the most banal grids of real estate developers is almost perfect—except the function of monetizing space fails. The form, antiquated and enormous, has taken the reason of the market too perfectly. The developer has maximized the surface area ratio of saleable development, but at the cost of actually surviving.

If it is one of the most commonly held faiths in media studies that the separation between form and content that serves as the infrastructure for mathematical theories of communication, and by extension digital media, is an engine for disembodiment and reductivism,[68] Songdo provokes a curious alternative understanding. The splits between the forms and contents at many levels—from the constraints in materials, capital, and engineering to the embrace of certain historical ideal types for urban planning—create spaces for action and imagination. On one hand there is something terrible about the connection between the infrastructure and the interface; on the other hand this split also allows different forms of materialization to potentially occur. If planners and designers had different concepts of urban form, and perhaps different genealogies of design, the city could look different, for example.

I might ask what imaginaries of planning and urban development have so affected the developers that they have failed, even at the cost of their own profit, to envision a different form of space and engagement with the environment? Are there other memories of information and space that might have been deployed here? Or has the quantification and monetization of space and sense reached so formal a point, the coupling between use and engineering so close, that alternative outcomes are impossible? The present is, of course, not known. History opens us to contemplating these spaces as assemblages of data layers and potentials, densities and probabilities that are never linear, causal, or inevitable.

Temporality is at stake. Songdo tells a story about the forces of history and powers of traces. This space is about the control of time and about breaking this control. In this space, older histories of architecture, economy, and modern vocabularies of planning have been both reified and destabilized. These folds in time incarnated in the landscape pressure the direct links between the past and future—and perhaps more pressing, which pasts are the grounds on which we speculate in the present? Is this the final fulfillment of avant-garde fantasy or corporate imaginaries? Were the earlier efforts at envisioning the global or sentient city capable of different visions than the contemporary instantiations? The links between counterculture and neoliberalism have now been well documented,[69] but the inevitability of this development also needs to be contested. Were there moments when technology and potentiality were more loosely coupled, when form was further from future? Would a knowledge of different efforts to imagine urban space in terms of computation, communication, and technology impact the visions of designers in the present? I certainly hope so. These are some of the questions that Songdo with its multiple temporalities—of capital, urban planning, technology, and politics—poses. I intend to demonstrate that there are many histories of our contemporary media-saturated environments, even within these most legible and obvious forms, even within spaces and objects we think we know. I excavate these genealogies and reveal these absurd, conflicting, and nondeterministic options for envisioning the future of how we sense and live in data-filled environments.

I have called attention to our present because it demonstrates so potently how our imaginings of the past get activated in envisioning the types of futures we would like to build. These sites of seeming extreme speed are also sites of accumulation and density. Their own futures are not known. While it is definitely possible that these infrastructures in our present are now making obsolete the many principles of design and communication that I will lay out in this book, it is also true that we do not know, and only the past can even begin to allow us to reflexively contemplate the present at a rate, and a scale, different from that being encouraged by the developers.[70]

In contemplating the historical treatment of vision and reason, the dominant methodology of this book is, therefore, to mime the practices of design and cybernetics to intervene and engage with these seemingly scale-free forms of calculation that underpin contemporary digital infrastructures. In our contemporary environments we tend to assume seamless mobility in moving from local to global, through interfaces like Google Earth and flexible mobile territories like Songdo. This mobility is, in fact, a critical element of our modes of perception in the present, a topological movement that offers the experience of an integrated global media system—a particular constellation of communication theories, data, design, and navigation.[71] However, while scale is often discussed, the logic of scale is very particular and historically situated. Scale in geography and planning often comes through two main approaches. On one hand, as the architect and theorist El Hadi Jazairy argues, scale is an "ontological fact" that "organizes matter in a Russian doll structure from infinitely small to infinitely large." On the other hand scale is also a method, a form of measurement that serves to manage data and reach conclusions within a defined space and time. Both of these models presume that scale is an ontological reality, and a stable entity to be used across locations. Events happen within the frame, and commensurabilities occur between different scales. Counter to these two approaches, Jazairy suggests a definition of scale "as the unfolding of events that produces a certain scale," which is to say an unfolding that creates conditions of possibilities. Scale is plastic because it is not stable, it is a matter of ongoing relations between technologies, objects, agents, subjects, and territories.[72] Scale becomes about relationships between surfaces, topological strata that are not automatically commensurate.

Another way to understand scale is from the perspective of cybernetics and the communication sciences and game theories that are the objects and subjects of this book. In cybernetics the fundamental epistemological quandary is how to relate the micro actions and macro systems. For the cybernetician Norbert Wiener, for example, the dominant epistemology was one of statistical mechanics, and the epistemological problem involved the incommensurability of translating between the actions of micro level phenomena and the behavior of systems. "Feedback" and "control" are the terms assigned to the practices negotiating these differences between "*actual* performance rather than its *expected* performance." The two are not deterministically linked; "in short, we are directed in time, and our relation to the future is different from our relation to the past. All our questions are conditioned by this asymmetry, and all our answers to these questions are equally conditioned by it." Wiener culminates his introduction by arguing that we now live in "Bergsonian time."[73] If in the nineteenth century concerns about time's discreteness and determinism continued to preoccupy scientists and philosophers, for cyberneticians this no longer held.[74] This incongruity between states and between times now sees itself encoded into the very infrastructural logics of cities like Songdo, economies based on data mining, and in financial and other speculative instruments that literally profit from and mechanize this asymmetry.

My intent here is to consider history as a matter of densities and probabilities rather than deterministic relations. This history operates like the logics of our contemporary data spaces between storage, memory, and interface. The book vacillates between demonstrating synchronic ideas of aesthetics and cognition at the time and diachronically exploring how mid-twentieth-century ideas of vision, knowledge, and recording were haunted and troubled in untimely ways by older nineteenth-century concepts borrowed from psychoanalysis, philosophy, mathematics, and physics. If there is a certain repetitive feature to this exercise, a performative stuttering that forces arguments to be returned to only to become new cyborg entities, then it deserves comment. The very nature of the phenomena of systems that use their past to predict the future in eternal loops mitigates against a linear or causal history. I have stayed true to my objects of study, and the book is organized thematically, not on a time line. It is also genealogical: the final chapter is an accumulation of those before it; an accumulation of densities. This feedback and looping is mimetic, and serves as a method to excavate the reformulation and reorganization of the senses through new infrastructures of knowledge and aesthetics that emerged through the merger of communication sciences, cognitive and psychological sciences, cybernetics, and design in the postwar period.

There is much at stake in the organization of such histories and how we

FIG. I.4__ "Central Park," New Songdo Smart City, Incheon Free Trade Zone, South Korea. Image: author, July 4, 2012.

wish to construct the answers to these questions I pose about the forms of vision and knowledge that now underpin our contemporary belief in data, visualization, and bandwidth as the very architectures for life. As the media theorist Jussi Parikka argues,

> what do we actually talk about when we address animals, insects, and media technologies? Do we think of them as predefined, discrete forms of reality. . . . Could [we not] approach things as intensive molecular flows, in which, for example, the notion of "media" was only the end result of connections, articulations of flows, affects, speeds, densities, discourses, and practices (namely, assemblages)? Could we see media as a contracting of sensations into a certain field of consistency—whether called an environment or a media ecology?[75]

If we were to consider media less stable, ontologically definable, or possessing particular necessary and defined attributes, would this contribute to rethinking our imaginations of technology? I focus on this "contracting of sensations into fields of consistency" that are the archival substrate of such technologies as entire sentient cities. This "contraction" is also an assembling of densities and forms; these are sites of accumulation in the way Bruno Latour speaks of producing "immutable" mobiles and accumulating agents into facts in actor networks, or the way Foucault speaks of visibility—those spaces where representation, practice, technology accumulate—to show things in the world, whether subjects or objects to power.[76] In unpacking these assemblages the present can become an unknown territory of accumulated densities rather than a natural and inevitable future.

This book is, therefore, a speculative endeavor, in all the ways "speculate" is defined—as both a matter of reflection and mediation, a matter of conjecture, and a matter of risk with possible gains and losses.[77] And, as in the case of the many speculators and corporations trying to bank in on these developments only to be frustrated in their financial ambitions, it is useful to be reminded that the present is often haunted by the past and the future is often cloudy and never predictable . . . is visualizable but not necessarily visible.

1_ARCHIVING

Temporality, Storage, and Interactivity in Cybernetics

 EW FIGURES ARE MORE PROMINENT IN THE HISTORIES OF digital media then the MIT-based mathematician Norbert Wiener. From 1950 until the late 1970s, his work was prominently featured across multiple fields from architecture to sociobiology. For almost thirty years after World War II, before the term "digital" gained prominence, cybernetics, a word he coined, was the language used to describe a transformation in life and a new technical condition related to, but not reducible to, digital computers. In the late 1990s with the advent of the Internet, his name returned in the effort to historically situate the rise of digital networks and the interactive interface.

Wiener almost appeared to anticipate his future popularity. His archive at MIT is a fascinating exemplar of a life turned into data. Carefully curated, every letter mimeographed and saved, it is as though Wiener was already preparing his life for transmission, assuming a seamless translation between personal experience and historical analysis. Fastidiously cataloguing his many failures in natural history and the sciences of empiricism and experiment, he turned to reformulating these experiences in service of another form of knowledge. Rather than speak of the value of personal experience or the specificity of his character, he sought to make that element of his innermost psychology—his character—the substrate for legitimating computation.

I wish to take up this turn away from an "external" world and the devolution inward, in this case to the very self, as a starting point to consider the relationship between the archive and the interface in digital systems. What might we make of this move from a concern with recording an external, perhaps "natural," world in its entirety to an obsession with processing the already recorded traces of memory? How do we wish to frame this shift to forms of representation whose reference is reflexive rather than indexical? Wiener was not naïvely recounting his failures in finding adventure, his inability to excel in the life sciences; rather he was articulating an aspiration for forms of technology—both of thought and machine, or perhaps of thought as a machine—that had not yet come into being when he spoke. In his work, and that of his many compatriots in the arts and sciences of the time, we hear similar statements that voiced a not-yet-realized aspiration to transform a world of ontology, description, and materiality to one of communication, prediction, and virtuality.[1] A world that, perhaps, speaks to our contemporary fantasies of a data-filled space where every screen is an interface, every diagram a process.

But if Wiener attempted to propagate the "new," it only came into being through the memory traces of the old. It was by way of Freud, the exemplar of a previous century's sciences, that Wiener implied the impossibility of describing a world in its totality, of ever rendering "reality" legible. Instead, he argued, we are faced with an "incomplete determinism," an operative lack that cannot enter description but can produce something else—a self-referential and probabilistic form of thought:

> one interesting change that has taken place is that in a probabilistic world we no longer deal with quantities and statements which concern a specific, real universe as a whole but ask instead questions which may find their answers in a large number of similar universes. . . . This recognition of an element of incomplete determinism, almost an irrationality in the world, is in a certain way parallel to Freud's admission of a deep irrational component in human conduct and thought.[2]

This form of probabilistic thought that emerged at the turn of the last century would now, in Wiener's work and that of his compatriots in the information sciences, be connected with theories of messages. Wiener was comfortable with acceding that the universe in its plurality might never be known. This accession, however, was only made to allow for the possibility that within far more localized situations, the future—chance—might yet be contained by way of technology.

But Wiener's invocation of Freud also complicated his own vision for technology and science. His statements posed the possibility that the contemporary systems he hoped to bring into being were not absolutely amnesic to their history. His statements would be, and still are, haunted by the residual problems of recording, translating, and transmitting information and associated concerns with indexicality, signification, and representation. Unconsciously, perhaps, even Wiener acceded to the possibility that not all forms of information could be similarly recorded and transmitted without loss, transformation, or change. It is precisely at this site, where the traces of older histories mark the desire for the production of the new, that I will excavate in this chapter.

Wiener's texts, and the work of his compatriots in cybernetics and the neurosciences, serve as useful vehicles, therefore, to begin investigating this historic attachment and displacement of older technical questions of documentation, inscription, and perception into terms of information and communication. The relationship, explicitly detailed in the work of many early cyberneticians, between the record, the diagram, and communication forms a bridge between our contemporary discourses about archiving, screens, and interactivity and historical concerns with memory, temporality, and representation. At this pivotal moment, demarcated by a catastrophic world war, these sciences were part of producing an aspiration for a new world constituted of information; but not without producing a novel set of conflicts, desires, and problems. I turn, then, to outlining what the conflicted relations between the archive and the screen might still have to say to our desire for "interaction" and communication with and through our machines.

Cybernetics: Communication and Control

The very definition of cybernetics already assumes a complex relationship to temporality and history—bridging the past with an obsessive interest in prediction, the future, and the virtual. As the etymology of the word suggests, cybernetics is a science of control or prediction of future action. In further adjoining control with communication, it is an endeavor that hopes to tame these futures events through the sending of messages.[3]

These rather abstract ideas of communication as the source of control consolidated themselves within the milieu of military research and development in antiaircraft defense systems during World War II. While scientific research has long been part of war, World War II is now widely heralded as marking a critical turning point in the organization of science both in scale (billions rather than millions allocated) and in the wholesale recruitment of both industry and academic researchers. To offer some substantive examples, major beneficiaries of the effort included MIT (an institution already active in recruiting military spending in the interwar period), which was awarded some $117 million in R & D contracts (approximately 1.6 billion in contemporary dollars) and remained among the top sixty defense contractors after the war); Caltech, with $83 million; and Harvard and Columbia, with about $30 million each; compared to $17 million for Western Electric (AT&T), $8 million for GE, and less than $6 million each for RCA, DuPont, and Westinghouse. Many other industrial groups also received assistance, including Bell Labs, IBM, and Norton. In sum total the U.S. Office of Scientific Research and Development spent $450 million (approximately $5.5 billion in inflation-adjusted contemporary dollars, although probably far more if other indexes of purchasing power or dollar value are used).[4]

Part of this reorganization toward what is now labeled "big science" was a transformation in disciplinary boundaries. Wiener worked in one corner of what was quite literally a very large complex. The Radiation Laboratory (commonly called RAD lab) at MIT was among the biggest of the American research installations. The space was costly at $1 million, and contained cutting-edge equipment in radar signaling, calculation, and communication technology. Housing at its pinnacle over four thousand different researchers from around the world, the site was host to engineers, physicists, mathematicians, physiologists, doctors, and individuals from a range of other fields, all investigating, in one way or another, signal processing and communication.

Built barracks style, out of wood, a haphazard block of extensions and convoluted hallways, it was, to quote one of its inhabitants, the future neuroscience and perception researcher Jerome Lettvin, "the womb of ideas. It is sort of what you might call the vagina of the institute. It doesn't smell very good, it is kind of messy, but by god it is procreative, and it doesn't make only replicas of itself, as other buildings do. It is sort of all-purpose."[5] As Stewart Brand, of Whole Earth Catalogue and Whole Earth 'Lectric Link (WELL) fame, writes, RAD lab was a sort of emergent architecture, a haphazard and growing set of wooden boxes and storage crates that simply grew and adapted, formlessly, to the different interactions of its inhabitants. Inhabitants reminisced about the way engagements would simply occur by accident and haphazardly in the hallways between the shed-like segments of the lab.[6]

But for all the purported camaraderie and creativity attributed nostalgically to the lab, the milieu was a high-stress space where the topic at hand was violence at a distance, particularly that coming from aerial warfare with its velocity and territorial reorganization. Within this context and under the imperative of rapid defense, Wiener, working with neurophysiologists and doctors and influenced by early work on computational machines (the differential analyzer), argued that human behavior could be mathematically modeled and predicted, particularly under stress; thereby articulating a new belief that both machines and humans could speak the same language of mathematics.

By reformulating the problem of shooting down planes in the terms of communication—between an airplane pilot and the antiaircraft gun—Wiener and his compatriots hoped to devise better defense systems. Wiener, working with the MIT-trained electrical engineer Julian Bigelow and the physiologist Arturo Rosenblueth, under the guidance of the director of the Applied Mathematics Panel at the Office of Scientific Research and Development, Warren Weaver,[7] decided to treat the pilot of a plane as a machine. These researchers postulated that under stress airplane pilots would act repetitively, and therefore have algorithmic behaviors amenable to mathematical modeling and analysis. To this end, they designed experiments to simulate flying through the turbulence, noise, and antiaircraft fire of a bombing run. Wiener and his colleagues set up a fake cockpit-type scenario, where a "pilot" was asked to guide a light beam through a particular set of invented obstacles. They then gathered data about the movements the individual made. They discovered that while different "pilots" appeared to have widely different patterns, for any one individual, there was a striking amount of redundancy and repetition in the movement. People do, it would appear, act quite mechanically under duress.[8]

Behavior, Purpose, Teleology

For a project so literally obsessed with problems of response time and the description of enemies, it is of little surprise that ontology (or the identification and definition of the enemy), representation, and temporality should be explicit and central concerns. What is somewhat less intuitive is how these questions of immediacy and identification became ones of storage and communication.

There was a series of moves by which this new world of perfect communication between networked entities emerged. The early efforts to rethink technological defense mandated rethinking what an enemy was, rethinking what communication was, and isolating the communicative relations between enemies in a manner that allowed them to be modeled and abstracted into computational methods. This wartime research engine therefore produced what the historian of science Peter Galison has defined as a new "ontology of the enemy," not the alien and animal opponent, not the distantiated space on the map of an airtime raid, but the "cold-blooded, machine-like opponent."[9] This move eliminated the enemy as visibly different and produced an imagined closed world of networked communications between informatic entities. This emergent assemblage would be codified later under the Cold War ideal of "c3i: command, control, communication, and information."[10] In this new imaginary, response time was the critical feature driving the system, and this produced a new understanding of teleology that was not about progress, linearity, or conscious humanist effort, but rather a new mode of technical thought. In fact, it is arguable that Galison's use of the term "ontology" is misplaced. Cyberneticians concerned themselves with process, not essence. Within this rubric, visually representing the enemy was not the core concern, but rather the training of the machine or person to communicate and anticipate further signals.[11]

The first major concept was to introduce the idea of black boxes. For the antiaircraft project, engineers and mathematicians, using early computing devices, began to view the gun and the plane behaviorally—which is to say that while the internal organization was opaque, unseen, and potentially different, the behavior or action was intelligible and predictable. The world, therefore, became one of black-boxed entities whose behavior or signals were intelligible to each other, but whose internal function or structure was opaque, and not of interest. We can say that this view of the world concentrated on process, not structure or difference. Julian Bigelow, who worked with Wiener to produce and articulate many of these ideas, put it succinctly when describing his fellow researcher: "Wiener thought as a physicist really. He thought in terms of process. He thought in terms of some kinds of physical models or some kinds of intuitive models."[12] These personal descriptions fit the larger project, as well, where precision was not valued, but rather the development of technologies that could be extracted, moved between locations, and do work. Wiener would argue, in the terms of gestalt, that the questions of perception, for example, were simply the abstraction of forms from the world, and that those forms were not unique to a particular sensory operation, "We have thus designated several actual or possible stages of the diagramatization of our visual impressions. We center our images around the focus of attention and reduce them more or less to outlines. We have now to compute them with one another, or at any rate with a standard impression stored in memory. . . . This is a general principle, not confined in its application to any particular sense and doubtless of much importance in the comparison of our more complicated experiences."[13] Behind the black box is the concept of a world that can be remade into exportable processes that point away from their initial sites of inception.

The second idea introduced by Wiener and his colleagues in their wartime research reworked the idea of communication and interaction into terms of statistical prediction and feedback. The capacity to predict a black box's action came to be known as feedback. Wiener would define feedback as "the property of being able to adjust future conduct by past performance."[14] It was argued that all behavior culminating in a goal was purposeful, teleological, and the result of negative feedback, or the need to correct the input signal. For example, a missile seeking a target swings afar and misses then adjusts its flight path in response to the signal given from its target. Negative feedback assumes that both entities are in relationship to each other, are both communicating, and are both changing their behaviors in relation to the other.

To appreciate what such an understanding of feedback implies mandates that we ask what, precisely, is being redefined in the ideas of "purpose" and "teleology." Purpose was now understood as any active behavior that was not random but rather "meant to denote that the act or behavior may be interpreted as directed to the attainment of a goal—i.e., to a final condition in which the behaving object reaches a definite correlation in time or in space with respect to another object or event."[15] This was quite explicitly delineated from, say, a watch, which has a given task—keeping the time—but no set or defined endpoint to this function. Teleology was a concept simply adjoined to any purposeful behavior of entities engaged in negative feedback (or dually responsive) interactions.

The result of this logic was that teleology was considered no longer "causal," and purpose no longer conscious, in that there is no given chain of events that consciously leads to a predetermined goal. In the seminal 1943 article consolidating these ideas—"Behavior, Purpose, and Teleology"—Wiener, Rosenblueth, and Bigelow argued that "the concept of teleology shares only one thing with the concept of causality: a time axis. But causality implies a one-way, relatively irreversible functional relationship, whereas teleology is concerned with behavior, not with functional relationships."[16] Which is to say that while both types of interactions occur within time, and are irreversible, one is simply assessed by any interaction that produces an effect, and the other—the functional and causal—is deterministic, it has only one possible outcome. In a functional interaction, A must cause B and only B to occur, whereas in this new "purposeful" and behaviorist approach there is a cause and effect, but the relationships between cause and effect are not predetermined; rather, they are a choice, a likelihood, the result of a series of possible interactions.[17] Change, which constitutes teleology in this system, occurs through a series of repeated and predictive actions. This is a space of encounter between two like objects communicating and responding to each other, forming a closed world of predictive encounters.

We may say, then, that three critical assumptions, or more appropriately aspirations—since very little that was theorized was actually operationally built or successfully implemented at that moment—emerged and crystallized in technical form in this wartime milieu. The first was that all entities—human, machine, animal—were intelligible to each other. The second was that all communication was a statistical and ultimately predictive endeavor. Third, feedback systems comprised a set of knowable and model-able entities in constant communication with each other, driving the system to "evolve" and even, in Wiener's terms, "learn." These three points are not important on their own, but when seen together are steps toward transforming the terrains of perception and representation.[18]

Prediction, Probability, and Algorithm

The interest of cyberneticians thus shifted in this research from describing in detail the mechanisms of actions to only considering the actions. Therefore, they refocused on the ability to calculate the probability that one set of interactions (the missile hitting the plane) will occur, over other sets perhaps less likely but possible. Rather than describe the world as it is, their interest was to predict what it would become, and to do it in terms of homogeneity instead of difference. This is a worldview composed of functionally similar entities—black boxes—described only by their algorithmic actions in constant conversation with each other producing a range of probabilistic scenarios.[19]

This obsession with communication as a question of potentiality and choice became the guiding framework for thinking about digital communication. Claude Shannon and Warren Weaver, Wiener's compatriots, working at Bell Labs, formalized this ideal of communication in the "mathematical theory of communication."[20] According to this pivotal work, which would influence information and communications theory for the next few decades, all communication was now digital; to argue for digitality was to argue for communication as a choice between discrete units. In the realm of digital communication, information does not denote meaning, only the choice between possibilities within a structured situation—"structure" denoting, in this case, a formally defined system where the range of possibilities for communication is designated (in this case by binary-encoded signs). Weaver summarized this emergent idea as referring "not so much to what you do say, as to what you could say.... The concept of information applies not to the individual message (as the concept of meaning would), but rather to the situation as a whole."[21] Information theory as emerging from cybernetics thus aspires to the future tense while existing in a heterogeneous temporal state where the control of this future comes through the abstraction of processes from historical data to produce preprogrammed, self-contained conditions.

This "situation" found its analogue across the social field. In the development of programmed computers, it was reflected in von Neumann digital computing architecture as a decision between on/off or 0/1 and "unit amount" as "bits." Computers came to be viewed as systems where the accumulation and rearrangement of basic decisions—"0/1"—would produce the conditions of possibility for a wide range of potential actions.[22] But it was not only in the realm of digital machines that information theory took hold. This aspiration for the perfect and unadulterated transmission of information as control of the future within a self-referential and contained space impacted everything from postwar architectural movements to genomics to politics. Cybernetics as a "science of form" would in many minds replace materialism, and relocate an earlier age of matter and diachronic descent, an age defined by Darwin, Hegel, and Marxian history, to "an age of form and the synchronic structure of information."[23]

The most literal exemplar of this emergent interest in communication and its impact on social science, policy, and economics was the series of Macy Conferences on cybernetics, initially titled "the Conferences on Circular and Causal Behavior," that were held during and after the war in New York City and sponsored by the Macy Foundation—an institution dedicated to healthcare research. Many of the central figures in computing, wartime research, communications and systems theory, anthropology, neuroscience, and psychology were involved. Participants ranged from the anthropologist Margaret Mead to the linguist Roman Jakobson and the mathematician John von Neumann. The historian-physicist Steven J. Heims has argued that these conferences served as an important site in producing a vision, and funding basis, for the postwar social sciences in the United States. These conferences not only produced a resource base for encouraging an emphasis on information and communication across many fields but also, more important, produced and affirmed a methodological emphasis on individuated enclosed systems, replacing other structural, historical and political economic approaches. Heims called this approach "atomism," by which sociological and political questions were reduced to studies of individual self-contained units. Heims argues that this work, produced within the context of McCarthyism and Cold War polarization, was part of a larger set of tactics affirming American political interests and diffusing political or economic critique, a viewpoint compatible with Wiener's own ethical writings in the wake of the nuclear bomb and subsequent proliferation. I return to this theme in chapters 3 and 4.[24]

The very continuation of the terms "cyber" and "cyborg" in our imaginings of digital technology, information networks, and human-machine interaction bears witness to this dissemination of cybernetics and information theory throughout the social field.[25] These words also remind us that in this transformation, the residues of historical questions and techniques reemerge—often with force.

Temporality and Communication

We might, then, seek to historically situate this relationship between older discourses and the present cybernetic one in order to ask what is at stake in such a movement, where we begin with an effort at documenting an external enemy and end with the question of prediction and communication?

The concept that statistics may have something to say to communication engineering did not, of course, appear out of nowhere. This work already rested on a longer history of feedback engineering, a modern concern with statistics, and Wiener's own work in Brownian mechanics in the 1930s. Wiener was already interested in introducing notions of statistical thought and probability to engineering.[26] This was an engineering that sought to be operational through a recognition of the impossibility of full objectivity or exteriority to the system—error and chance became the very platforms from which technology emerged.

Wiener, and his colleagues, thus took a modern concern with the "taming of chance"[27] and the emergence of statistics and attached it to the possibility of prediction and communication in engineering. Wiener frames his own project to apply a statistical form of thought to mechanics as a fundamental reworking of older modern dualisms in the interest of overturning and replacing historical questions with new ones. Wiener, but also his compatriots, such as Warren McCulloch, who pioneered neural nets (and to whom I return in chapter 3),[28] were trained in and responded to longer traditions in philosophy. Wiener specifically invokes the relationship between Bergsonian "vitalist" and Newtonian "mechanical" and deterministic temporalities. Wiener argued in *Cybernetics* that our age of complex automata and feedback systems exists in an active Bergsonian "vitalist" time, which is to say a temporality that is nonreversible and probabilistic. (The past and the future are always interpenetrated through conditions of potentiality.) Wiener further correlates Bergsonian attitudes to temporality with the lifelike, or evolutionary, potential of cybernetic systems.[29] In articulating this understanding of time as irreversible but not necessarily progressive, Wiener is reflecting and advancing a discourse of temporality that had already become dominant in many fields by the turn of the century.[30]

Wiener, however, also sought to signal a break from these previous histories. He had already explained that there are methods outside of meticulous documenting from observation that are meritorious of our attention. He argued that we live in a universe of "process," where "possibility" can be created through the recourse to action.[31] In focusing on complexity, and aspiring to produce new forms of organization, cybernetics sought to displace the question of referentiality and insert that of prediction.

The disavowal of the present in the interest of prediction recalls a previous historical move by which "presence" and the present emerged as formal sites of articulation and concern—the historical moment from which Wiener and his colleagues seek to separate their form of thought and formally introduce an age of complexity or postmodernity.[32] This effort to produce history recalls the comments of the French literary critic Roland Barthes on the emergence of both reality and history in the nineteenth century as a cardinal site in the production of the "modern." For Barthes, this is a modernity produced through recourse to an idealized external nature, what he labeled "reality effects," in literature and other mass mediums. This concept, itself, of a textual and mediated device that produces both new subjects and worlds, bears an intimate relationship to Barthes's own contemporary relationship with theories of communication, semiotics, and linguistics.

To use Barthes's framework, cybernetics disavowed "reality effects," in which "reality" is produced through speech acts that seek to resemble the quality of history or "having been there." The presentation of reality, a feature of nineteenth-century literature, emerged, according to Barthes, at the moment when human experience was increasingly mediated—through new techniques of writing, reading, and recording. Barthes tells us that reality effects are descriptive, and produce a historical time through recourse to "a referential (and not merely discursive) temporality."[33]

In opposition to this referential temporality of language, Barthes places the predictive communicative temporality of the honeybee; a language of noncognition and pure form that, incidentally, a few decades later would align socio-biology with information theory and computation through numerous donors and research agendas in genetics, social science, and cognitive psychology.[34]

> This is an opposition which, anthropologically, has its importance: when, under the influence of von Frisch's experiments, it was assumed that bees had a language, it had to be realized that, while these insects possessed a predictive system of dances (in order to collect their food), nothing in it approached a description. Thus description appears as a kind of characteristic of the so-called higher languages . . . to the apparently paradoxical degree that it is justified by no finality or action or of communication.[35]

This comparison between languages rests on Barthes's own belief that reality, as embodied by the ideal of lived experience, was being destroyed through the emergent industrialization in the nineteenth century. The French critic wrote that the modern obsession with description enacted through interruptions in the diegesis produced a relation of presence, or "what is," that was resistant to meaning; a resistance that supported the ideological belief of a (still existent) reality or experience outside of mediated representation and yet (paradoxically) available for transmission in the text.[36] He is also arguing that the purely predictive and action-oriented "language" of the honeybee cannot denote presence. Such a representational schema cannot speak of experience, for it contains no grammar by which to problematize its abstraction of space and time.

This conceptualization is another way of articulating the notion that the modernist grammars deployed the ideal of experience, paraded now as an illegible "desire," in Barthes's words, to replace the fact of mediation; obscuring that in fact the reader is not there, and neither is the author—they are separated by the technology of inscription, in this case writing. Barthes does not, however, restrict such effects solely to literature; this is an entire discursive network that includes other media technologies such as photography and tourism. Most significant, therefore, Barthes signals to us that the idea of exteriority and interiority, or real and represented, are modern conventions, and that for the nineteenth century the displacement or "destruction" of reality became a site of problematization that produced new techniques of representation.

Wiener, however, signals to us precisely the disavowal of the "problem" of mediation in favor of a new set of questions. By extension of Barthes's argument, we can argue that cybernetics, like Von Frisch's bees, was invested in developing a universal language temporally uninterested in referentiality through description, producing instead a statistical grammar of prediction. Mediation, which has long been the foundation for the idea of "representation," was therefore no longer a site of problematization or obfuscation. Rather it became the site of potential and probability. We are no longer focused on the "meaning" or origin of the signal but rather on its transmission.[37]

I call attention to Barthes, therefore, because he marks both the emergence of a new model of representation in the nineteenth century and the space that this history of representation has in producing our contemporary critique of mediation and mass media, a critique, one could argue, that only emerged in relation to a new set of questions—those of information and virtuality. Cybernetics is thus complicit in producing both the ideal of an older concept of representation and the turn toward mediation as a site as potential—for both critique and technology. Wiener, himself, self-consciously sought to mark the passing of an age, and the emergence of a new one, by taking from what was, to produce what may become. Wiener wrote that "cybernetics is bound to affect the philosophy of science itself, particularly in the fields of scientific method and epistemology, or the theory of knowledge." Wiener points out to us that we are now interested in process, not rigidity, description, or stasis.[38] It is useful to hark back to his initial comments on diagrams and descriptions, where the question of "reality" is not so much gone as displaced, no longer the site of intellectual or technical interest.

Teleology and Action

We might ask, however, what residual relations persist between these modernist representational technologies and computational thought. Immersed in larger technical projects to build weapons, decode enemy tactics and messages, and later, produce multipurpose technologies for the inscription, organization, retrieval, and communication of data, cybernetics contended explicitly and implicitly with older questions involving mechanical reproduction and mediated communication in large networks. Researchers of cybernetics and information theory called on previous modern heritages and discourses involving temporality, inscription, and representation in photography, the cinema, psychoanalysis, and psychophysiology. Wiener, as an example, specifically invoked Henri Bergson, Sigmund Freud, and other philosophers and theorists of both time and cinema. Cybernetic ideas emerged from and operated through a reframing and reattachment of these older concepts and practices in modern thought to produce the conditions of possibility for an interest in multimedia computational machines.

Bergson already heralded an emergent form of philosophy that anticipated cybernetic ambitions when he announced in *Matter and Memory* that "I call matter the aggregate of images, and perception of matter these same images referred to the eventual action of one particular image, my body,"[39] collapsing clear demarcations between the psychological recollection or image and the external "reality" of movement. Bergson's overriding ambitions were the production of process philosophy, with its heterogeneous temporality, emphasizing the elements of becoming, change, and novelty in experienced reality. Perception, in this view, is durational, simultaneously encompassing both an emergent past and a future. Philosophically, Bergson attempted to produce forms of thought that did not remain static and always combined the memory of an event with its future, producing possibility out of the synaptic, or embodied space, that merges historical temporality and sensation with its processing and response. The separation between the body and the representation cease to exist in the interest of movement. It was in the interest of recuperating this affective kernel of thought in Bergson that Wiener deliberately turned to Bergson in producing his own cybernetic approach.

Bergson, of course, did not answer to the same questions in science or technology as Wiener; but they both struggled with problems of mediation, recall, and action, and each marks the emergence of new historical forms of assemblages and statements. Bergson's philosophy sought to respond to an emergent psychological field in which reality and consciousness were opposed; where knowledge of the world was always subjective and mediated, essentially bifurcated from "nature"; and in which temporality had become probabilistic and synchronous—inexorably linking the past with future potentials through an inaccessible and mediated present. Bergson, and many of his compatriots, most significantly Freud, therefore answered specifically to problems of mediation in a transforming social field.

For both Bergson and Freud, the rise of new mass mediums with their technologies of recording and inscription, and the increasing stimulation of the sensorium, were sites of vexing problematization and possibility. Freud himself, in producing his conceptions of memory, wrote that modernity was producing many "forms of auxiliary apparatus . . . invented for the improvement or intensification of our sensory functions." Among these apparatus he mentions the camera and notes that of the perceptive functions, it is memory to which these apparatus are most "imperfect."[40] It was this imperfection that, one can say, drew Freud toward the problems of inscription and storage, and to the management of stimuli to which the unconscious and consciousness respond. The film theorist Mary Ann Doane has pointed out the historical relationship between Freud's problematization of storage, which is in her words a question of "representation and its failure," and the emergence of a larger mediated landscape that produced new questions about temporality and the storage of time as it threatened older symbolic systems.[41] Wiener also would return to psychoanalysis as a template for thinking about the relations between the record, storage, and communication, as we shall see.

One central modern concern, therefore, was how one might remember, recall, and distinguish moments of experiential meaning from an endless flow of stimuli. Bergson's refutation of the past as an unconscious reservoir of stored stimuli, a direct response to Freud, in whose view perception precedes recollection, was in the ontological interests of, in Bergson's words, "becoming." This rescripting of divisions between the psyche and action was arguably a move that could make thought itself an ontological object and actor. Bergson's effort to reconcile consciousness and reality might appear antithetical to psychoanalytic concerns, but both intellectual projects mark an emergent historical transformation in thought whereby "reality" and temporality are inextricably linked to problematize representation and experience as vexing, but productive, sites of inquiry.[42]

More specifically, both Bergson and Freud provided processes for operationalizing and activating thought that would be explicitly seized on, and unmoored, by cyberneticians. While often conflicting, this ongoing discursive engagement between Freud and Bergson also revealed the shared investments of these two modern projects. Both provided tools for creating systems that were self-referential and in which the temporal frames of recording the past and producing the future became compressed—in psychoanalysis in the production of a psyche through analysis, and in Bergson's work through process philosophy. Despite a disagreement over time and the unconscious, therefore, both projects can arguably be seen to anticipate the emergence of new technical systems, where the process of analysis, the operationalizing of memory, and the emphasis on affective possibilities became the core tenets. One could contend, taking Freud's and Bergson's concerns with stimulus management, recording, memory, and recall to their extreme, that they both shared in making the psyche both an abstraction and a material actor, often counter to their own aspirations—whether the disciplinary ideals of Freud or the metaphysical dreams of Bergson.[43] As programs, therefore, both psychoanalysis and Bergsonism contributed to the very possibility of technicizing perception, and even thought, as later efforts in artificial intelligence and cognitive psychology would demonstrate. Wiener already anticipated this possibility by invoking Bergsonian "time" for his automata.

It was in the interest of ontology that Bergson was, however, specific in his condemnation both of Freud and of the emergent cinematic apparatus for recording and externalizing sensation, when he argued that temporal moments were rendered equivalent, and therefore meaningless or insignificant, through a "cinematographical mechanism" that constituted "ordinary knowledge."[44] At the legendary (though inconsistent) sixteen frames a second, the early cinema made all moments equivalent, with no moment that could enter the realm of "experience," since every moment was the same, producing no grounds for differentiating temporal states.[45] For the cinematic apparatus, difference ceased to exist. Out of still, and technically equivalent, frames erupts movement, but it is only a false illusion; the lapses, the cuts, the overlaps of event and time in between the frames is obscured by the projection apparatus. The cinema gave a sense of infinite recording capacity, but its true operation was the spatialization and leveling of time.

Bergson, does not, however, reduce this critique solely to the apparatus of the cinema; he calls attention to the fact that this form of spatialized and dedifferentiated temporality is a larger mechanism of thought, ingrained within philosophical and psychological determinism, the physical sciences, and the organization of industrial space and production. Bergson goes so far as to imply that such a mechanism is internalized, projecting within the psyche a false belief in perception preceding thought. The cinema thus parades as movement coming into being, when it is only stasis. Static, because the cinema, as apparatus, produces and supports the illusion, which separates an external and recordable world from the production of the image. The desire to represent temporality, one might say, presents for Bergson a larger ethical dilemma entailing the possibility of thought coming into being at all. But more specifically, modern problems with temporality and memory also produced both new forms of subjectivity and technologies of thought that would continue to inform later media systems.

Bergson's earlier work in *Matter and Memory* proposes one such possibility, seized on by Wiener, through a more ambiguous relationship to mediation in the reunification of the image and movement. Bergson produced a philosophy in which thought and movement were merged, the movement-image, and put a full emphasis on the operational or affective capacity of thought, over and beyond the space between thought and action. For Bergson, perception is theoretically lodged within the real, within the referent, and is external to the subject. This perception is only theoretical, however, because it is inaccessible; "in concrete perception, memory intervenes, and the subjectivity of sensible qualities is due precisely to the fact that our consciousness, which begins by being only memory, prolongs a plurality of moments into each other, contracting them into a single intuition."[46] Certain forms of "memory" are therefore continually active, producing the future. In the process of "recalling," we induce action.

The fundamental "error" in psychological and physiological conceptions of temporality, according to this reading of Bergson, is the idea that there is a clear separation between sensory perception and representation (or thought); that we first perceive an external world, store that perception as recollection, and then retrieve it, in a set order with clear differentiation between on the one hand the material and embodied actions and on the other the cognitive thought processes. Bergson, it can be assumed, meant nothing of the sort. Rather, as Deleuze explains, "we do not move from the present to the past, from perception to recollection, but from the past to the present, from recollection to perception." One might say that the mind does not simply respond to and synthesize a series of abstractions gathered from the external senses, which precede the thought, or the image; rather, both action and thought are coconstituted. The experience of perception for the human being is always, therefore, produced in the lag between the stimulus and its recollection; it is inextricability linked to both recording and recalling. Gilles Deleuze summarizes Bergson's argument as defined "less by succession than by coexistence."[47] The temporality of the organism is not diachronic, the simple unidirectional and stable procession of cause-event, but is, rather in keeping with Wiener's new understanding of automata and cybernetic time, probabilistic and conditional. Deleuze explicitly identifies the historical and methodological possibility of Bergson's thought when he argues that Bergsonism speaks to our contemporary, post-information theory biology and to our understanding of post-electronic media and cinema. For Deleuze, rethinking the cinema involves reviewing it as a closed system that produces a new reality, a possibility that he recuperates from *Matter and Memory* and that bears an uncanny resemblance to the notions of cybernetic productivity.[48] Taken to its logical extreme, we may, as Wiener (and later Deleuze) did, argue that Bergson's formulations of memory, duration, and perception hint at a world with no exterior—an internally and self-realized one.

Nervous nets and thoughts are conflated. What was once abstract— "thought"—is now material in its capacity to act and its literal performative instantiation as a net that works like a machine. This possibility appears literally manifested in cognitive psychology and neural nets. The psychiatrist Warren McCulloch's innovations in neural nets, developed with the inspiration of Wiener's article "Behavior, Purpose, Teleology," reflects this comprehension of an organism as system, or series of processes—now, however (and in this we mark historical change), assessed through measurable outputs. I will return to this model more extensively in chapter 3. For now, it is important to note that McCulloch, and his mathematician colleague Pitts, demonstrated an approach to the mind, answering Kant's idea that "the schema of the triangle can exist nowhere but in thought" with the idea that "a schema for a universal" could exist in the brain, in specific actions of neural circuitry, and not as a priori abstract thoughts.

For cognitive psychology and neuroscience, viewing complex actions out of accumulated systematic behavior of networked neurons puts the emphasis on the process, or algorithmic pattern that facilitated processing, and viewed this processing as productive in itself. Processing was a thing in itself, not only an intermediate stage toward a more complete and final state or a representation of some external thought or reality.[49] Such ideas could be forwarded, however, by making the site of processing—the thought—indifferentiable from the animal-machine body. The image's production is the movement, the operation so to speak, of the network. We can, perhaps as Bergson already signaled, no longer truly speak of representation in its static and referential form.[50] Modern separations between ontology and epistemology, or reality and mediation, may no longer act as organizing principles.

Bergson's critique of the mediated and "spatialized" landscape of his world may have, therefore, both described and produced the future of the very technologies he condemned. Deleuze, who speaks as a contemporary of Barthes, and in some sense of Wiener, already argued that Bergson's very notion of the movement-image anticipates the future of perception, and of media-thought. For Deleuze the apparatus of cinema, taken as an entirety, and including the spectator, was most capable not of abstraction in the interest of stable representation but of "emancipating" movement in a manner congruent with Bergson's thought project. Deleuze goes on to explain that this is not merely an abstraction, because this new mode of "being," or thought, does not claim reference to an external "real." He argues that Bergson, even in his critique, was close to the cinema—an intimacy emerging from within his philosophy of active systems without exteriority or interiority. Deleuze explains Bergson's formulation: "my body is an image, hence a set of actions and reactions. My eye, my brain, are images, parts of my body. How could my brain contain images since it is one image among others? External images act on me, transmit movement to me, and I return movement: how could images be in my consciousness since I am myself image, that is, movement?"[51] From within this formulation can emerge a self-contained universe, which is constantly producing new forms of existence. "This is not mechanism," Deleuze argues; "it is machinism. The material universe, the plane of immanence, is the machine assemblage of movement-images. Here Bergson is startlingly ahead of his time: it is the universe itself, a metacinema."[52] The universe has become a space of productive enclosure.

In thinking about the cinema in such a manner, Deleuze suggests that the cinema shares an integral impulse with the efforts of cyberneticians, and later computer science, for abstraction through the formalization of process through programming; where abstractions of processes—whether images, algorithms, interfaces—always produce actions, and refer not to exterior spaces but to the production of new worlds. In cybernetic understandings, descriptions of processes always become sites for the further production of new techniques of production rather than static descriptions; materiality, action, and concept are inseparable. Deleuze, as though influenced by cognitive psychology and cybernetics, can find this possibility in Bergson because he is interested not in the external mechanism of projection (the projector) but in the spectator, or more specifically the internal relations between the image and thought. He produces a self-referential world that is relational, and where "abstraction" does not provide a static representation of moment but constantly produces new processes and communicative exchanges. Deleuze already views the cinema as interface.

It is worth noting, however, that for Deleuze this is a strategic "misreading" of Bergson. Bergson's concepts can also be read, as the Futurists, for example, in his own time, did, or as contemporary media theorists such as Mark Hansen do, as a specific invocation of the live body as a necessary framing device for image flows. Hansen argues that Bergson "empowers" the body by placing it at the center of the perceptual act: "Bergson's theorization of perception as an act of subtraction installs the affective body smack in the center of the general deduction of perception. . . . What is more, Bergson places his emphasis on the body as a source of action; it is the action of the body that subtracts the relevant image from the universal flux of images."[53] Within this matrix, in an age of digital information, the body is bequeathed new powers to be the only entity to lend structure to information flows.

Locating Time

There is much at stake between these two understandings. Deleuze, I argue, is concerned with the virtual, or that which is to come, and therefore does not seek to locate either matter or memory in set entities like the biological body or a metaphysical concept of the "present" or an external, "natural," and eternal temporality. I argue that theorists like Hansen overvalue, determine, and essentialize the body in a manner that replicates Bergson's own sentiments.

Counter to this reification of the present and the body, Deleuze is strategically rereading Bergson to excavate an ethical potential within the emerging electronic, cybernetic, and computational mediums of his time that are transforming cinema, and philosophy. For this reason, I read Deleuze as putting temporality within the system. It emanates from within the time-image, from the irrational cuts of the cinema. Time is not outside, and neither is the body, in Deleuze's articulation of the image throughout *Cinema 2: The Time-Image*. Systems emerge out of their communicative interactions, not from forces external to them.[54]

Wiener himself marks this historical mutation to the Bergsonian heritage. He demonstrates a loss of concern, or perhaps deferral, of the problem of presence or representing time, nature, or an external "reality." He argues, as we may recall, that "the whole mechanist-vitalist controversy has been relegated to the limbo of a badly posed question." It is a badly posed question because in the languages of cybernetics, there is no desire to articulate a form of life outside industrial capitalism, society, or technology. What starts with a structural similarity to older questions—for example, the separation between vitalism and mechanism, between determinism and probability, or between ourselves and the enemy, ontological questions that seek to explore and represent what exists—becomes research endeavors without ontological interests.

I argue, therefore, that this error is committed on purpose by Deleuze. The philosopher is invested in a radical experiment to rethink the relationship between biology and mechanism, and between the mind and body. To hold the body as immanent, for him, does not permit such an analysis. For certain versions of cybernetic discourse, as well, there are many forms of matter, and abstractions also act. The idea of an immanent, or particular, biology separable from mechanism is untenable.

Of course, many figures, particularly Wiener, rejected this possibility by attempting to return to older epistemologies of authority and objectivity. I would argue that Deleuze is, however, involved in an ethical act of excavating this possibility and repeating this cybernetic displacement of ontology in the interest of producing new opportunities for thought, or "becoming," in his language. Perhaps Deleuze is completing what Bergson himself was never ready to do—abandon the real and the present as prime points of reference. Irrespective of whether we choose to embrace or disavow Deleuze, his reformulation of Bergson makes clear the epistemological shift that occurs after the mid-twentieth century, and the manner by which older ideas are retro-fitted within a context no longer concerned with the same questions. But we are also forced to ask about the sudden vogue for affect and materiality framed in rather reactionary terms of antagonism to cybernetics. What is at stake in continuing to resurrect the terms of materiality and abstraction, mind and body in our present?

Historically speaking, this form of statement, this aspiration for an emanci-pation from the separation between mind and matter, body and consciousness, or reality and representation, speaks to a transformation in the very grammar of philosophy and, more specifically, of a new site of unfulfilled desire, no longer for "life" or the present but for this new active form of abstraction—in both philosophy and engineering. One could argue for this correlation in the fact that the first decade or so after World War II saw an eruption of such state-ments in philosophy, science, and education. What this aspiration is produc-ing is not yet clear. What could be said is that the realm of the "lived," in the Bergsonian sense, that realm outside of but always desirous of representation, was transformed into an operative site that permitted a new form of techni-cized perception. The inability to render the present representable was trans-formed into the condition of possibility for cybernetics. In the martial envi-ronments of the World War II laboratory, the critique of modernity—and its horrors—became literally realized as a technical possibility. A strange irony.

Retrospectively, it appears that to make the machine, the animal, and the human compatible, so as to build a not-yet-existent sensorium through a sys-tem, necessitated a foundational transformation in points of reference. To comprehend the profound potential transformation in our perceptual field that might allow such a statement to be uttered, it might be useful to con-sider what is at stake in relocating Bergson's questions to the machine. Self-consciously, cybernetics positioned itself as the inheritor of a previous mo-ment in critical thought. Wiener seized on the Bergsonian reworking of matter and perception, indeed their collapse. While the implications of such thought, if any, is as yet in a state of becoming, what this move may herald is the his-torical possibility of speaking without concern for ontology; the utilization of probability does not come aligned with the problem of indexicality. In relocat-ing the methods of statistics to computing and communications, the project that could be said to have started much earlier in the nineteenth century found a material form in digital media technologies.

But Wiener explains to us that we no longer answer to the same questions, that in fact they are obsolete. What starts with a structural similarity to older questions—the separation between vitalism and mechanism, for example, between determinism and probability, or between ourselves and the enemy—ontological questions that seek to explore and represent what exists, become research endeavors obsessed with process. Bergson's "innovation" in Wiener's estimation was a reconsideration of temporality, and an operationalizing of perception through memory; a philosophical project amenable to technicization and literalization in computing, as we shall see.

Bergson, however, was obsessed with, and still answered to, questions of ontology and of "life." His critique of spatialized time, and his production of a metaphysical thought to accompany an emergent disciplinary science, produced transferable techniques, but his points of reference—disciplinary knowledge, the idea of an "external reality" (even if to critique it), being, interiority, and consciousness—no longer framed the cybernetic endeavor.

Certainly neither Wiener nor his compatriots were directly concerned with problems of experience, mediation, or even explicitly representation. They were concerned, however, with effecting actions in the world—an obsession with "purpose." The implications of the deliberate rescripting of older ideas of psychology, perception, thought and consciousness toward the realm of communication, control, and feedback was a radical reframing in an entirely new grammar; a language now concerned with prediction as its dominant interest, and temporal mode. For cyberneticians, the initial questions that had precipitated the modern rethinking of both perception and its relations to temporality—mainly the increased consciousness of the mediation of the sensorium and the temporal nature of perception—had ceased to be points of reference. The subjective and produced nature of perception could now become the source of an unrealized, but productive, aspiration to model thought—the site of a new dream for freedom and the exercise of will. We return to the start as Wiener aspires to a form of thought no longer aligned with the archive and its ontological and taxonomic orders. Mediation was therefore no longer a problem for Wiener and other cyberneticians.

Eliminating this problem, however, produced new sites of interest. By virtue of all these steps cybernetics took to produce feedback—which can be viewed as a nascent form of interactivity—two new areas of investigation or problematization emerged. The first was perception, and the second was memory, and these two now had a new relationship, historically speaking. For Wiener, memory itself would become the space tying the past action with the future one; bridging an older concern of presence with a newer problem of transmission and communication. Memory would become the fantasized space of processing; a space where the trace of a stimulus could be utilized to dispense with the totality of the original in order to utilize this abstracted form—this "essence" of the object—for future operations. This reformulation spoke to a post–World War II moment marked by a new set of statements that saw perception, and ultimately even thought, as technical projects. In the course of rethinking communication, older questions of memory and perception were operationalized toward new goals.

Redefining Perception

While cybernetics was invested in facilitating perfect, and rapid, communicative exchanges, the residue of these interactions was a vast cumulative space of data and information. The desire to predict from past behavior called into being some form of recording, storing, and retrieving information. This same process could become the terminal failure point in perfect transmission of a communicative message; too much noise would interfere with the signal. Systems not capable of erasing the excesses of stored material or sufficiently "damping" or time-lagging their impulses would be prone to error—error always being a failure to effectively transmit a message.[55] Within closed systems, too much information could overwhelm the system's stability. In short, faced with an excess of information/stimuli the system may lose its capacity to manage and respond. A continually nagging "problem" of overabundant information that is the preoccupying obsession of distributed and networked systems.

The selection of information became a pressing problem for cyberneticians, as it had been for the theorists of perception before them. Failure to adequately sort and sift through stimulus, and allowing too much "noise" into a system, results in the loss of homeostasis, and excesses of oscillation and instability. Cyberneticians used, for example, models of servomechanisms responding too rapidly to a moving target, or of "functional disease," such as mental disorders or blood clotting, to make this point.[56] For Wiener, such losses of stability were possibilities not only located within the human subject but also operating at the level of large systems, particularly possible in nations at the brink of atomic disaster.[57]

In cybernetics, avoiding transmission failure but still producing viable feedback systems mandated, therefore, not only storage, but a process of selection by which only that information necessary for response would be stored. Perception came to be defined by the ability to respond, and memory as the site of processing. Abstraction could facilitate transmission. Wiener, and other adherents of cybernetic theory, including psychologists, became obsessed by nonconscious acts of abstraction that permit negative feedback to commence without error.

Vision was one prominent site of this reworking of perception into a modelable and technical project. In vision, the eye would take up another function as a machine for abstracting the world—a black box. Vision, that sense that collects so much information, must, in the minds of cyberneticians, have some sort of dampening or straining process to facilitate the flow of information toward a more abstract state where what is stored is not everything seen. This process was imagined as a series of steps, in which each step of the process brings visual information "one step nearer to the form in which it is used and is preserved in memory." Wiener posited that the most "plausible" explanation of vision is as a process where outlines are emphasized. The eye, starting with the retina, must begin filtering the information, otherwise it will lose its ability to transmit the stimulus onward, being bombarded as it is with constant stimulus. Wiener argued that alterations in "storage elements" are necessary for transmission.[58] For Wiener, the eye consisted of processes working to filter and arrange information. Information entering the eye is at every moment both an index and a command for future action.

Perception came to be defined as the ability to respond and merged with cognition on the basis of efficacy under the duress of information overload. The nerves, as in Deutsch's formulation, discussed in the introduction, whether of government, televisions, or humans, begin to scan and find "outlines" far before the brain is involved.

This assumption of a nonconscious and material perception that operated on principles of pattern seeking out of data-rich fields was the substrate for producing computational approaches to the senses. In the territory of communications engineering, as the historian of science Mara Mills has demonstrated, cybernetic work contributed to the "industrialization" of speech.[59]

But cybernetic approaches to perception went far beyond telephony, or any simple definitions of reduction or abstraction. By the late 1950s and early 1960s, for example, Béla Julesz, a Hungarian émigré during the Soviet occupation, now working at Bell Labs, researched what he labeled "cyclopean" perception:[60] an ode to the mythic monster who stands in for the ability of the nervous system to produce the perceptual myth of a unified visual field. These landmark studies produced new ways to model human and machine vision.

Julesz was an innovator in applying machines to the study of human vision and worked closely with cybernetic, cognitive, and communication theories. His pathbreaking work separated human vision into two vectors—depth perception and form recognition. Using the nineteenth-century model of stereoscopic vision refined by David Brewster and Charles Wheatstone, Julesz sought to resolve the one question unknown at the time and unresolved since—

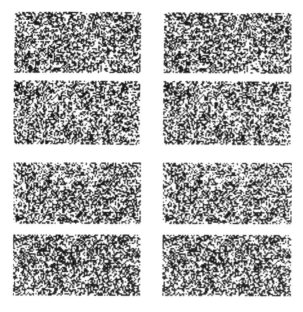

FIG. 1.1__Examples of random dot stereograms, computer generated to prompt pattern matching without identification. From "Cyclopean Contours and Closure Phenomena," in Julesz, *Foundations of Cyclopean Perception*, 257. © 2006 MIT, by permission of The MIT Press.

whether depth perception resulted from perspective (geometric "reality") or from the subject's recognition of relationships between objects, as a result of recognizing forms.[61]

In his studies, Julesz postulated that vision was a matter of finding repetitive patterns out of stimuli received by the eyes. The experiments involved showing test subjects abstract patterns, randomly generated through computers (in fact necessitating the machines in order to generate such levels of patterned complexity without indexicality), and incredibly dense in data points. These image fields were totally illegible to most humans and outside of the lived experience of any of the test subjects. These computer-generated stereoscopic blocks appeared one-dimensional and lacking any formal organization or pattern when seen monocularly, but when put together and seen through red-green glasses, patterns emerged. Encountering these strange forms, test subjects would begin to see in three dimensions after staring at the pattern for a time (see fig. 1.1).

Julesz noted that while subjects did not immediately "see" depth in the pattern, after looking for some time, the field became three-dimensional. The visual apparatus, so to speak, was learning; it could autopoietically produce depth in relationship to the computer-generated blocks. The visual process was pattern matching and distilling depth out of the data by filtering information and finding patterns out of dense data fields.

This ability to produce three-dimensional sensation occurred without any conscious recognition of relationships between objects, identification of form, or even gestalts. In short, depth perception could occur without entering the realm of conscious representation. Viewers shown incoherent computer-generated patterns, impossible to witness in the "real" world (at least at the time, 1960), could still find patterns, and begin to see in three dimensions (this led to the autostereogram, and to Salvador Dalí producing a painting in 1978 as an ode to Julesz titled "Cybernetic Odalisque—Homage to Béla Julesz").[62]

This temporal delay in producing visual experience, and the bifurcation of vision into depth and form perception, made vision a process amenable to simulation and computational modeling—in both humans and machines. Vision no longer had a stable referent in the world but rather was a self-generated process from within a system.

While the use of random patterns, such as Rorschach tests, had been common in psychology, Julesz suggested that these previous methods had now achieved a new era of autonomy and self-regulation. As Julesz pointed out, the "era of cybernetics and information theory is marked by the study of complex systems with stochastic input signals and the measuring of stochastic output parameters."[63] What marked this change was the capacity to probe the central nervous system through complex generated images with precisely defined statistical and geometrical properties. Most critically, only computers allowed the measure of and generation of such complex random patterns.[64] Finally, psychologists using tests such as the Rorschach always sought to return to representation. Patients or subjects were asked to identify a relatable form in the image, a form they recognized and verbally described, and psychologists read this response as a diagnostic of subjectivity. In Julesz's examples we have pure perception without representation or subjectivity.

Perception became a probabilistic channel whose capacities were variable, and capable of being engineered, enhanced, and modified. This is a vision that stretches into the nervous system and out to the computationally generated blocks. As a result the process of seeing could be simulated, making human vision amenable to computation, and in reverse making computers, perhaps, see, as vision became an increasingly algorithmic process whose complex operations could be broken down into discrete elements, and reproduced, materialized, and circulated as technology.[65] For Julesz, depth perception was a process in vision untethered from any specific site or landscape, or from any situated relationship to experience, consciousness, or unconsciousness. However, this is not a concept of perception that returns to the idea that subjectivity obscures an objective and external reality. Vision came to be understood as an autopoietic process, emerging from within a system of interactions, amenable to algorithmic treatment, and materializable in experimental method in relation to machines.

Perception became an informatic entity assumed to operate according to some set of algorithmic and communicative principles that could be cordoned off and isolated. Cybernetics would become a mode of operations interested not in representing the world but in understanding what templates, approximations, agglomerations of information facilitated generalized productions of universal concepts that allowed the eye, now an independent set of processes not attached to conscious reason, personal history, or specific situations, to perceive and act on the world.[66]

This reformulation of what had been a vexing debate in the nineteenth century between Wheatstone and Brewster had now become the source of an entirely new epistemology in psychology, and a method for integrating machines and humans into novel processes for making vision modelable and psychology experimental.[67] For such cybernetically and computationally informed research in vision, perception was a thing in the world, a process emerging from interactions between the observer and the stimulus, a process that began already at the site of input, through the screening and comparison of inputs. The patterns that machines generated to produce these autostereograms and stereograms were inseparable from the perception.

If perception and mediation throughout the nineteenth and twentieth centuries was an on-going and vexing site of interest, speculation, investigation, and problematization across the social field from philosophy to behavioral and social sciences and to cinema, cybernetics succeeded in suppressing these previous questions in favor of a modular and literally technical approach. One can say that we really rarely speak of perception anymore except as a medical technology.[68] Instead of attempting to overcome mediation, the process between stimulus, perception, and analysis becomes the site of technical potential.

In a turn of historical irony, I might say then that cybernetic research makes technical the Bergsonian concept that perception is now "in things." As the literary critic Temenuga Trifonova summarizes the Bergsonian position, "perception is material just as matter is already perception."[69] Process is a thing in the world, even producing the world, and therefore amenable to being materialized, circulated, and technically reproduced.

In relation, therefore, to the predecessor philosophy-psychology that already had prepared us for an operational sensorium, the older question of recuperating "life" or "will" became a new problematic of producing action and creating technology. Perception was temporally marked by past elements, it was a space between information reception, recollection, and reaction, but this was the opening for the possibility of processing, not a vexing problem of the historical record, or of the representability of time.[70]

Activating Memory

Isolating the terrain of perception into the terms of interactive or feedback exchanges still left a second problem—that of memory and storage. As Wiener redefined it, the question was how to cordon off, yet communicate between, the vast realms of raw, and largely useless, information and the sites of processing and translating this information.

Memory continued to be a nagging residue for Wiener and other cyberneticians. To deal with this problem of system stability and the demand for a nonconscious abstraction, a new functionalist idea of memory emerged throughout the cybernetic fields. Memory would facilitate the systems' capacity to predict by acting as a repository of only the necessary information for functional reaction. While memory, in cybernetics, is an undertheorized aside to the central concern of communication and transmission—a persistent reminder that it is hard to process or transmit something that has not been recorded or stored—it would come to play a central role in the dreams, aspirations, and structures of future digital systems.

Wiener would define memory as "the ability to preserve the record of past operations for use in the future." Memory, itself, either in men or machines, in Wiener's discussions is not represented; it does not possess some metaphoric analogue, such as the Mystic Writing Pad of Freud. Memory exists through the absence of a direct metaphoric machinic equivalent. Memory is only a set of discussions concerning where it ceases to function, and needs to be repaired—or some new mechanism constructed. There is no absolute representation of the memory ideal. There is, however, a functional ideal. Memory was now fantasized as a severed entity with a series of layers or levels of storage. The first form of memory is "short term," which serves the functional necessity of carrying out the current processes, processes that do not need to be stored but that themselves mandate the implementation of some stored information, such as an algorithm, but whose immediate results are of no use. This memory can record quickly (and of course perfectly), can be read quickly, and can be erased quickly. The second form of memory is one "intended to be part of the files, the permanent record of the machine, or the brain and to contribute to the basis of its future behavior, at least during a single run of the machine."[71] If these notions appear at all familiar, it is because they seem, from this vantage point, to reflect and advance themselves in the compartmentalization of contemporary random access memory and hard-drive systems.[72]

Memory, however, had to be reconnected, now that it had been separated from the realm of communication. Cyberneticians would consciously refer to, and reappropriate older disciplinary sciences and representational practices to substantiate, this transformation toward a process- or "form"-oriented memory. Wiener would call on the analogues of photography and psychoanalysis to quickly transfer between the older disciplinary scientific and taxonomic model of the world, where description and classification of a theoretically external and "real world" preside, and another site—one concerned with conditioning perception and producing thought.

Communication failure was, in his terms, the analogue to mental illness.[73] Wiener, and others, understood disorders such as manic depression or schizophrenia as functional failures to conduct a chain of operations without disruption. These were diseases "of memory," the results of circulating information accumulating in the brain and unable to be discharged. This excess of unstable circular processes over time would result in loss of stability of the system as the storage "space" in the organism ran out, signals interfering and deforming each other.[74] To produce equivalence between the apparent complexity of the organism and the basic recall modes of the machine, time, in terms of runs, became equivalent to space in an organism. As Wiener wrote in "Behavior, Purpose, and Teleology," "scope and flexibility are achieved in machines largely by temporal multiplication of effects; frequencies of one million per second or more are readily obtained and utilized. In organisms, spatial multiplication, rather than temporal, is the rule."[75] A machine could run through numerous operations and basic sets of decisions to approximate what an organism must do through a more complex physiological structure. Time for a machine was what space was for an organism. Memory, therefore, came to be viewed not as an endless static repository or archive of stored information but as an active site for the management and execution of these operations. The limitations in storage, in fact, meant that what was stored was an abstraction sufficient to be used to execute an action, not a perfect image or representation of an external sensation. In this sense it recalls the ideal of an active memory from Bergson, while extending the belief that memory and perception are now the same thing.

While Wiener had no clear representational schema for memory, he was clearly obsessed throughout his personal writings with the sciences of memory, psychoanalysis the most predominant of them.[76] Psychoanalysis, or psychotherapy, could be the solution to mental illness rather than the more violent interventions of the day, such as shock therapy. His adherence to psychoanalysis lies in the parallel in psychoanalysis to an emergent notion of processing. Psychoanalysis works because of the concept that "the function of psychoanalysis in this case becomes one of processing, a perfectly consistent point of view with cybernetics. The technique of the analyst consists of tactics by which to mobilize these hidden memories, to accept, and to modify them . . . and in this lies the success of the therapy."[77] Psychoanalysis becomes a process of moving information, not unearthing meaning.

Psychoanalysis, already a terrain of automata and unconsciousness, could mutate into formal technical strategy. In this move, a science that sought to disciplinarily derive knowledge of individuals becomes a question not of truth claims but of perceptual training, a technique to move information and train communicative forms. The process of psychoanalysis could almost be said to permutate toward more contemporary understandings of an "interface"—a zone where messages could be processed and translated in order to continue the seamless movement of information between different areas—of the brain and between the individual and other entities. This concept of interface as translation zone was further developed immediately after the war in both psychology and in computing in association with ideas about personal computers and windows interfaces. Wiener defined cybernetics as the ability to communicate through a controlling device—a steering mechanism, as the term "cybernetics" suggests—between different entities. This insight, as a contemporary (2001) text on multimedia argues, "is the premise behind all human-computer interactivity and interface design."[78] The controller, which for Wiener could be a psychoanalytic session, a screen, or a steering wheel, continues to operate as that space where otherwise increasingly disorganized, entropic, and differentiated messages could be organized and assembled into communication.[79]

In this imaginary, the role of psychoanalysis was not to specify the content of all the stored memories in the mind but rather to define the forms, the "affective tone," in Wiener's words, of the stored information. This affective tone would condition future behavior, and the necessity of the therapeutic encounter was to discover the general patterns and modify them, in short to translate and abstract them into a form compatible with "normal" functioning.[80] The therapeutic encounter, for Wiener, was not about unearthing what was within a patient, but rather about producing a space where the patient and the therapist, and perhaps later the patient and my computer, could communicate and effect future interactions. The main purposes being the conditioning of future exchanges. Friedrich Kittler has already suggested that this was, indeed, the initial effect of psychoanalysis: the externalization of the psyche and its incorporation into larger discursive networks. In distinguishing the "discourse network of 1800" from the "discourse network of 1900," Kittler specifies the latter as being concerned with an obsession with the minute, unimportant, and indiscriminately recorded, which characterized the nascent media technologies of the time—most specifically, for our purposes, the camera. But this is a camera no longer aligned with photography but rather with cinema and the phonograph, technologies that "can record and reproduce the very time flow of acoustic and optical data."[81] For Kittler the realization of McLuhan's argument that "the 'content' of any medium is always just another medium" has the implication of making the senses autonomous. Through a variety of new recording apparatus, the senses can be separated and stored—offering the possibility, which Kittler aligns with the necessary condition for cybernetics and computation, of a sovereign perceptual field where memory, now a technical operation, becomes merely the site of storage for the further circulation and remediation of signals into other mediums. This capacity of "discourse networks" makes all forms of inscription interchangeable and mobile, and facilitates, for Kittler, the possibility that memory is merely an operative form of storage for further transmission and operations without any alignment to meaning. This process, however, is located not solely at the level of any one technology but rather at the level of "network," of which psychoanalysis is but one part. Kittler extends this notion to make psychoanalysis subordinate to, and supportive of, the positivist science of psychophysics.[82] Therefore, Freud's obsessive concern not with obviously scripted "events" but with slips of the tongue, minute details, and so forth advances a larger technical assemblage obsessed with delivering recorded and stored events from any clear referential relation to an external, and meaningful, "reality." In this formulation, psychoanalysis sits much more comfortably with the isolation of perception as a modelable entity as posited by cognitive psychology. We may now add that not only does psychoanalysis externalize the psyche, but, in the project of cybernetic thought, it has become an explicitly and directly technical project aligned with computational machines.

Wiener's explicit use of psychoanalysis is, however, a subtle inversion of the Freudian concerns that disrupts Kittler's excessively seamless and automatic extension of the 1900 network into the electronic one. As Mary Ann Doane, for example, has framed it, Freud was interested in accumulation. The unconscious, in his essay "The Interpretation of Dreams," is, according to Doane, "a vast storehouse of contents and processes that are immune to the corrosive effects of temporality."[83] Memory records everything. Psychoanalysis' disciplinary fantasy is the total representation and catalogue of this repository, a task that drove the disciplinary enterprise but that even Freud understood as impossible.

The residue of this vast archive of experience was consciousness. Consciousness was simply the visible and articulatable residue or symptom of this memory; a memory that is now "out there" and outside of representation but whose excesses of information could overwhelm and incapacitate the subject. Consciousness was the filter, the translation zone, that allowed the organism to function and produced a teleological and functional temporality—a temporality not of infinitude and flow but of marked events and history. Consciousness and its related measurable and historical time was thus antithetical to memory, although protective of the organism. Consciousness was a barrier to accessing the moment of impression—the present, or (again) the "real." As a discipline, psychoanalysis wants to overcome this barrier to representation . . . but cannot.[84]

Wiener fundamentally sought to displace these questions by problematizing accumulation and not memory's inaccessibility and representability. Despite this, the older heritages continued to plague the new fantasy. Storage continued to be "problematized" as an issue of system stability, entropy, disorganization, and noise. The excess accumulation of stored or extraneous information always produced the problem of lowering the statistical probability for communication, and forcing the system into "oscillations" and instability. These problems of information management now emerged without recourse to the taxonomical, objective, and static terms historically associated with the archive or the ideal of an external "reality" or "temporality" to be brought into representation.[85] This "problem" emerged most visibly and consciously in conversation over storage and recording mediums.

Early theories of computing and interface engaged heavily with the fantasies of film and photography as graphical and recording apparatus. Wiener specifically refers to photography, and more specifically film's, problem as a storage medium, due to its indexical heritage and the demands of erasure and mutability. The problem, as framed by Wiener, with photography, was its slowness, its lack of efficiency, its inability to keep up with the autonomous recording and circulating computing systems. Photography, Wiener repeatedly argued, could be ideal for its perfection in recording and documenting perfectly. The problem, however, was with slow development and nonrapid erasure.[86] In cybernetics, photography's heritage as a perfect record, indexical, and archival was seized on to define, and now make problematic, the question of recording and storing information. For early computer and information theory, photography seemed resistant to that feature of the cybernetic material world—abstraction. The image was solely a storage mechanism, it was static, it did not store processes, or forms that would create future functions. In cybernetic terms, so was film, because it was only a medium of representation, not inscription, it could not respond, react or change within the temporal structures of real time and prediction. In doing so, these mediums were no longer the site of training a perceptual field that was now lent autonomy.[87]

This problem with representation continued to drive interests in designing storage structures not based on tape or the literal medium of film. Wiener would argue for a dream of photography in which the very production of the record through alteration in the storage element (the film) could already also inform the further transmission of the message. In short, that the record and its recollection or memory could become more closely synonymous, if not the same. "We have already seen in the case of photography and similar processes," Wiener said, "that it is possible to store a message in the form of a permanent alteration. . . . In reinserting this information into the system, it is necessary to cause these changes to affect the message going through the system."[88] The potential of mechanical reproduction for distortion and abstraction, so central to modernist concerns about recording, now became the core aspiration for cybernetics and information theory. Wiener signals to us an emergent hope that the acts of recording, storing, and recalling may no longer be held separate.

However, the very things that made film a problematic storage device also made it the perfect recording device. Control demanded the indexical document. To make a perfect prediction would, it is assumed, demand perfect data. Photography's mechanical nature and indexicality was both its problem and the solution to systems that sought simultaneously to build a novel perceptual field that was totally fabricated, referential only to the system and no longer located on the plane of human observation but still seeking to visualize and record the world in order to respond to it.

This new relationship between the record and its recognition became a new site of discussion and debate. In the Sixth Macy Conference, where many central figures in information theory, computing, and social science assembled, the arguments in attempting to understand memory were specific about trying to distill the difference between the processing elements that facilitate recognition, or perception, and the process by which information is recalled. A tense relationship thus continued between the role of memory as a site for storing stable records and the aspiration for memory as an operative function, the site of actual processing, and the seat of an active perceptual field. The psychiatrist Warren McCulloch (who will return prominently in chapter 3) struggled to separate memory from learning within a matrix of his own work on neural nets and networked cognition, which had made the separation between memory, cognitive processing, perception and sensation permeable. McCulloch argued that human memory has at least, "three kinds of things that are distinguishable from the curves whereby one learns." He laid out an ontology of memory, similar to Wiener's, except that there is now a second form of physiologically produced memory that is hardwired, through training, to perform set acts, such as playing piano, and a third, and ultimately problematic, memory:

> there is some kind of a process which is involved in skilled acts and is obviously different from the kind of memory that takes snapshots of the world and files them away for future reference, whether or not [they are of] any importance at the moment. We do have a memory of the third kind [snapshot], that is not immediately accessible, that has certain different properties . . . but this third kind of memory, which I strongly suspect is more important in neuroses than the rest, I think first needs an examination of the mechanism of recall.[89]

In this formulation the problem memory takes the photographic position, and in this position it is, ultimately, pathological—neurotic. And like all neurosis, it can facilitate the subversion and repression of one desire to allow another set of actions or possibilities to commence. It becomes a site that McCulloch and indeed all the participants of this segment of the conference—the psychoanalyst Lawrence Kubie, the neurophysiologist Ralph Gerard, the ethnographer Gregory Bateson, Wiener, and others—view with great interest as a research agenda, and as a core issue in the production of both cybernetic models of both minds and machines. The a priori stability and uniformity of the record becomes a form of memory that requires further theorization and investigation. It is precisely this residue of an older vision of autonomous, perhaps "mechanically objective"[90] recording that births a new aspiration toward a hyper-recall that can so perfectly analyze the known and perfectly recorded world as to be able to produce something new out of its own documentary practices.

Later, in his important essay "As We May Think," Vannevar Bush, the head of the American scientific war effort, would formalize this ideal in the fantasy of a not-yet-built machine, the Memex—the memory extender. His idea of the Memex was that a user could access a perfect and total database of information, all recorded on microfiche, and be able to bring up information on numerous screens for comparison, and for the production of new relation-ships. The importance was that the machine would break the taxonomic and stable structure of the archive, and would work "as we may think," by cre-ating rhizomatic linkages and nonlinear associations between different pieces of information. The hope was that a fully recorded world waited to simply be reaccessed and analyzed. The scientist of the future, Bush hoped, would auto-matically and constantly be recording the world; there would be faster, and better, and more autonomous forms of recording and picturing, and more automatic and novel forms of indexing. Whatever one wanted to know, one could access through the scientists' interface to the networked libraries of the future. Once a discovery was made, then one would "photograph" one's own research path, the trails one followed to get to a result, and send this "pic-ture" of thought to one's colleagues: in short, an endless world of recordings, of which the main goal is not the documentation of an external world—it will have been done, it is a given, as Wiener already signaled, and the military cer-tainly took as a serious goal—but rather the production of a new world. The screen for Bush, and also Wiener, was not a representation of an outside reality but a dynamic space to encourage the production of new associations, and further interactions—between people and between people and machines. The screen referred to further modes of interaction, not to anything outside the system, per se; it was not a representational display of a world "out there" but a translation zone aimed at inducing new modes of thought. The Memex was never built, and computers went digital, but nevertheless the essay is consid-ered an important contribution to the dream of networked, hyperlinked, and personal computing machines.[91]

While photography and film would continually emerge as theoretically ideal mediums for storage in the imaginings of future technologies, their per-sistence and inadequacy belies this problematic—that of total and perfect recording, on the basis of which to make the most accurate predictions, while simultaneously posing an older set of conventions involving storage and time. A tense, but productive, relationship would come to exist between representa-tion and archiving on one pole and perception, interaction, and immediacy on the other. This relationship would be the site for the development of a new and as yet unrealized aspiration by which storage and memory could become one.

To sum up, what can we say about this world of predictive but nonteleo-logical temporality? Of the effort to reframe the terms of memory, represen-tation, and archiving as a matter of storage, behavior, function, and trans-mission? And finally, of the effort to produce these questions in the form of myriad technologies (and I do not just speak of computers here), which split interaction and communication off from realms of processing and storing information?

Wiener, and others, deliberately seized on older forms of recording, storing, and transmitting information and sought to produce new points of reference—a new language to encode the very ideal of thought. But this was not done without creating new sites of failure and problematization, although now no longer around questions of referentiality and historical temporality but rather around those of interactivity and mediation—the relentless encouragement of future communications. We are forced to ask, however, whether we find the separation between the archive and the interface, between the storage system and the screen, a site of desire and potentiality or a technological failure to be overcome?

Counter to dominant narratives of digital media and computation, I argue that this desire still lurks in our networks. It was not, and is still not, clear that speed and programmability are the only sites of interest for culture or technology at the cost of complexity, history, and the space between the archive and the interface. In our contemporary adoration of data, the structure of computing machines, and the sensory prosthetics of cyberneticians, temporal multiplicity still played a critical role in producing imagination and emergence. As the psychoanalyst Jacques Lacan commented about cybernetics in his famous 1955 lectures on the ego in psychoanalysis, "cybernetics [is about] chance . . . that it seeks to contain by way of laws of the binary order." But, he reminds us, "in keeping on this frontier the originality of what appears in our world in the form of cybernetics, I am tying it to man's waiting . . . to the chance of the unconscious." And he continues, "at this point we come upon a precious fact revealed to us by cybernetics—there is something in the symbolic function of human discourse that cannot be eliminated, and that is the role played in it by the imaginary."[92] There is something about the history of representation and expression that is both revealed and repressed in cybernetics. By deduction, it is arguable that the role of postcybernetic psychoanalytic practices is to make evident this possibility lying latent in the lag that builds our machines and makes our data beautiful. The work of poststructuralism was (and perhaps is) to not only give us an analysis of cybernetics, and perhaps its associated forms of governmentality, but also to realize this latent potential of "man's waiting"; perhaps the potential to finally become human through our media in realizing the temporal process between recording, storing, and analyzing data.

Theorizing the Perceptual Future

Earlier in this chapter, I used Roland Barthes to argue that "reality" was produced at the very moment that the world became a mediated one. Does cybernetics, and its affiliates across the social field, mark another such turn? The transformation that Wiener, in his memoir, at the start of this book, describes as the natural progression of one age to another. These questions cannot be divorced from the larger concerns they seek to advance—a series of concerns in which a wide-open world—one that must be described, one whose "truth" mandated our response—no longer holds us at bay.

I opened this chapter arguing that cybernetics may aspire to an elimination of difference in the name of perfect communication—a perfection of transmission that would obliterate the separation between the archive and the interface, or, to return to Barthes, between the sign and its referent. What Barthes gave voice to in the early 1980s, already well within an age of electronic communications, was the consummate danger that in the aspiration for a "real time," the possibility for signification, and by extension thought, would be eliminated in order to meet the demands of an immediate, and immediately effective, form of interaction. Therefore, the example of the small cybernetic honeybee engaged in thoughtless, but communicative, actions.

If our contemporary media field fulfills the relentless desire for abstraction, and the absolute interchangeability and manipulation of all symbols in the demands for automation, we are left with the question of what ostensible desires are left to be fulfilled on the screen. Do we, as Barthes implied, become little satiated automata?

This question returns us to the dilemma between the demand for infinite storage and archiving and the command to interact in real-time that still preoccupies the design and architecture of digital systems. Derrida, Barthes's contemporary both philosophically and historically, contributed his own electronically informed response to this problem. In his interrogation of the fate of the archive and memory, dedicated in 1994, not incidentally, to Freud, Derrida wrote that in the very process of recording, in the act of seeking to represent the world, we make it. We read in his text *Archive Fever* of the "technical structure of the archivable content even in its very coming into existence and in its relationship to the future. The archivization produces as much as it records the event."[93] Derrida preceded these statements by arguing in relation to the new technologies of memory, of which he names computing and electronicization, that "the upheavals in progress affected the very structure of the psychic apparatus, for example in their spatial architecture and in their economy of speed, in their processing of spacing and temporalization, it would be a question no longer of simple continuous progress in representation, in the representative value of the model, but rather of an entirely different logic."[94] The Derridean critical project in many ways repeats some of the ideals of diagrams and virtuality that Wiener wrote of, as discussed at the start of this chapter. Derrida himself is invested in producing this logic. He argues for the possibility of ethics, indeed love, through the structuring of communication, and in the failed collusion between the sign and the signifier. It is, in fact, in the impossibility of such collusions that much of poststructuralism has found its imagination. He shares epistemically, therefore, in a field of thought that is no longer invested in description, origins, ontology, or the present. To seek both life and love through mediation is one implication of this form of thought, and it comes situated (both historically and philosophically) within a relationship to these new machines of inscription, recording, and communicating and their related epistemologies.

This is not an unproblematic relationship. Derrida, in writing of the (now) electronic technologies of inscription, recording, and archiving, also poses to us the ethicopolitical problems of systems that seek to record in order to destroy that which is being recorded. Which is to say, that we record to produce the capacity to forget, on the condition of forgetting, in the name of "real-time" as the only time; a technological fantasy of accessing the present in the name of immediacy, and erasing the lags and resistances in translating and transmitting information. In recording we have destroyed the need to remember and, Derrida hints, we have mechanized that loss, made it no longer a pain to be felt but a site for further technical projects; at its extreme horizon, devoid of anything but its own technical imperative, this drive becomes "radical evil." This "radical evil" is, of course, the failure to imagine a future through the loss of all points of reference; an automation of recording that facilitates death.[95]

This comment returns us to the original question Wiener posed, as discussed at the start of this book: what are the stakes between a world based on referential representational schema and a world of complex diagrams and ongoing communicative exchanges parading under the guise of "control"? Writing in response to World War II, Wiener asks a moral question as to the application of science to human welfare. Wiener marks a moment where our subjective perception becomes a site of possibility; but this also marks an effort to eliminate phenomenology, "reality," and exteriority as dominant concerns for culture. Engaged in the midst of the Cold War, Wiener is, therefore, forced to ask an ethical question as to the possibility of human survival and the fate of humanism in the midst of this strategic conflict couched in the terms of "game theory," strategy, and information, with an endpoint in nuclear confrontation. The specter of Marx haunts Wiener—both literally, as he writes for new possibilities in the aftermath of Auschwitz and the Bomb, and figuratively, in his theory obsessed with the productivity of abstraction and symbolic manipulation.

Marx observed that "ideas . . . first have to be translated out of their mother tongue into a foreign tongue in order to circulate,"[96] and therefore the analogy between money and language, their similarities as arbitrary systems for the production of value, exists only insofar as the latter is understood as translation. Rosalind Morris argues that "the bourgeois and the structuralist response to this observation has, of course, been one that fantasizes the possibility of total commensurability or translatability."[97] This is a fantasy we could extrapolate into the obsession with "real time," reality television, and immediacy at the interface—all markers of a dream world composed of perfectly homogenous, commensurable, and convergent entities. This is a dream that was once also articulated in the military lab, where all entities became behavioral black boxes. If we are to believe Marx's older dictums on circulation and translation, we might then lose the possibility for freedom, or even a future, to the dream of perfect communication. Everything can be remediated or translated without change. Thought itself becomes a technical project.

We must avoid this fate. As I have attempted to demonstrate by way of cybernetics, our contemporary digital multimedia field also responds to older heritages of media, psychology, and philosophy not solely grounded in the mid-twentieth-century war machine. The turn away from ontology as a referent to an external reality and its transformation into a malleable category in computer science, the emphasis on affect and prediction in the interest of a pragmatic performativity, the movement away from metaphysics and phenomenology have all also informed much of contemporary critical theory, filmic, and artistic practice. The semiotic and semiological portions of poststructuralism, which Friedrich Kittler already posited as historically specific forms of already electronic or postcybernetic thought, are therefore aligned with certain tendencies in digital media interested in seeking possibility through mediation.

It is to this possibility that I turn—we may call it an estrangement or "foreignness" from representation—that is the core site of possibility for poststructural and other critical theories. This incapability of, or resistance to, perfect translation is the "source of a complex," and of usually unrealized, yet possible, freedom. "It is a freedom (both the 'difference' that Marx posited as the internal contradiction of the commodity between use-value and exchange value, and the temporally defined dimension identified by Derrida's term différance) that is the necessary condition, or possibility for revolution."[98] How we define and maintain the temporal and spatial separation between the archive and the interface is part of this struggle. Does communication and translation automatically assume homogeneity and convergence between all mediums and entities? Cybernetics both posed this aspiration as control and opened us to its impossibility. Calling on Freud, Bergson, and many others, cybernetics also hinted at the possibility that within our imaginaries for computational and (later) digital media lay far more complex heritages than simply the demand for another uninterrupted and automatic interactive exchange. Despite the rhetoric of convergence—the faith in perfect translation to the point of absolute commensurability that has become the dominant fantasy for media conglomerates and technologists of our day—it has not yet become a new reality, and need not.

Our questions today are, of course, no longer Wiener's. We are no longer forced to respond to the immediate demands of fighting a world war against Fascism or a Cold War against a mythic Communism, or contending with crisis in industrial capitalism. We may not even remember these events. But we do respond to the technical legacies and forms of the time period. We are left to ask: in what directions will we try to push our technical imaginaries? I resurrect these relics from the archive, "blasted" from the past,[99] so to speak, in the attempt to excavate other possibilities and tensions for the future. In the numerous tensions, eruptions, and resistances posed from within the cybernetic ideal, we confront the implications and possibilities that an archeology of our own thought produces. The cybernetic concern with human possibility operated through a dream of interaction based in ideals of complexity and emphasis on process. This fantasy has incarnated itself as a desire to turn inward, to a myopic obsession with our instruments, where the "archival fever" operates at a technological level. What to do with this obsession has now become a defining feature of our present relationship with, and future imagination of, interactivity; permeating our screens, occupying our networks, and feeding our data centers, as both a daydream and a nightmare about the beauty of data as prophecy and history of the future.

2_VISUALIZING

Design, Communicative Objectivity, and the Interface

 HE AERIAL OBSERVER FOR WHOM CAMOUFLAGE HAS TO BE largely considered today is a mobile observer. Every factor involved in his vision is in continuous movement. His eye is moving, the light conditions are changing and the landscape is moving."[1] With these words written in 1942 the prominent designer and artist Gyorgy Kepes inaugurated a new concept of visual perception. Writing for an issue of *Civilian Defense*, Kepes described a course on designing camouflage that he taught at the School of Design with László Moholy-Nagy as part of the New Bauhaus in Chicago. Working for the U.S. Department of Defense, the designer took flights above the city, where his perspective was transformed. He wrote of an eye no longer moored in a single space or time. He was trained to trust instrument panels streaming data from radar and radio transmissions, to rely on the guidance of machines and the recordings of surveillance teams. Calling on this experience, Kepes described a new form of vision, one that was

FIG. 2.1__**Landscape, Oscillation, Survey.** Images from Kepes, *New Landscape*: patterns emerging from charged particles; aerial survey of Chicago; oscilloscope patterns of an Analogue Computer (1956).

mobile, relative, nomadic, and autonomous. He began to consider designing for information flows coming from communicating machines.

Developing this theme of technically transformed vision, Kepes later insisted, in a text fittingly titled *The New Landscape of Art and Science*, that "precise observation" and a new armory of sensory devices coming from nuclear physics and electronic computing that no longer operated at the levels of the human sensorium had produced a "new foundation for our material existence." This material existence converted what he had previously labeled the "language of vision" to a "landscape" of "forms"; the terms "language" and "vision" mutated into environment and process by way of a new computational sense.[2]

What are we to make of this turn from language to landscape? Arguably, Kepes demonstrates a midcentury reconfiguration of cognition, perception, and sense into algorithm, pattern, and process. In his work, and that of his many colleagues in the computational, communication, and design fields, we witness a subtle hope that a world of static objects and pictures might become one of interactive images and pattern recognition.

In chapter 1, I demonstrated how cybernetic attitudes to storage and time heralded a new epistemology of data inundation. These attitudes to archiving and recording made perception autonomous and material and produced a discourse of storage and interactivity that continues to inform our relationship to the interface and data. This chapter traces how designers, like Kepes, encoded this epistemology into environment. In design practices, ideas of cybernetic temporality, data inundation, and process transformed the nature of the image, perception, and observation.

To offer further sustenance to this argument, let us consider another example. In 1953, the prominent designers Charles Eames and George Nelson stood before a crowded hall at UCLA and proceeded to run an experiment. The purpose of the course was to see "how much information could be given to a class." The function of the class was to combine and apply the latest theories at the time in communication science, cognitive and behavioral science, and design to the training of budding engineers and business management students.[3]

Titled "A Sample Lesson," Eames and Nelson implied that this course would be but one element of a process, a "sample" of something larger. In the optimistic tone that personified Eames design, the brochure announced that "something new" was happening, "bringing down the barriers between fields of learning." This something was not, or perhaps could not be described, because "a *sample* lesson is more of an experience than a tangible solid . . . because it is more of an emotion than an action . . . it is difficult to explain. . . . A sample lesson must be seen and heard and felt and smelled."[4] Implying that language itself was behind the times, Eames and Nelson appear to suggest a pedagogy of sense. Emotion described as active, and pedagogy as affective and material (touchable, feelable, seeable), it was not clear what the course would teach, except perception itself.

Coming at a moment of massive change in higher education spawned by the war and perpetuated by the GI Bill, the work of Eames and Kepes marks a historical shift in the relationship between knowledge and vision. In this span between the training of the camoufleur in design to the education of the future manager of the information economy, wartime imperatives of surviving by means of the identification and evasion of the enemy became autonomous and self-referential technologies of perception. Vision and cognition were rendered equivalent, a "process," to repeat Kepes, and envisioned as part of a single communicative channel that could be algorithmically represented, materialized as technology, and circulated autonomously, separate from content.

This algorithmic optic did not terminate in the hallowed halls of special seminars run for elite students. Arguably, what had begun as the aerial surveillance of cities became a new measure of environment and territory.

Scaling from the personal vision of the designer to the perception of the city, one of the most famous urban planners and policy-makers in American history, Kevin Lynch, began a study at MIT and Harvard, under Kepes's tutelage, titled "The Image of the City," also in 1953. One of the single most influential studies on urban space in the postwar era, it was a landmark in challenging policy at a moment when the city was rapidly being transformed by way of new technologies and economies.

While the title of the study utilized the term "image," Lynch initially equated urban space with the rhythm and cadence of music. He wrote that the city is composed of "sequences" and that urban planning is a "temporal art . . . like music." Continuing in this vein, Lynch argued that vision is musical because of its interactive qualities; "nothing is experienced by itself, but always in relation to its surroundings, the sequence of events leading up to it, the memory of past experiences."[5] Lynch repeated the cybernetic insistence, perhaps inherited from his mentors, on translation between sensorial forms, and chose to reinterpret topics of space, structure, and environment in terms of sensation and affect. Transforming the study of cities, Lynch forwarded the idea that it is through sense and cognition that the urban could be planned. "The purpose of this study . . . will [be to] consider the visual quality of the American city by studying the mental image of that city which is held by its citizens." Psychology and memory are conduits to reinterpret and ultimately reconstruct built environments. Lynch's assessment of urban life both straddled the long-running idea of the city dweller as inundated by information and overwhelmed by stimulus while offering a new type of research methodology by which to contain that deluge, unearth its patterned sequences, and reconstruct space as a mental process. Lynch, like Kepes, signals a transformed attitude to vision and images: perception collapsed with cognition and memory, and used as a conduit to scale between individual subjects and vast territories.

I open with these three case studies because they offer, at different scales, evidence of an emergent form of observer and a nascent concept of environment or landscape. Each of these practitioners was closely related and important to late 1950s design and planning and was central to popularizing ideas of communication and cybernetics throughout culture. Each of these practices, in important ways, offers insight into a newly bequeathed autonomy given to vision as a material process in this immediate post-war period, and each gestures to the reconceptualization of space as an interface. This channeling of the divide between the object and subject redefined aesthetic practice and human perception not in terms of surfaces, screens, or mediating bodies obscuring fantasized political or natural realities, but rather as conduits for communicative exchanges. In the course of this chapter I will move from the training of designers, planners, and engineers to the application of ideas of communication into the structure of urban space, culminating with a classic example of how new models of attention and knowledge negotiated changing economies, racial tensions, and urban formations in the 1964 New York World's Fair. The chapter scales from within the classroom to the organization of corporations and territories. In these movements ideas of perception, cognition, and environment were reformulated and were contested in a myriad of manners.

Counter to our standard assumptions of information theories as disembodied or abstract, this move to give vision autonomy and to turn language into an environment was not a return to some mythic Cartesian perspective. Rather, this move was, in Kepes's words, an "experiential" form of vision, even as it was grounded in nascent concepts of information and communication. In postwar design practices, cognition and perception were rendered equivalent, and both took on new forms of materiality that could be technically and aesthetically manipulated—objectivity was redefined as subjective. To be objective, Kepes wrote, was to learn the "basis of the language of vision," a "basis" that was a process and a technology to be designed. Objectivity, Kepes and his colleagues in the information and communication sciences and psychologies intimated, was no longer about documenting an external truth or reality about the world, nor was it about taxonomy or ontology (describing the essential characteristics of objects). Instead, to be objective would now require producing the most effective and affective method or process to induce, if not replicate, conscious experience.[6]

The infrastructure for this transformation in the practice of urban planning or the training of managers was an epistemology of informational surfeit. Assuming a world of abundant data, designers, planners, and social scientists focused on methodology. Information was redefined as apprehension; a measure not of content but of the way the observer would process data. Designers and urban planners began to view their work in terms of communication, focusing on interactions between agents and concentrating their efforts on producing replicable methods and processes that could be transferred into any environment. Behind these changes lie fundamental shifts in the treatment of archiving, documentation, and objectivity as related to aesthetic practice. For the figures I portray here, the practices of storage came to focus not on documenting individual data points but in storing the traces of method, in making process a material and archivable object.

While much scholarship attends to the relationship between the military, communication sciences, computation, cybernetics, and fields ranging from design to the social and the life sciences, what has not received much attention in the historiography of postwar and Cold War science and its relationship to art and visual culture are the attendant forms of epistemology and knowledge that condition and accompany such aesthetic transformations.

Architectural and design historians have often returned these studies to discussions about built space or home (domesticity) and have largely framed these materials within the context of debates over modernism and its heritages. At the same time historians of science have largely ignored the aesthetics of truth and the centrality of method as an autonomous and central feature of Cold War discourse, with a few major exceptions. Most important, neither group has asked: what are the stakes attendant to making representation a question of process and environment instead of meaning and identity? How do these new tactics and strategies enter the lived field of history, and to what effects? How do older histories of archives, power, and knowledge intersect with these newer modalities of technicized vision?[7]

As the nature of the observer was reconceived, knowledge claims were also transformed. As cognition, perception, and the body (both social and individual) came to be redefined in terms of feedback and patterned interactions *between* objects and subjects (as a communication process), what it meant to produce a truthful account of the world (or a product) shifted, coming to be no longer about hidden truths, invisible elements, or psychological depths but rather about affect and behavior. This transformed idea of truth found itself embedded in an entire new set of tools for the measurement and analysis in the social sciences and behavioral sciences, and a new set of tactics with which to train the observer.

This section excavates this epistemology that links the way we might think to the design of a new type of screen—the interface. This reorganization of knowledge and perception produced new machineries of computing, social research, and marketing. In the course of this section, I will trace the relationship between the emergence of a new form of observer, one both radically individuated and simultaneously networked, and a novel form of knowledge production based on assumptions of informational infinitude, a "communicative objectivity." In the immediate postwar period one can document a shift from modern normative and disciplinary concerns with documentation, objectivity, indexicality, and archiving to a new set of investments in process, communication, and circulation, now encoded into built environments, machines, and attention spans. In these many movements and translations between different sites and practices, however, we can also witness bifurcations and multiplications in how cybernetics, cognition, and communication were understood in relationship to human perception and life. It is as important to examine irreducible differences in interpreting shared epistemologies as homogenizing similarities.

Perhaps most important, in these design and urban planning projects, environment came to replace discourses of structure, class, and race, as the observer was conceived as the subject of a personal and reflexive data space both radically isolated and always networked into a broader ecology. While it may appear obvious from today's vantage point, at the time it was not automatic that transformations in economy, and the increasing changes in urban space, stratification, and racial and class relations would be negotiated through design and a turn to aesthetics and personalization. One might say that these designers were part of a move to produce the world as an interface, making attention itself a material and scalable technology. This shift between structure and landscape did not occur, however, without creating a new set of tensions and possibilities that continue to inflect themselves in our contemporary media environments and urban forms.

Learning to See: The Algorithm of Design

In 1951, Gyorgy Kepes wrote to Norbert Wiener thanking him for his contribution to *The New Landscape of Art and Science*: "after reading your essay I saw that your contribution could be the focal point of my book and that gave me the courage to ask you more than I originally dared."[8] This interchange between the cybernetician and the designer prompts a more global question: what was this "focal point" on which the education of vision was to rest? Kepes answered—method—and introduced a pedagogy to train artists, designers, and engineers by which this "focus" might now rest.

As one of the foremost design and arts pedagogues in America at the time, Kepes had much to say about the future of media and education. With a biography that traversed many legacies of modern design and art, Kepes's life mirrored his recombinant, archivally dense practice—merging influences from multiple genealogies in design and art and remixing aesthetics of nation, identity, and class. A Hungarian émigré who had fled Fascism, he had been born into an aristocratic family in the final years of the Austro-Hungarian Empire. In his memoirs he recalled turning to art at the age of eighteen in order to address "the inhumane conditions of the Hungarian peasantry," some of whom lived on his father's estate. World War I furthered his concern with finding ways to address the suffering of human beings within a technological world, turning at the time to new techniques such as film. He wrote, "only film could bring into a single focus my joy in the visual world and the social goals to be realized in this world."[9] This interest in social welfare apparently inspired his obsessive desire to reconcile art and science throughout his life.[10]

Seeking an ideal with which to negotiate this industrial modernity and the human being, he seemingly found sustenance and inspiration from the ideas of both Soviet constructivism and the Bauhaus (although he never formally joined either). He left Hungary to study art in Berlin, befriending there his fellow Hungarian Moholy-Nagy in the late 1920s.

At the time, Moholy-Nagy was part of the emerging design and arts movement the Bauhaus, which had been started by the architect Walter Gropius in Weimar in 1919. Dedicated to modern approaches to art and design, the school sought to integrate all the arts, craft, and technology in the interest of improving industrial design. The school embraced the machine and technology and never taught history, as design should be taught according to principles, not precedent. It was a large and far-reaching movement whose complexities cannot be fully interrogated here. Most of its practitioners were forced to disperse when the Fascists closed the school in 1933. Many of its leaders found homes in the most prestigious American art and design schools. Tel Aviv, Israel, is another center of Bauhaus design. Kepes himself fled Germany with Moholy-Nagy for London in 1935 and arrived in Chicago to teach in the New Bauhaus in 1937.[11]

Kepes's career, however, was largely marked by his time at MIT, where he taught for upward of thirty years. Invited in 1945 by the then president of MIT, James Killian, to begin a program in visual arts and design, Kepes went on to start the Center for Advanced Visual Culture (CAVS) at MIT and became a central figure in revising the architecture, design, and urban planning programs at the university. Working at one of the central institutions for reenvisioning architectural, planning, and design practice after the war, Kepes had great influence in American (and global) design.[12] In the United States his colleagues and interlocutors included figures such as George Nelson, Buckminster Fuller, and Charles and Ray Eames. His students included figures such as Kevin Lynch, whose work would go on to pioneer environmental psychology and reconfigure urban planning through psychological models of feedback between subjects and surroundings.[13]

Kepes opened the book *The New Landscape of Art and Science*, to which his aforementioned letter was dedicated, with the following words, devoted to his methodology: "the method . . . has served as a kind of laboratory experiment—fuses visual images and verbal communication in a common structure. The visual images . . . are the content. The verbal statements . . . are illustrations. They do not constitute a connected systematic account. The quotations touch the subject from one angle, the comments from another, with the visual images forming the basis of the interrelated structure that alone tells a connected story."[14] The designer implied that verbal statements are illustrations, and images serve as grammars, syntaxes, or structures generating "stories." His focus, however, was not a specific image or text but the "systemic account" and the "common structure" organizing the "experiment." Kepes implied that this text operated to create a story, or meaning, if the reader could create connections between mediums and objects. The fact that Kepes labeled this an "experiment" implied that the concept underpinning the book was not to train individuals in a style, or single practice, but as scientists in a method of conducting inquiry.

In his treatment of vision, Kepes was therefore translating one history of visual practice and psychology into the postwar American milieu. Of great influence on his work was *The New Vision*, put out by Moholy-Nagy in 1929. In this text Moholy-Nagy sought to merge the "physiological experience" of vision with the psychological and cultural aspects of life to produce a form of pedagogy that would encounter the changing technical experience of life, and address the "ABC of expression itself."[15] Moholy-Nagy, like Kepes after him, was seeking the patterns that organized perception, and attempting to formulate a visual pedagogy based on this concept of an expressive abstraction.

The foundation for this design was a "new structure-order," in Kepes's language, that emerged from the recombination of vast data fields. These were archives of images that the designer had compulsively collated through a constant outreach campaign to corporate and academic labs, art museums, and a vast range of colleagues in almost every field imaginable. Kepes's personal archive at the Smithsonian is largely constituted of a network of correspondences concerning the imaging techniques and image acquisition of various institutions and labs in the physical, human, and life sciences.[16]

Archive Frenzy

Of what, however, were these archives composed? What did Kepes seek when he collected images, texts, diagrams, charts, and mathematical equations? The texts are highly idiosyncratic. These books published for the purpose of engaging the design and art community with topics in aesthetics and science, are not organized through taxonomies of historical periods or content (see figs. 2.1–2.4. There is no set organization of history in the display of images or artifacts. Chapters in the books, and sections of his courses at MIT, were not organized around a material or medium or a method. Syllabi left in Kepes's papers list, for example, guest lectures crosscutting neuroscientific ideas of vision and perception with urban planning studies of space.[17]

Kepes did insist, however, that both courses and books should be organized around three terms—"pattern," "problem," and "scale." Kepes collected different varieties of ways to capture processes and to compare seemingly disparate phenomena as linked. So for example, in his book *The New Landscape*, he arrayed aerial photographs of a city through the advanced fish-eye lenses of geographical surveillance teams alongside the famous intimate images shot by Julius Shulman of Mrs. Kaufmann before the pool of her Richard Neutra–designed home in Palm Springs, California, an architectural rendering of the Crystal Palace from the 1851 Great Exhibition in London, aerial images of crop planting, and short texts by William James, Coleridge, and Kafka. Whether town and country are automatically linked, and suburbia or bourgeois sensibility hovers in between, is a matter of ongoing debate by urban planners, but from a pedagogical standpoint, these many forms of life were linked through tactics of scaling and a focus on line and movement—the graphic line of a blueprint, the traces of crop tilling, the curvature of a skyscraper from the lens, the cadencing of language—these elements compose a process where the focus is not on the representation of the world but on the many modes that may be used to do so. Kepes would repeat such exercises with many physical and social phenomena, such as recombining photomicrography and crystallography with aerial examples of earth patterns or geological formations and patterns of urban development.

Authorship had become a curatorial project, and learning came to be about scanning. The student of this course, being inundated with data without context, or linear historical or spatial organization, presumably was urged to examine the relationship between images, to attempt to extract relations between objects. This was, therefore, a matter of collecting not artifacts but methods. This was a collection, and celebration, of the many (particularly technologically) available modes of apprehending the world. Kepes's diagrams, collections, and notes evoke the idea of a course that has no medium specificity but is focused entirely on producing "structures" for vision, or perhaps equivalent to vision. Kepes proposed the terms "dynamic," and "effective action" to describe seeing as part of a new method. The conjunction of "action" and vision denoted both a materiality to the perceptual process and its autonomy as an object of study and as an independent actor. The visual process equated with method, and made material. Kepes's attitude to vision anticipates later attitudes in conceptual art and in cinema where the medium or the artistic process, becomes itself an object to collect, curate, and reorganize.[18] As the archi-

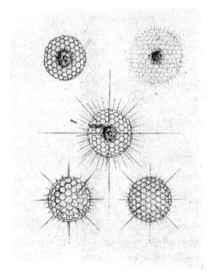

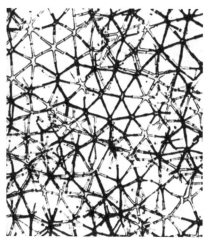

FIG. 2.2__**Structure, Network, Environment.** "The notion of *structure* is taking an ever greater Importance in the field of human knowledge, even with immaterial and abstract concepts; we hear, for instance, of: Structure of thoughts, structure of mathematics, and so on. . . . Seeing in space is not as we believe merely having a keen sense of the occupation of space by some physical object, but rather being able to grasp the notion of combinatorial arrangements in view of obtaining certain peculiar conditions." Le Cicolais, *Contributions to Space Structures. Radiolaria*, from Haeckel, *Report of the Scientific Results of the Voyage H.M.S. Challenger* (*London, 1887*); Geodesic dome, by R. Buckminster Fuller. "Discontinuous three-way grid which stresses its members equally and acts almost as a membrane in absorbing and distributing load." Photograph: Fuller Research Foundation; "Stellate Cells in Pith of Juncus (rush)," photomicrograph by Carl Strüwe. Preceding text and images from Kepes, *New Landscape*, 365.

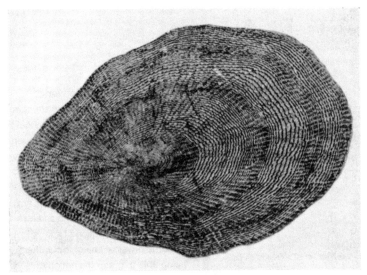

255
Haddock Scale, 9 Years Old
U.S. Department of Interior.
Fish and Wild Life Service
Photograph: Howard A. Schuck

256   257
Beech   *Schwan Glacier Tasnuna Valley*
Photograph: Tet A. von Borsig   Photograph: Bradford Washburn

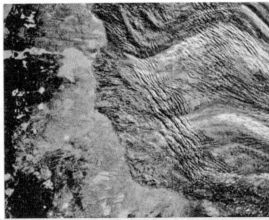

FIGS. 2.3 (LEFT) AND 2.4 (RIGHT)__Pages depicting patterns,
scale, and time. From Kepes, *New Landscape*, 214–15.

258
Strip Cropping and Contour Cultivation
Aerial Photograph: Mitchell
U.S. Soil Conservation Service

259
Shield. Australian Aboriginal. National Museum of Victoria

215

tectural historian Rheinhold Martin has noted, Kepes inaugurated a discourse of patterns and organization in design and architecture.[19]

Archival Truths

This was a most curious pedagogy, therefore, predicated on a strange imperative to agglomerate data. While it has perhaps always been the nature of design to deal with generalizable approaches to the production of objects, Kepes demonstrated a particular reconfiguration of the relationship between process and material. He wrote, "what is called technical education, the mastery of a particular skill or a particular habit of visual representation, should be put off as long as one learns the objective basis of the language of vision."[20] Situated historically, this comment marks a break from an object-oriented or utopian design practice. For Walter Gropius, the famous architect and former head of the Bauhaus from 1919 to 1928 and later professor at the Harvard Graduate School of Design, different elements and mediums were to be taught separately and then brought together into a utopian synthesis identified, for him, with architecture.[21] Johannes Itten who designed an image of the Bauhaus curriculum in 1923 demonstrates this principle in his diagram (see fig. 2.5). Kepes, however, articulated no such linear and progressive structures for the student to work toward; all mediums became media. Kepes took from the Bauhaus and from his previous work with Moholy-Nagy but reformulated many of the same methods toward an expansive use of mediums, and away from concerns with architectural space, or specific forms.

Kepes also had had an extensive interest in the science of Gestalt psychology and had been exposed to the Theoretical Biology Club during his time in London while in exile before coming to the United States.[22] As Donna Haraway has noted in her work on twentieth-century developmental biology, the Theoretical Biology Club was part of a greater movement, engaging gestalt, forwarding a new idea of structure as encompassing interrelationality, wholeness, change, and self-regulation. These ideas in the sciences, Haraway notes, developed before the war, emerged with force and prevalence after the war, becoming the very basis for life and computational sciences.[23] Leigh Ann Roach, in her work on Kepes, has made visible this close intimacy between Kepes's work and thinking and these earlier movements in developmental biology.[24]

Kepes applied these ideas liberally. He took from gestalt and previous modern design movements, but often modified their tenets. He demonstrated little interest in ideal forms, or in finding static biological or immutable structures

FIG. 2.5__ "Images of Pedagogy," diagram of the Bauhaus Curriculum by Johannes Itten (1923). From Lupton and Miller, *ABC's of Triangle Circle Square*, 5. Courtesy Princeton Architectural Press; Plant stalk cross-section, polycrystalline aggregates, and the structure of a leaf. Kepes was interested in linking structure to growth through finding patterns in the natural and scientific world. Images from the essay by Wiener, "Pure Patterns in a Natural World," in Kepes, *New Landscape* (1956), 274.

that underpin perception. His was a nonnormative vision; a pragmatic and empirically driven idea of practice. For Kepes, the designer was supposed to reflexively reproduce and mime this method of algorithmic seeing. He or she was supposed to simulate how *any* image or data set could be integrated into new scenarios to produce an affective response in the viewer. The ideal notion of design, for Kepes, was not about medium specificity and ideal forms but rather ecology, process, and interactivity.[25]

In the courses, the student was explicitly encouraged to concentrate on the method before developing expertise in either any single medium or technique. Vision itself, understood as an algorithmic method or a logical pattern, could be extracted and made the object and "objective" of education, from which designers must form "experience."²⁶

Experience was based on data inundation as a form of truth and moral virtue. "These observations [made by sensory technology, computational, and electronic devices]," Kepes would come to write, "have made a new order of objectivity possible when looking at the world. The permanent record that the camera provides gives us the opportunity for sustaining visual experience as long as we wish, long enough to overcome the errors that the eye makes because of our impatience, prejudice and inability to recall."²⁷ Objectivity here was associated with the power of recall. The human eye might possess "errors" as a result of its archival limitation, but the machinic eye does not. The perfectly extended memory of the machine affords a perfect recall that facilitates a "sustaining visual experience.²⁸ Ignoring or dismissing problems of recording, Kepes focused on memory and access, like the Memex mentioned earlier, as the fantasized sites defining both objectivity and the objective of science and design. The designer being trained in these courses was building a process and "landscape" of vision, not producing individual and isolated objects to be seen. Correlating memory with objectivity, Kepes argued that error was *not* the result of subjectivity, embodiment, or mediation but was, rather, based on a failure to recall and store data. He depicted an informationally dense world, where it was access, not the recording of data, that would be the future challenge. Objectivity was the ability to produce different forms of subjectivity. To be objective, Kepes implied, was to be able to "sustain" and modulate an experience for as long as one "wishes."

The new form of truth being valorized was, thus, a claim for manipulation and mediation, not documentation, as the site of value, goodness, and aspiration. Objectivity was redefined in terms of the production of algorithms, methods, and processes that facilitate interaction, based on the assumption of an infinitude of stored information always/already readily available. The best, which is to say most objective, system for Kepes was the one that allowed the most conditions of possibility for seeing to emerge from recombining data. His work gestured to a wholesale relocation of objectivity away from unearthing a perfect record to the management and organization of patterns and the construction of dynamic structures out of vast data fields in the most effective manner. He wants the designer to produce these autopoietic structures— to mime, perhaps materialize, the process of seeing itself. Kepes, as I already mentioned, explicitly and repeatedly told students not to focus on any one style or one medium or one image.

Design practice is not imagined as answering to existing problems or operating by prescribed principles. Rather, designers, artists, and engineers are encouraged to agglomerate information and retroactively discover patterns. This "communicative" objectivity was data driven, nonstructural, and relational. Turning away from ideals of medium and form and assuming the world as informationally dense, designers could focus on process as material and method. This process was equated with perception and cognition simultaneously—to see and to think being analogized into a single channel.

Reconfiguring the Practice of
Art and the Experience of Space

This production of entirely novel environments found its focus in Kepes's dedication to producing new spaces for artistic practice. In the course of his career he moved increasingly from visual design to spatial and environmental design, and to creating new institutional spaces for art practice. While he was always an artist, his greatest contribution and enduring legacy may be the CAVS, started in 1967 after decades of increased interest in humanizing engineering and science.[29]

Kepes's vision for the future of the arts was unique at the time. As Anne Goodyear has noted in her work on CAVS, the 1960s saw an intensified artistic interest in science and technology. This interest, however, did not always take the same form. Artists and designers coming from pre–World War II Europe often had conflicting concerns about technology as a Faustian bargain between hell and heaven. The work of art, for many of these figures, including Kepes, was to make technology more "human." New forms of visualization should offer proof of the interrelatedness of life to assist in averting future conflicts. Art's work was to use the informational surfeit available through the technical optic of the day, but never to subscribe to it. However, for many of the engineers and artists whose attitudes were forged after the war, technology was often seen as a route to economic and political success. The focus was not on art's place in challenging or forming technology but rather art's subservience to science and technology. Art should be defined by using the latest technologies, rather than challenging or reenvisioning them.[30]

Critical to Kepes was less the application of technology than using art to enhance science. "The scientific-technical enterprise needs schooling by the artistic sensibilities. . . . One of our most urgent and significant educational tasks is the fleshing out of our atrophied sensibilities," he wrote in a 1967 letter to a potential donor. This filling out of sensibilities, an enhancement of capacities for seeing, feeling, and, in his language, attaining "the fullest and richest human use of our opportunities," would not come through exploring the deficiencies or lacks in the human senses, psychology, or capability. His was not a discourse of insufficiency, disability, or lack. Rather, he viewed CAVS as a "testing ground" of new ideas, tools, and media that could "explore the creative forms engendered by technology";[31] CAVS would be the embodiment of his ideals of research and experiment.

This focus on art as actually abetting, challenging, and improving science separated its advocates radically from the engineering-oriented artists of other art-technology groups of the time, such as Experiment in Art and Technology (E.A.T.), a New York–based group fronted by the engineer Billy Klüver and the pop artist Robert Rauschenberg, and closely associated with Bell Labs.[32] Art, for Klüver, was about testing materials to produce new affective experiences. Klüver spoke regularly and often of bringing artists to Bell Labs in order to *do* "technical" things. The focus of E.A.T., arguably, was on transforming engineering into a playful endeavor, and merging artistic practice into engineering and business. The organization of E.A.T. mimed that of a corporation, with boards of directors, and a vision not of sustainability or institutionalization but of applicability. Kepes's vision for CAVS, which was founded in 1967, on the other hand was collaborative, scientific, grounded in the academy, and separated, at least in concept, from industry (despite his work in advertising). His main concerns were with the ethical impact of art in (re)imagining technology and science in relation to society. He was invested in arts of environment and connection.[33]

Both Kepes and E.A.T., however, participated in a data-driven and performative practice. Both groups also assumed a world where emotion was to be carved out of information, but each organization hoped to attach this new epistemology of information to different affects. For E.A.T. it was the improvement of technology through artful engineering; for Kepes it was about returning technology to signification.

In an unintuitive merger between Romanticism and cybernetics, this humanity and "meaning," in Kepes's words, would arrive through the broadening of perception. So dedicated was Kepes to pure affect and this materiality and autonomy of perception that by the 1960s he had turned entirely to producing light sculptures, films, and environmentally focused design projects. Major projected works emerging from his design studios at MIT, never built, included, for example, ephemeral and mobile lighting sculptures for Boston Harbor linked to plans for cleaning the water. This project sought to use the perceptual reorganization through light of the space of the city to induce a hoped-for transformation of the ecology of its waterways. Kepes described this project as an orchestration of the urban nightscape by means of "developing simulation devices of light patterns coupled to a computer," in order to achieve "creative use of kinetic light designs on an environmental scale."[34] The images of the model show beaming lights looming upward on a grid, imposing a surreal alternative landscape on the space.

These light sculptures were indicative of a new form of observation—one that was networked and interrelated to the environment. Kepes wrote that "art . . . without loss of personal vision—in point of fact through the expansion of such vision—is fast approaching the environmental scale and by its own inner dynamics as a craft becoming a collaborative enterprise involving science and engineering. . . . The focal expression of our corporate existence . . . may well be the shape of gigantic luminous forms celebrating our civic pride in our knowledge and high technological achievement, fountains of light, produced by projected sources of powerful artificial light."[35] Light could bridge the individual observer with a civic network founded on a materiality of sense that linked computers and control systems that he envisioned modulating these experiences in response to changes in the environment and in utilization, producing what he theorized as new forms of relationality and identity (fig. 2.6). For Kepes this was not merely a question of vision but truly a modulation of sense itself. His favorite reference was to "color music," to contemplating an integration of senses in these mobile environments enhanced by artificial lighting.[36]

Kepes proposed a number of other projects, including installing mirroring buoys in the Charles River and creating mile-long programmed luminous walls. As Otto Piene, later head of CAVS and among its first fellows, recalled, this was to "compensate for the lost pageantry of nature,"[37] and was meant to produce new concepts of identity and affiliation in urban landscapes through an exploration of previously unexplored aesthetic and psychological dimensions that focused on the interactions between people and between people and space through raw sensation.[38] Kepes would also build interactive light installations for corporate lobbies, of which the most prominent is an interactive light sculpture in the lobby of the KLM headquarters in New York. The mural is a large set of screens with areas of fluid lights and sections of discrete lights that modulate. The ephemerality and interactive quality of the medium refracted the client's global ambitions.[39]

This move toward light as a pure medium was partially in the legacy of Kepes's earlier work with figures like Moholy-Nagy, who also was deeply concerned with light itself as a structure for artistic perception and as a site for the exploration of this new-found materiality of perception. Kepes scaled this

FIG. 2.6__Simulated effects of a proposed mile-long programmed luminous wall, suggested for the Boston Harbor Bicentennial, 1964–1965. From Kepes, *MIT Years, 1945–1977*, 69. Image courtesy of Imre Kepes and Juliet Stone Kepes.

concern with pure opticality to the landscape of the city and into civic engagement in an effort to reunify art and science in the interest of social relevance. His sensorial city, abounding with light, provides an interesting genealogical challenge to my opening scene of sentient space in today's global smart city spatial products. Central to his pedagogy throughout the late 1950s, and particularly later to art practice at CAVS, therefore, were environmental planning and psychology, where the landscape of the city came to be treated as an interface for aesthetic experimentation and as a reflection of both social and individual psychology.[40]

Perception took material form as a channel scaling between the microscopic to the macro systemic. This materiality of process is made evident through the fact that few of his projected works were built and that his greatest success may have been CAVS, an institutional site that formalized this "experiment" and method in using art as abetting science in reconfiguring vision.

Kepes thus demonstrated an aesthetics dedicated to producing a world out of the recombination of images and forms. These texts and projects do not explore an existing world but aspire to produce a self-referential one from the detritus of a world assumed always already fully recorded. As he would write in an interview late in 1971, "today we are in a critical stage of human evolution. Evolution is becoming self-conscious. Our future relies upon how clearly we understand and how well we control the self-regulating dynamic pattern of our common existence. . . . To agree on objectives it is necessary to reach a better common understanding of 'reality.' What I call reality here is neither absolute nor final."[41] In this model of self-referential evolution, the future was about cybernetically regulating patterns—perception becoming material and subject of design and art, circulative as an autonomous process; knowledge redefined as the subjective ability to modulate, enhance, and manipulate perception.

For Kepes the idea of a world fully recorded pushed design toward materializing process and focusing on the relationship between subjects. Perception itself became a form of thought and created new challenges for design, science, and art; not to reveal some truth of form, of nature, of society but rather to organize the interactions between users. The focus of design turned toward the structure of organizations, systems, and environments. The remaining ethical question was what shapes these networks would take. Evolution could now be "self-conscious" if designed appropriately.

Found Educations: The Pedagogy of Communication

Kepes's work, therefore, should not be seen in isolation but should be understood as a broader effort to revise concepts of knowledge and the practice of business, science, and design. In a reflective postwar moment, when concern with the ethical and humanistic impact of technology was high, Kepes's initiatives were hardly unusual. Many universities introduced new art, technology, and design pedagogies. This reform impulse was spurred further by C. P. Snow's famous discussion of "two cultures" and his critique implying the degraded moral and ethical effects of such a condition. Over the next twenty years MIT, UCLA, Bell Labs, and many other labs, corporations (even Rand Corporation),[42] and universities developed programs to integrate (successfully or not) arts training into the engineering curriculum. What was guiding this reform impulse, however, may have been less a concern with ethics and morals and more a steadfast belief that knowledge (and business) was being transformed, and central to this transformation was the paradigm of information, communication, and computing.[43]

Propagated by the designers and university administrators was a regularly articulated faith that vision, above all, could provide a tool to reconcile the humanities, sciences, and arts; providing a universal language not only within the university but for what was increasingly understood to be a global and interconnected planet. This faith in vision kept repeating itself, ad nauseam, whether in urban planning and design or in architecture and design. The imperative, therefore, was not only to teach art appreciation but also to focus on

FIG. 2.7__Lecture hall during sample lesson at UCLA, May 1953. The Work of Charles and Ray Eames, Negative no. LC-E12, CE 492-B, lot 13181, no. 2, Print and Photography Division, Library of Congress. © 2013 Eames Office, LLC.

visual communication as a location where students might be taught new forms of production and experimentation.[44]

A course taught by the preeminent postwar designers and intimate colleagues of Kepes, Charles Eames, and George Nelson for business school and engineering students at UCLA in 1953 is demonstrative (fig. 2.7). Previous versions of the course were taught at Georgia Tech as well.

Designers whose clients included IBM, Herman Miller, and the U.S. Information Agency (USIA), these are the individuals most commonly associated in contemporary culture, and historiography, with the public face of government, the aesthetics of the Cold War consumer lifestyle, and the office space and style of the postwar (modern) corporation. The course was intended to teach the idea of communication to the uninitiated.

The brochure intimated that the course had to be "experience," and that it was "new." It was so new and experiential, in fact, that the class could not be described.[45] Language seemingly replaced by sensation or affect.

Considering the interactive and contemporary language of the brochure, it is somewhat surprising to witness Eames arguing that this "something new" must come from something old. He spoke, repeatedly in the course of his career and in reference to this course, of a "found education," suggesting that education should be about recombining data. Reflecting on this 1953 course in an article also titled "The Language of Vision," he argued that the main concern for designers and managers was to reduce "discontinuity" between disciplines in an information age. The inverse and implicit corollary was that such a reduction in information flow would result in an increase of the capacity of the individual or the institution to process data.[46]

To produce this smooth space between disciplines, Eames felt "vision" was the best tool. To teach students to see, however, must come through exposure to an excess of data. For Eames only through information *inundation* could learning commence. He was very specific on this point; the purpose of the class was to experiment with "how much information *could* be given to a class." Data overload as pedagogical principle.

This principle was a long-running one. The work of the Eames Office in education was not contained to the courses Charles taught. Charles and Ray Eames were deeply involved in many exhibitions and in developing educational materials for the public. The Eames Office designed brochures (fig. 2.8) for Science Research Associates, Inc., an educational products company owned by IBM, a major client of Eames. Science Research Associates introduced behavioral research to education, and its learning system was influential in postwar elementary education.

The brochures introduce a learning system and new forms of teaching math in lower grades. Central to this "new" math is the teaching of set theory, a central component for programming and computers, at far earlier ages and before university. The courses also emphasize group and collaborative work, modular teaching, and constant assessment and measurement of outcomes. Science Research Associates also emphasized speed reading and analytic capacity, and was a major provider of grade school educational materials throughout the 1950s and 1960s, as part of Cold War concerns about educational achievement (SRA was purchased by IBM in 1964). These methods and training systems were the childhood equivalent to the engineering courses, a system designed to help students deal with large amounts of data, find patterns or sets, and analyze quickly.

Charles Eames believed that design should *amplify* and accelerate the availability of information. In what, from the vantage point of contemporary de-

bates over attention deficit disorders, now appears as anathema, Eames spoke of distraction and overstimulation as an education. Rather than worrying about information overload, Eames thought more data offered more "choice," giving the spectator a freedom to "choose" from and produce his (or her) own patterns and combinations. He would speak throughout his career about "connections," and making connections possible was the purpose of good design.[47]

The Vision of Communication

This was not any form of choice, but a specific type—choice as creatively interpreted from digital communication. As the premier science educators in the United States, the Eames Office maintained close contact with individuals in cybernetics, psychology, computing, and communications engineering. In this "sample" lesson, Eames and Nelson specifically sought to teach both cybernetic and communication principles involving information and feedback.[48]

This course was structured around, and dedicated to, the *Mathematical Theory of Communication*, first formally introduced by Claude Shannon and Warren Weaver in 1949, and defining information as a probability (also discussed in chapter 1). This model, emerging from the multiple influence of telephony, antiaircraft defense, cybernetics, computing, and radar war research, famously split form from content. In the realm of communication theory, information does not denote meaning, only the choice between possibilities within a *structured* situation. Repeating Weaver's summation, we may recall that the idea of communication "not so much to what you *do* say, as to what you *could* say. . . . The concept of information applies not to the individual message (as the concept of meaning would), but rather to the situation as a

FIG. 2.8__Science Research Associates brochure (1964). The brochures state: "in March 1964, Science Research Associates, Inc. became a wholly owned subsidiary of the International Business Machine Corporation, operating with its own management and board of directors. Under this arrangement, SRA [Science Research Associates] will continue to develop new materials based on research in the behavioral sciences as it relates to the learning process." Works of Charles and Ray Eames, Print and Photography Division, lot 13195–395/396, Library of Congress. © 2013 Eames Office, LLC.

whole."[49] Redefining information not as an index of a past or present event but as the *potential* for future actions (not what you *say* but what you *could* say) encouraged engineers to defer concern about particular messages and to refocus on interactions between sources of signals, in a move that mirrors the turn in design to complexity, process, and connection.

In the *Mathematical Theory of Communication*, under the joint pressures of military and economic interest, building channels or communication systems was recoded in terms of efficiency, compression, and logic, and came to be understood as engineering the most efficient (least-choice) possibility for transmitting the most potential types of data; leading to the decision to use a binary numeral system—a choice between only 0/1—as the basis for digital communication systems. In communications science there is no ontology or interest in presence, the present, or the index, there is only the modeling of a situation or a relationship that delimits the types of future exchanges to be conducted.

The Eameses, however, were not interested in efficiency as understood by engineers. They were interested in the most choices that could be produced,

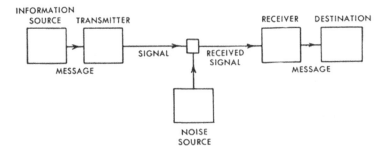

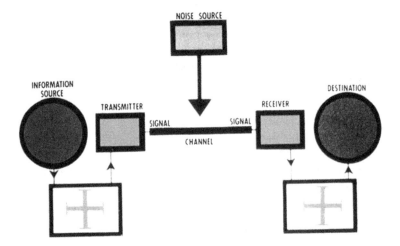

FIG. 2.9__Interpretations on a theme: (top) diagram by Claude Shannon, from Shannon and Weaver, *Mathematical Theory of Communication* (1949); (bottom) diagram by Charles Eames from Eames, *Communication Primer* (1953).The Eameses add an extra step in the diagram for (translation or corruption of message) the little crossed boxes, and interpret the theory more as a concept of interactions with the environment than as a strictly functional idea for information transfer. Particularly salient is their idea of redundancy, which for them can be virtuous by allowing clarity, versus engineers, who often want to remove redundancies, as in audio or video compression algorithms. Film stills from Prelinger Archive, http://archive.org/details /communications_primer.

all the while seeking to make useful and pleasureful things (fig. 2.9). Eames's and Nelson's classroom method was one experiment in attempting to produce a new type of informatic spectator and designer—both the consumer and the producer of such communication systems. Under the influence of these theories of communication, but also influenced by ideas of feedback and behavioral psychology, the team, repeating Kepes, defined the class as a methodological experiment, where students would be exposed to a vast amount of data and asked to distill this data into a single, coherent visual presentation. Anticipating our contemporary multimedia perceptual field, vision should not, however, be understood as a singularly optical register. In the class students were inundated with sensory stimuli, "a live narrator, a long board of printed visual information, and complementary smells . . . piped into a chemistry lecture theater." The principle, Eames argued, was not to produce a "far-out" experience but to develop the concept of the class. This concept, they argued, was that of information theory as it applied to design and architecture.[50]

The purpose was for students to unearth a distilled logic or pattern within a non-media-specific data field constituted of sound, smell, and image that they could then isolate and reenact through a slide show or short film. Decades before the age of PowerPoint presentations, the Eameses trained budding engineers, architects, scientists, and business managers to present their information in the logic of visual language and succinct bullet points. Eames wrote that the best tactic was "to put it on film because through the medium the central idea can be supported by images which give substance and liveliness to it. This reduction of one idea to its essence, using the support of visual images, is the core of several films we made on mathematical topics." He wanted students to distill the "essence" of the data, and he felt that vision offered the best tool to do so.[51] The image, here, is lively and substantial while also reductive and essential. Eames hints that this form of image is a process and pattern that embodies "ideas." This durational image is also historical. It was through the legacy of a different medium, film, that Eames hoped to move to computation. This was not a discourse about the end of cinema but simply the consumption of one medium into another in the recombinant and archival logic we now understand as underpinning computing.

The remaining slides from the course, now in the Eames Office in Santa Monica demonstrate some of what was shown. The slides comprise of hundreds of images. They are arranged in such a way that one detail or a repetitive pattern links what are otherwise discontinuous and unrelated phenomena. So, for example, an eagle's claw turns through its circular shape into a staircase that slowly, through a sequence of shots, becomes a dinner party. In the film that resulted from the course, *A Communication Primer*, which teaches the mathematical theory of communication, a red light becomes an on/off message, a logic gate, and then a mathematical theory of communication. What stays stable is the structure of presentation, the ratios, the patterns and speed of image delivery, the setup of the screen or interface in relation to the observer. This stability in the organization of vision and sense makes the content malleable in scale and meaning. These slide presentations and films established examples of ideas the Eameses would later encode in their large-scale installations and educational films.[52]

For example, in the images from the *Communication Primer* (fig. 2.10) the Eameses demonstrate how the expression "I love you" can carry no information if uttered within a context in which there is no possibility for the message to entropically degrade to "hate." If one is "in love" and is told by one's partner "I love you," there is no information in the exchange, since the receiver has no choices (no degrees of freedom in selection). A message that will definitely arrive to the receiver contains no information, even if it has great meaning. Only if we are not sure of the love, or if there is a choice to not love, can there be information in the system. The goal of the exercise is to give an example that is "meaningful"—love—and turn it into an example that bears information by making it equivalent to communication and action (maybe affect) instead of an identifiable emotion.

Making "love" unsentimental and emotionally unidentifiable demands some creative modulation of spectatorship on the part of the Eameses. As the voiceover of a man intones the definitions of a mathematical theory in highly standard 1950s monotone documentary style, we see the movements of grass, a man's mouth grazing the ear of a woman, and then a set of flash cards flipping from love to hate. This irrational set of images is lent coherence through repetition, cadencing, and voice. The images themselves possess no stable coordinates in space and time, roaming between locations, scales, and examples. Repetition is key, therefore, to producing this equivalence.

The movie demonstrates the same concept of symbol distortion repeatedly throughout the film. Before love, the definition of communication is repeated four times in the example of reading, the transfer of money between banks, a diagram of the mathematical theory of communication, and finally in the telephone "game" in which individuals whisper to each other. Love is the fifth example. The movie re-performs, literally, the idea of redundancy, the production of noise, and the concept of information, by giving multiple examples of the same idea. In the example of love, which is the apex of the exercise, this redundancy is demonstrated through the compulsive repetition of icons of love

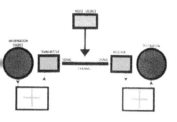

FIG. 2.10__Stills from Eames, *Communication Primer* (1953). Film stills from Prelinger Archive, http://archive.org/details /communications_primer.

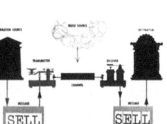
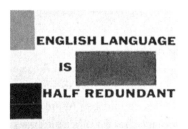

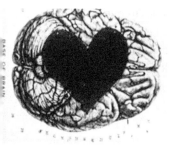
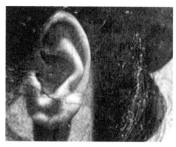

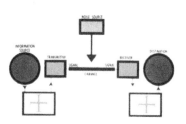

again and again from art, including Romantic eighteenth-century paintings, Chagall images, and hearts. There is no mise-en-scène of the love scene, and therefore there is no way for us, as spectators, to identify with the emotional or subjective nature of the scene. What we are permitted to do is recognize our familiarity with such scenarios generically. We see love as a technology equivalent to communication channels, finance, and computers. Spectatorship, finally, is not about scopophilia, or identification or desire with images, but rather about deducing this pattern from the data field.

The Remains of the Archive

Eames did not leave this archival vision in the classroom. The Eames Office, as a design practice, largely operated through recombination, and Charles Eames argued that films should be "found films," implying that the films should recombine footage shot for other purposes in science, art, marketing, or design to produce new patterns and messages. With literally everything being stored—from production stills, mockups, and slides to numerous objects regularly collected in world travels—the work of the Eameses regularly repeats images and conventions from other pieces. Their archive at the Library of Congress is extremely extensive, and was well catalogued, by themselves, upon donation, and their office still contains endless rows of cabinets where Ray Eames would store every button, doll, piece of cloth, yarn, string, and other objects—often toys from around the world—used in their work. Almost all their movies have a scene using footage, just like their found education, from earlier work. Arguably the very distinguishing feature of Eames design was their sense of play, the ability to bring in toys, buttons, games, representations of childhood to the presentation of topics like computers and math, recombining and making nonlinear the very human life cycle.

These are archives, however, that are organized not through a logic of classifying objects but through modes of apprehending objects. It is perhaps no longer even appropriate to label these practices "archival" in the sense of the nineteenth-century interest in taxonomizing and organizing objects, artifacts, and peoples into stable relationships with one another. In the histories of cinema and photography it is the endless efforts of Étienne-Jules Marey to capture the present, or the documentary zeal of Alphonse Bertillon or Francis Galton, that has laid the foundation for theorizing archiving. While these individuals all worried about informatic excess, they just as assiduously worried about capturing, recording, and organizing data; the archive was not always already there, it had to be built through complex assemblages of many instruments, machines, and techniques.[53] If the archive has always been about its structure and the associations facilitated through its organization, this had now become not the implicit and repressed function of the storage endeavor but the *explicit*, and openly advertised, purpose—"found educations," "found films"—a spectacular discourse of remainders, recycling, and storage. This discourse was made more marked by the concomitant absolute disregard for, and total silence about, organization or taxonomy.

This is a subtle but important mutation in the Eameses' own history in design. Originally trained at Cranbrook Academy, and familiar with the Arts and Crafts tradition, both Charles and his colleague and partner, Ray, had come from a space where automation was not aesthetically desired and was eschewed. They took what they understood to be the "human" element of this training, the immanence that the movement had attached to craftsmanship and the people-centered, to rephrase Charles's term, approach and reattached it to a new type of machine. The Eameses sought to balance a desire for mass consumption, information technology, and reproducibility with a fantasy of quality, immanence, and "humanity" in design. What separated their practice from this previous history of design lay in bequeathing the value of craft, essence, and immanence onto information patterns, with these patterns taking the place of the well-crafted and singular object. Even their objects, like the infamous Eames chairs, cannot be seen in isolation from an entire system for reconfiguring the human perceptual system, lifestyle, and work.[54] Innovation, beauty, objectivity, and moral value were all reassigned to a new site—the production of methods. Methodology, process, and, in Charles's words, "connections" assigned the value of craft and object-hood. This, again, may have always been the implicit idea animating many schools of modern design and industrial manufacturing, but now it became the explicit and central goal, separated from any clear-cut endpoint in the single design object or utopian ideal of a pure aesthetic (or even a school) and reformulated to apply to everything from architecture and design to strategy and management in business and engineering.

The Eameses took seriously the idea of process as a material for storage. Their archive is not only full of found materials, which they recycled continuously, replacing redundancy for specificity and signification. They also assiduously documented the actual production of every project. From the first sketches to the action of their assistants arranging images on light walls, the Eames Office was comprehensively obsessed with collecting, collating, and showing not the content of their exhibitions but the process of designing itself. Their understanding of storing information appeared to be imbricated with the concept of apprehension—how to catalogue and store the tactics and strategies that induce perception and action in the client or observer? Good information design was about figuring out the communication structure of the interaction. The retrospective of their work in 1969 at the Louvre is demonstrative. The Eameses show not the final products, or the exhibitions, but the process of building these exhibitions and the details that come together in their designs. Organizing the show is a famous chart by Charles Eames demonstrating how design must be a dynamic process, by showing the constantly changing relationships between clients, designers, and the public. Central to the office was collecting visions of process itself, and focusing on the ways that design could form attention.

These students were thus being trained to become both the consumers of data, capable of choosing patterns, *and* the designers of vision itself, the managers of this consuming algorithm. Charles Eames regularly implied in his writings and with his diagrams and exhibition designs (figs. 2.11 and 12) that to observe was to scan, to find "connections," with the implication that the cognitive process of learning to see would become equivalent to his self-defined "language of vision"—to see, to think, and to speak all becoming part of a single channel.

The Ecology of Sense: Urban Planning, Information, and Affect

Eames's and Nelson's course offers evidence of a new technique of management by way of an algorithmic optic that made attention and cognition synonymous and could be extracted, and perhaps measured and managed, a theme I will continue in the next section. The move between an organism's perceptual field and the organization can perhaps be no better demonstrated than in the work of Kepes's most famous student and perhaps one of the greatest and most influential of the postwar American urban planners, Kevin Lynch.

Lynch initially hailed from Chicago, where he was born in 1918. He was trained in Catholic schools that were under the influence of John Dewey's educational philosophies. This philosophy's faith in democracy, participation, and pragmatism inflected his later design practice. While his high school experience was during the Great Depression, Lynch in his interviews mentioned the Spanish Civil War as the "first real political influence" in his life, stirring an interest in socialism, communism, and contemporary political matters.[55] On reaching university age, he opted for architecture and was advised to attend the very conservative Beaux-Arts program at Yale. By his sophomore year the

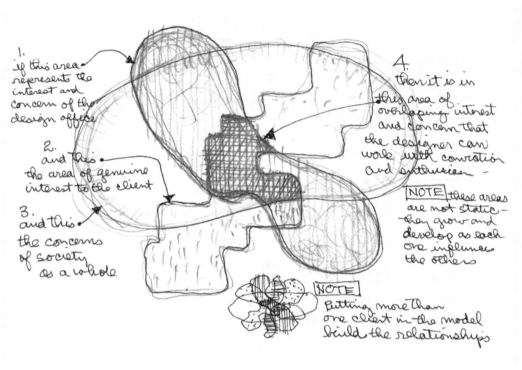

FIG. 2.11_ "What Is Design," diagram by Charles Eames (October-December 1969), sketch for the exhibition "What Is Design," Musée des Arts Décoratifs, Paris. The diagram represents the interest and concern of the design office, the client, and the society: "Then it is in this area of overlapping interest and concern that the designer can work with conviction and enthusiasm. NOTE: these areas are not static; they grow and develop as each one influences the others." The Work of Charles and Ray Eames, box 173, folder 9, Manuscript Collection, Library of Congress © 2013 Eames Office, LLC.

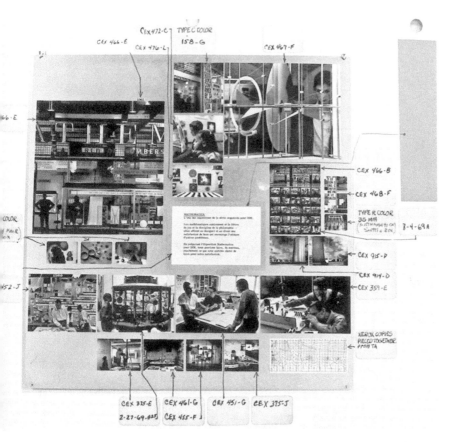

FIG. 2.12__Images from the exhibition "What Is Design" (October–December 1969), Musée des Arts Décoratifs, Paris, depicting the show "Mathematica." There are panels depicting the process of selecting images, their layout, the mockups, the piecing together of Xerox copies collating mathematical equations and so forth. The exhibition itself was another Eames project in educating the public in mathematics and practices of calculation. Work of Charles and Ray Eames, box 173, folder 10, Manuscript Collection, Library of Congress. © 2013 Eames Office, LLC.

curriculum had frustrated him, and on reading about Frank Lloyd Wright, Lynch wrote to him at Taliesin, "filled with enthusiasm to go to a place with such an atmosphere of freedom and creation." Wright responded that Yale could in no way foster creativity and invited Lynch to Taliesin. There Lynch worked for a year and half, during the school's move to Arizona.

Lynch, while inspired by Wright, found his social philosophy backward, however, and very "arts and crafts" (which put him at an interesting disjunction with the Eameses). He left to study at Rensselaer Polytechnical Institute to gain engineering experience, and was shortly thereafter drafted into the army; he served in the Army Corps of Engineers in Palau Islands, the Philippines, and Japan. On returning, he decided to attend MIT on the GI Bill and complete a bachelor's degree in urban planning. He was hired at MIT with only a BA, on the strength of his thesis and work, and he continued to teach there until the early 1980s.[56]

In Lynch's studies of urban space, landscape, environment, and pattern took the discursive place of structure or function by way of a new epistemology. To understand Lynch's work, and its innovations, demands some recognition of the history of planning. In the United States, planning in the early twentieth century was heavily influenced by men like Daniel Burnham (who planned Manila after the Spanish-American War and Washington, DC), Fredrick Law Olmsted (of Central Park in New York fame), and Ebenezer Howard, the British planner who conceived of the "garden city." For these men, the city was the producer of and mirror to social order and health. Cities were massive instruments of normative force shaping populations into the best forms for society. Lynch had also been exposed to the hyperindividualistic and mobile ideals of Frank Lloyd Wright, whose Broadacre city spread out across the plains as a diffuse space, networked by roads, focused on individuals owning homesteads. Wright was symptomatic of modern utopian planners who extended, in many ways, the late nineteenth-century concept that the city was an organism, or machine, for producing ideal social orders. For Wright this order was about rugged individualism and the motorcar; for other influential architects of the time, such as Le Corbusier, and his Ville Radieuse, it was about the city as a machine for industrial organization.[57] All these ideals of planning shared in a structuralist assumption that sought to impose an ideal social order, rather than generating it through longer term processes of interactions between agents in the environment. The citizens of these cities were envisioned as normative. The work of architecture and planning was to organize these populations into stable hierarchies, locations, and spaces. Kevin Lynch would separate from this history.

In 1960 he published a study titled *The Image of the City* that continues to be one of the most influential texts in planning, transforming modern visions of utopian cityscapes. Dedicated to Gyorgy Kepes, the study had emerged under Kepes's supervision at the Center for Urban and Regional Studies at MIT. It was part of a joint research project between Harvard and MIT that involved examining the relationship between the changing nature of cities at a time of increasing suburbanization, eroding tax bases, and deindustrialization (certainly true of Boston by the early 1960s) and the "look," in Lynch's words, of the space. "Giving visual form to the city is a special kind of design problem," Lynch wrote, "and a rather new one at that."[58] Lynch's hope was that in studying this "look" new answers might be found to solving the problems of what even in 1960 was clearly a rapidly changing urban environment increasingly confronted by racial and class division and clearly degrading in physical infrastructure.

Lynch later recollected in a UN policy paper dealing with the future of the city in the 1970s that the purpose of this research was to link perception and cognition to policy.

> Research into environmental perception began about twenty years ago. Its original purpose was to change the way in which urban areas were being designed. These first efforts have had continuing consequences in many fields but only a superficial effect on public planning at least until rather recently. . . . We began to inquire into the way people perceived their city surroundings in the late 1950s. The aim was to clarify some vague notions about the visual qualities of large environments, and particularly to show that you cannot evaluate a place, and should not plan for it, until you know how its residents see it and how they value it. The very first trials were concerned with how people locate themselves in a city, and find their way through it, but these experiments quickly escalated to considering the entire mental image of a place.[59]

By implication, Lynch was among the first to produce a psychological account of urban space, making individual cognition and perception a dominant discourse in urban planning over the course of his thirty years as an active educator, policy-maker, and planner in Boston and at MIT. While in the 1950s this was merely a side note in planning, public health, and urban public policy, by the mid-1970s sociological discourse concerning urban problems was fully invested in psychological explanations. Lynch's particular account of vision therefore linked the urban form at a large scale to the individual account of space.[60]

The initial studies followed three cities—Boston, Los Angeles, and Jersey City. The city was mapped at different scales: from the air, through standard cartographic methods, street maps, traffic maps, maps of infrastructure, and at the street level. Lynch worked to produce an optic that could scale from the vast to the personal.

Cognition and Communication in Planning

As part of this new perceptual form there was also a novel experimental and methodological apparatus, aimed at measuring affect and sentiment. Lynch produced a taxonomy of process to define and categorize perceptual processes. The first was *imageability*, "that quality in a physical object which gives it a high probability of evoking a strong image. . . . It is that arrangement which facilitates the making of vividly identified, powerfully structured, highly useful mental images of the environment. It might be called *legibility*, or perhaps *visibility* in a heightened sense." The best city is highly imageable, according to Lynch; it produces distinct sensations and well-formed mental maps.

Lynch's concepts of "imageability" and cognition were tightly derived from cybernetics, the rising cognitive sciences, and computing. In formulating the study, Lynch, for example, wrote regularly to J. C. R. Licklider, who helped found Lincoln Labs and was one of the pioneers in computer development.[61] His famous 1960 essay *Man-Computer Symbiosis* reframed cybernetic concepts of feedback into ideas of an evolutionary coupling between humans and machines that would be prosthetic and augmentative rather than competitive. He applied ideas of time-sharing to the relationship between humans and machines, where each would do separate tasks to free time for the other.[62]

Lynch's discussion with Licklider is illustrative of how ideas of cognition and subjectivity were transforming planning, but also the liberties that urban planners were taking with communication theories and computation. Licklider was concerned about the "normative" elements of the study in Lynch's preferences for more "imageable" cities. He was also interested in applying standard tools like Fourier transforms to define form in the city and compare it to human perception. Finally, Licklider worried about the black-boxing of "meaning" in Lynch and Kepes's proposal.[63]

Lynch responded with interest with regard to the normative concerns, but unlike the computer scientist, Lynch was ready to relinquish objectivity and programmability in either defining form or "meaning," arguing that while Licklider's concerns were understandable, the intent of the study was quite different:

in other words, instead of saying that "cats are black" and being led to a direct test of whether they are indeed black, we are saying, "it would be best for all of us if cats were black." Since we are not prepared to prove this vague statement, we are really saying, "since we assume that we would all be healthier and happier in a black-cat world, then one of the most important things we could learn about cats is how to make them black." It gives us a set of values as guide posts in a complicated beginning [from which future policy and research will emerge].[64]

Rather than defining the goals ahead of time, Lynch works in a fantasy world that accedes to its own illogical assumptions but uses them to change the direction of planning. Instead of asking *how do people see or feel or think*? Or *what is wrong with the city*? Lynch wants to push for an anticipatory design that gathers data first in order to then retroactively unearth the characteristics that might be desired. On one level, Lynch still wants to create "happier" cities, but in terms of method, his focus is on the process—discovering the nature of black cats—rather than on defining terms and definitively proving or disproving facts about cities that have been clearly delineated before the start of the study. In keeping with more contemporary methods in building responsive environments, Lynch prefigures such attitudes by substituting fuzzy logic and data gathering for structural and hierarchical approaches to planning. For Lynch the best social science focuses on post-data-gathering analytics and the production of instrumentation rather than the generation of structural ideals or types for urban forms.

Lynch also pragmatically argued against a reduction of language or signification, including what both he and Kepes labeled "visual" language, to strict logical computation. For Lynch and Kepes, signification might be returned to planning by way of reflexively using the archive of individual memory and history.

Lynch on one hand shared a sympathy with communication sciences in attempting to develop standardized methods for comparing human cognitive understanding of an environment with other maps and plans of the same space. Computing, cybernetics, and cognitive science offered Lynch an armory of metaphors and ideas, but on the other hand he resisted attempting to pretend that the techniques of computer science to represent reality were a scientific, or objective, approach. Lynch saw a diversity of patterns and rhythms emerging from comparison, not a single standard approach, such as Fourier transform, to make discrete data points appear continuous. He was even more antagonistic to the idea that human perception operated in one such standard manner.

For Lynch the disjuncture between what is measured, visible, and logical and the complexity and meaning of the space of the city was precisely the disjuncture driving planning interest. Like the cyberneticians discussed in chapter 1, Lynch found a differential between the archive and interface, or the "image" of the city. But unlike many in the computer and artificial intelligence projects of his day, his interest was in multiplying that disjuncture rather than suturing it. Lynch did not seek to suture the space between the imaginative and meaningful "image" of cities in inhabitants' minds and the measurable logical forms that planning could produce, control, and anticipate.

If Licklider was still interested in precise definitions and clear-cut mathematical approaches that were predetermined, Lynch was interested in the reverse—in studying the small differentials between representational approaches to produce a more complex idea of the city. In many ways, Lynch's approach allowed greater penetration of ideas of communication, method, and process than the failing efforts that were being made in the computer science of the time to clearly define language or thought. I draw attention to this feature of the study because it signifies a broader movement in the social sciences and design practices at the time to translate cybernetic and communication science concepts of feedback and communication into practice by repressing the constraints of formal logic and positivism.

The Measure of Perception

In order to study this psychological and perceptual interaction with space, Lynch proceeded over the course of five years, 1954–1959, to rigorously interview numerous subjects (fig. 2.13). Participants in the study were asked to draw mental maps of significant landmarks, to sketch the paths they navigated every day, and to draw and classify elements of their environment from memory. They were also asked to classify, from memory, types of buildings and types of landscapes. They were asked to describe the segments of the space they found most memorable, remarkable, or affective. The drawn maps were then compared with standard maps made by geographical and aerial surveys of the space, including the actual buildings, parks, and other environmental features occupying the space, and were also compared with the perspective of planners on the environment. To get this other perspective, that of the planner, the studies also involved the systematic field reconnaissance of the area by a trained observer who mapped various elements, their visibility, their "image strength or weakness, and their connections, disconnections, and other interrelations."

FIG. 2.13__Figures from Kevin Lynch (1960), demonstrating the link between aerial views, street mapping, and perceptual mapping. From Lynch, *Image of the City*, 18 and 24. © 1960 MIT, by permission of The MIT Press; Personal "Images" of the "City" drawn from memory by study participants. From Lynch, studies of *Image of the City*, MC 208 available at the Dome website of MIT Libraries, http://libraries.mit.edu /archives/exhibits/kepes-lynch/index1.html. Courtesy of MIT Libraries, Institute Archives and Special Collections, Cambridge, MA, Kevin Lynch Papers. All rights reserved.

Lynch openly admitted that all these methods were "subjective." In saying so, he valorized subjectivity as a tool, an approach to measurement, and a standard for truth in the sciences of urban planning and administration. Subjectivity was no longer a problem but an operation that could be used to rethink and represent the city if planners, designers, and policy-makers but examined the spaces between different types of mapping rather than the maps themselves. The city emerged by comparing different forms of "subjective" maps rather than through a structural approach that assumed an ideal or normative form before the planning process and evaluation commenced.

In these studies, therefore, perception was conceived of as a channel or a process, if we will, with a temporality. The study states that cities are "of vast scale, a thing perceived only in the course of time." City design, the study concludes, is a "temporal art," and experience is a matter of pattern recognition and manipulation. The new function of urban planning is to deduce these relations and to mime them—solving the problems of urban blight reframed as a perceptual and psychological problem rather than, say, a problem of class or structure.[65] Urban space could be conceived as a channel that could always be improved and enhanced, made more "imageable" and more conducive to circulating sensorial information.

The importance of these studies cannot be overstated. Lynch's work was critical to the imaginary of urban redevelopment and politics emerging at the time. The running of these studies served as an important training ground for many students at MIT and Harvard who served as these observers who conducted these studies and were trained in these methods. From these early studies of the 1950s grew an entire methodology and a series of tactics for urban redevelopment. More significant, Lynch's cognitivist approaches in-

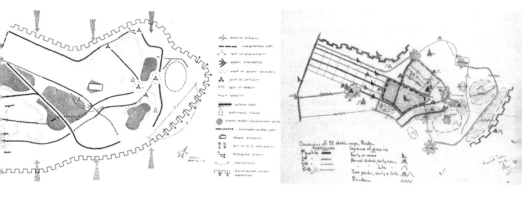

spired and underpinned Nicholas Negroponte's approach and were closely affiliated with the rise of Media Lab and Negroponte's work on simulating urban space and transforming architecture.[66] Lynch also spearheaded initiatives in community gardening, for example, and he is considered central to the formal rise of environmental psychology, a field that owes its existence in part to his work. While today environmental psychology finds itself embedded within organizational psychology or relegated to a side note in public health, psychiatry, and sociology, it was a prominent discourse throughout the 1970s and 1980s, perhaps making a return in our present with the ubiquity of interactive and sensorial collection devices in our lived environments.

When situated within a broader discussion about the future of the city coming at the very threshold of the urban renewal programs of the late 1950s and 1960s, Lynch's work marked a discourse of psychology and space that both critics and advocates of the urban policies toward office space and business development would implement. Major works such as Jane Jacobs's *Death and Life of Great American Cities* and community-based research such as that of Herbert Gans reflected a broader understanding in sociology and urban studies of the importance of psychology and spatial arrangement to the subject, and heralded an affective turn, so to speak, in the social sciences, where the alienated subject of urban life, first posited by such figures like George Simmel in the early twentieth century, could be reconciled with the urban space through a reconceived and newly introduced discourse of environment.[67]

More important, Lynch and Kepes also demonstrated a bifurcation in the social sciences. As Rebecca Lemov has shown, the 1920s and 1930s already saw the emergence of a social science interested in "systems."[68] Integrated with cybernetics and communication theories through the Macy Conferences in the mid-1940s, these sciences consolidated an approach toward systems that relied heavily on a similar transmission of methods. However, the systems approach came to view cities as exceedingly complex, and planning therefore as anachronistic.[69] Planning projects were viewed as impossible, modern, utopian ideals with no relationship to the society into which they were being imported. Lynch was among the pioneers of a new strategy by which to reconcile the understanding of systems with individuals by using cognitive approaches as a conduit by which to reformulate both the individual and the individual's relationship to populations and territories, making all these entities more flexible, and allowing localized interventions with systemic implications without assuming the need to act globally.

Lynch's work, fostered and encouraged by Kepes, demonstrates a strange folding. Discourses of information inundation, pattern seeking, and communication became those of communal psychology, environment, and community. These new concepts of planning and territory linked subjects to scalable territories through the standardization of method applied through organs such as the UN.

Planning, psychology, health, and space had long been linked throughout the nineteenth and into the twentieth century. We have but to think the relationship between psychoanalysis and Secession art in fin de siècle Vienna, Ebenezer Howard's notion of health and spirit in his garden cities, or the projected plans published in the issue of *Architectural Forum* titled *194x*, which articulated ideals of postwar cities that would manage and control veterans' shock while creating environments for physically and psychically healthy families.[70] All these utopias assumed that the perfect plan would foster spiritual and physical health in inhabitants. However, unlike these earlier efforts, Lynch's planning linked perception and cognition without recourse to utopian forms, and without a preexisting or ideal structure to organize the plan. Furthermore, the single normative notion of subjectivity tied to a specific space disappeared in favor of the notion of a population produced through interactions and data analysis. The population and the individual are linked in Lynch's work through the optic of perception. The nerves of the subject were integrated, at least in theory, with the design of the city and the territory.

Lynch had definite biases toward producing spaces with "strong" images, but no clearly defined ideas of class, economics, or community to mandate a set pattern for designs to follow. The plan should emerge from the relationship between cognition, perception, and the actual structure of the space. For Lynch perception became a channel that was material, and spatial, and could always be modulated and enhanced. Cities could always be made better;

their denizens had perceptual capacities that could be modulated and shifted through design.

More important, Lynch provided a method by which to gather and document psychic processes at an organizational level. Unlike his predecessors, he produced a truly novel methodological apparatus for data gathering, and sociological investigation, that while clearly related to the efforts of other sociologists at the time was also particular and novel for planning. Lynch took older histories of anthropology and psychology, of which he mentions individuals like Bronislaw Malinowski and sciences like gestalt and psychoanalysis, and pushed their methodologies into a new territory with a different type of truth claim. Lynch displaced concerns about what could not be recorded with an infinite faith in the study of subjective perception in and of itself as a tool. Rather than promoting an ideal or normative city, Lynch promoted the subjective space, a city built by means of the playback between individual memory and perception. Lynch put method ahead of end goals or idealized forms in the work of the planner.[71]

Lynch's measurements make visible the emergence of another phenomenon, the displacement of measures of risk and actuarial concern by "unknowability" and uncertainty.[72] Human beings and environments are envisioned here as constantly modulating without clear endpoint. Cities become systems with an endless capacity for change, interaction, and intervention, and problems of urban blight, decay, and structural readjustment have no clear definitive endpoint, just as their solutions become constantly extensive. Total risks, with defined endpoints such as nuclear destruction, were slowly consumed in the course of the 1960s toward a very different model of metabolic change and a methodological imperative for data gathering.[73] With no final ideal prototype or disaster in mind, for these planners, clear-cut endpoints and goals were relocated into variances between perception and action.

This method could be globally applicable while simultaneously capable of engaging the specificities of individual situations and environments. The image of the city vacillated between views from above and views from within the mind. The boundaries of the subject, the environment, and the interviewer compressed into one channel. Subjectivity turned into the very site of study. The space between the mental map, the trained observer's map, and the aerial map became the site of intervention. Lynch, like his mentor Gyorgy Kepes, treated perception as a pattern or algorithm that could be made material, visible, and instrumental. And like Kepes, Lynch also participated in a pedagogical and epistemological effort to turn data flow and memory into a capacity for design and a tool for social improvement. Interestingly enough, however, it was the very discourse of society that disappeared beneath the discourse of images and perception, seemingly consumed into the interactive space between the individual and the environment. Ideas of structures that defined urban forms—like capital, modes of production, or even social psychologies—were relocated to ideas of personalization, perception, and local networked action.

Territories of Communication and Affective Spectacles

If we were to seek then the implications and effects of the emergence of this algorithmic optic, and the folding of identity and space into environment, what would we find? Linking together the training of the observer with the production of space, it might be helpful to take a final example.

One of the central spectacles of urban redevelopment in the early 1960s was New York City's ambitious and expensive plan to host a world's fair in Corona, Queens. The New York World's Fair is today famous for restructuring New York's racial and urban landscape through Robert Moses's aggressive redevelopment and highway production. The fair has also been cited as a final moment of universalism and technical optimism before the Vietnam War and the emergent countercultural revolutions. The mottos of the event, "Peace Through Understanding" and "Man's Achievement on a Shrinking Globe in an Expanding Universe," reflect this techno-fetishized and rosy picture of human improvement by way of better mobility, technology, and communication. This mix ended up proving to be popular—some 51 million people visited—making it one of the largest live events ever recorded.[74]

This popularity was by no means assured. Repressed beneath this emergent and optimistic discourse of scalability and globalization, increasing conflicts brewed. At the start of the fair, which came at the height of the civil rights movement, there were concerns that protests, even riots, critiquing discrimination in transportation, economic opportunity, and employment, would disrupt attendance. It is rumored by historians that initial figures for visitation were very low on account of concern about racial violence.

Critiques of the fair's discriminatory geography and development abounded. Robert Moses, New York's master builder and the manager behind this event for the city, had successfully barred any development of public transport to the site, instead encouraging the building of roads. His vision of New York at the time largely consisted of high-rises and highways, an attitude to urban form that has been highly criticized for its encouragement of suburbanization and the deliberate race and class segregation that resulted from that highway construction that cut up the continuity of urban space and made mobility impossible for people with lesser means and without cars. Eventually, after lobbying by local city counselors, cheap buses were added to enable residents of boroughs such as Brooklyn to get to the Queens fairgrounds.[75]

The segregated geography of the expanding highway systems was not the only charge brought against Moses's and the city's practices in developing the site. The Urban League of Greater New York (ULGNY) in 1961 began to take the fair organizers to task for discrimination in hiring practices, resulting from the fact that Moses probably hired no African Americans, or Puerto Ricans for that matter, to work in any of the professional or technical jobs associated with planning or designing the fair. Throughout 1962 and 1963, and concordant with increased lobbying for the Civil Rights Act to be passed in 1964, the ULGNY, and the Joint Committee on Equal Employment Opportunity picketed the UN, with the fair as target, carrying slogans such as "End Apartheid at the Fair," and "African Pavilions Built with Lily White Labor," drawing parallels between American and colonial/neo-colonial European racism.[76] In the context of decolonization and the Cold War, the governor of New York State, Nelson Rockefeller, stepped in to attempt to resolve issues of employment discrimination and to ensure that concerns about race were addressed. Only days before the fair's opening, however, the NAACP, the Congress on Racial Equality (CORE), and other organizations were preparing to protest the content of some of the displays mounted by southern states as well as to use the event to focus more national and, perhaps more important, international attention on the demands for racial equality in the United States. Protests on the first day did indeed garner sizable national and international news coverage.[77]

While the city obsessed with finding the funds, building the infrastructure, and negotiating the turgid mix of national and international politics emerging from a combination of Cold War logistics and civil rights agitation in order to successfully host this event, the pavilions on the grounds, perhaps unsurprisingly, appeared to reflect little of this context. Science fiction, not current events, was the reigning topic and aesthetic in the attractions that garnered the most attention. The GM pavilion, for example, was the most popular, showing a "World of Tomorrow," complete with amazing gadgets, multimedia rides, and a curious absence of human beings.

The other centerpiece was the IBM pavilion, which contained an installation called "The Information Machine," consisting of an oval building designed by the architect Eero Saarinen and a twenty-two-channel installation designed by Charles Eames titled "Think." The entire space was dedicated to introducing analogies between how machines and people think. If today,

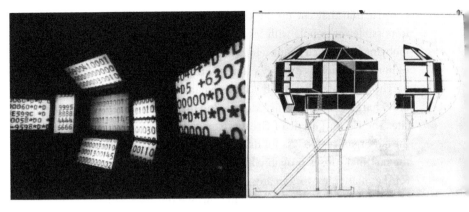

FIG. 2.14__Images from the exhibition and sketches of "Think" screens for the "Information Machine" Pavilion, New York World's Fair, 1964, designed by the Eames Office. Screen shots show the number of screens and varieties of information from numerical abstractions to highway infrastructure; the middle panel shows Eames interior architecture for the twenty-two-panel installation. The Work of Charles

then, Apple urges its users to "Think Differently," it is only in negation of the history of this slogan's predecessor in computer science and corporations like IBM, who urged us simply to "Think." But what form of thought was it that would be introduced to the public at the dawn of an age of ubiquitous computing?

The ascendant computer company took the concept of thought quite literally. Viewers were lifted from the ground into a closed oval, and entered a space of multiple screens, filling the field of vision (fig. 2.14). On these screens of different shapes were projected a filmic flow of images from different sources and in different formats, including animation. The intention was to get viewers to make connections, by will, between the images; to actively navigate the informational field and, through choice, create new connections and thoughts.[78] The speed of the film, according to the architectural historian Beatriz Colomina, was intended to replicate the processes of thought. The multiple-sized screens arrayed in different locations and projecting varied types of discontinuous information forced the eye (at least in theory) to move rhizomatically, making unexpected and nonlinear connections.[79] The "host" welcomed the audience to the IBM Information Machine with the words "a machine designed to help me give you a lot of information in a very short time. . . . the machine brings you information in much the same way as your mind gets it—in fragments and glimpses . . . like making toast in the morning."[80] Inside this machine, which was now no different from the human self, the spectator was ex-

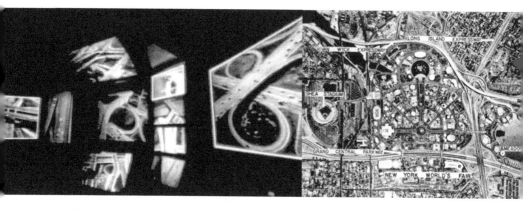

and Ray Eames, lot 13195, no. 187, Print and Photography Division, Library of Congress. © 2013 Eames Office, LLC; images from *IBM at the Fair* (1965), video transfer from *The Films of Charles and Ray Eames*, vol. 5, Image Entertainment (2005); World's Fair aerial view from September 2, 1963, courtesy of New York City Parks Photo Archive.

posed not to any singular piece of content but to a perceptual field. From the myriad screens, sequences of football games, train exchanges, dinner parties, and mathematical equations all flashed.

Reperforming the logic of Charles Eames' engineering courses, this installation, which was a "machine," brought to you "a lot of information in a short time" to model exactly how your brain worked—perhaps even to rewire your brain. Like Kepes's algorithmic design, this was a new design where finding the pathway through an overrecorded and infinitely available archive was what constituted "thought." It was not identification or recognition of subjects or objects that mobilized the eye and the brain of the spectator in this environment. The installation could be understood as a mimetic reenactment of the process of perceiving data in an information economy; a process that was now so automated as to be like making toast in the morning.[81]

Computational Space and Design

The pavilion, it must be mentioned, was designed in tandem with a radical reconfiguration of IBM itself as a corporation. Throughout the late 1950s IBM had been creating a more decentralized managerial bureaucracy, transforming the nature of its products and moving to selling systems, not machines, and constructing new laboratory facilities designed by the premier architects of the time and organized to facilitate interaction between scientists and managers.

The corporation relocated its headquarters to the suburbs of New York—Armonk—and built numerous large, stylish research complexes outside the urban center. The decentralization of management was seemingly followed by a networking and dispersal of research and operations. If New York City was contending with an increasing problem of dispersal, fueled by the policies of Robert Moses and a focus on highway construction, this was accompanied by a transformation in economy of which IBM was symptomatic[82]

Central to this reconception of IBM was the use of design. The 1973 lecture, at University of Pennsylvania, titled "Good Design is Good Business," by IBM chairman James Watson Jr. reflected back on this history of IBM and design. In a subsequent article, he wrote that it was a "story of a company's increased sense of design." Watson discusses the pivotal place that design had in rethinking IBM. He recalled that he "came to see design become one of the major reasons for the success of the IBM company." Initially inspired by a Fifth Avenue Olivetti showroom, Watson became convinced that the "Olivetti material fitted together like a beautiful picture puzzle." At a moment when IBM was hoping to expand into numerous services and markets, Watson desired a coherent aesthetic to join the machines, the services, the research, and the manufacturing of computational products.[83] He went on to hire many of the top names in modern design, including Eliot Noyes, Paul Rand, Charles Eames, Eero Saarinen, and many other designers (figs. 2.15 and 2.16).[84]

Design was particularly vital when one considers that IBM was moving from selling individual machines to entire enterprise solutions and systems. The machine to which we may assume that "think" was, in spirit, dedicated was the 360 series computer, introduced in 1964. This was the first machine built by IBM that separated the architecture of the hardware from the specific use of the machine. This separation allowed IBM to sell the series for different scales of businesses—from very large to small, and for different functions from commercial to scientific—since the system could be customized. It is the first time that a customer could change the configuration of the system without having to rewrite or discard older software. The system cost IBM billions of dollars and took at least a decade of effort (if not longer) to develop. As the former IBMer and historian Emerson Pugh notes, the 360 demanded entirely new manufacturing systems and a reorganization of how computers were designed. IBM released the machine not as a single unit but a "family," an innovation in computer retailing at the time, such that customers could purchase a smaller, cheaper configuration, knowing that in the future it could be expanded. The series was modular and scalable.

FIG. 2.15__IBM Laboratories, Yorktown, NY, designed by Eero Saarinen. From "Research in the Round," *Architectural Forum* 114 (June 1961): 80, 83. Note the windows on the outside and the laboratories pushed inward. The original prospectus is adjacent.

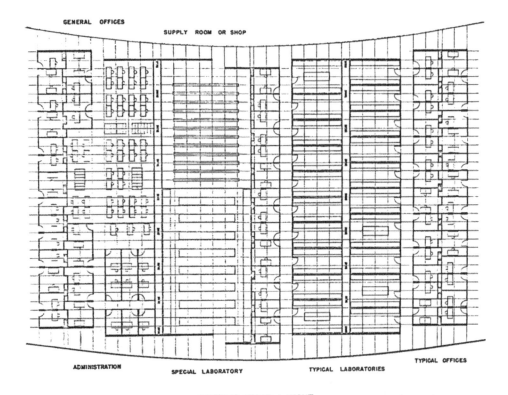

FIG. 2.16_ "Possible Space Plan for Yorktown Laboratory Facility." From "The New IBM Research Center in Yorktown," project prospectus (1959–60). Courtesy of IBM Corporate Archives.

The company itself refocused on selling accompanying management strategies, control systems and supply chain solutions as part of an integrated package with its machines. In corporate architecture and management, as in the perceptual city of urban planners, environment and ecology replaced the focus on objects and discrete commodities. The IBM man with his classic suit and tie, with all IBM men dressed alike, imparted to clients a sense of coherence between many aspects of computing—from the machine, to the technical support, to the control and logistics softwares—that IBM would provide.[85]

The company's redesign of its laboratories followed suit (fig. 2.16). Note that the elements of the plan are often organized as "typical," assuming potential conversion in functions, and that labs are flanked by office and administrative facilities, assuming an integration between the research and the administrative operations of the company. The corridors and windows are on the outside, to allow flow and contact between the labs inside the building.

Saarinen described the space as a self-contained world: "labs and offices today depend on air conditioning and fluorescent lighting for their air and light—not windows. Windows are like fireplaces, nowadays: they are nice to have but rarely used for their original purpose." The partitions between offices are a combination of glass and steel, innovated for the building, with the attribute that it is "translucent for visual privacy but it retains the psychological advantages of the glass." A walk between labs, however, is, in the words of *Architectural Forum*, "as spectacular as a Cinerama travelogue." If there is an "outside" in this architecture, it is already framed as being produced from a media mechanism, as though emanating from the technologies being made "inside" the building.[86]

The data-filled worlds of the IBM pavilion refracted this changing nature of knowledge and commerce. Inside the pavilion, human perception was also treated as a channel and capacity to be extended, increased, and circulated. This "expanded" media (as later labeled by Gene Youngblood, discussed in chapter 4) not only reformulated perception but linked this form of seeing to forms of thinking. For simultaneously with these new machines also came a reformulation of what constituted knowledge, even truth, in engineering and science, as well as in design and business education, a reformulation whose principles were being developed aesthetically as well as technically.

What are we to make, then, of the suppression of one set of concerns over the role of the fair and its space for another set of interests in redundancy, reverberation, and information inundation? This affective education for the eye and, now, the mind negotiated the transformation of economies and spaces through the modulation of attention by way of a novel self-enclosed and self-referential space of data inundation framed as "choice."

The Exhibitionary Complex and the Training of the Observer

The promotional materials for the IBM exhibit strongly confirm the idea that this was a new form of entertainment and a new mode of spectatorship. The materials given to journalists announced: "here from 22 screens, a new kind of motion picture entertainment leaps out at the viewer. . . . The aim of Charles Eames and Eero Saarinen Associates, co-designers of the pavilion, was to create an entire environment." The purpose of this environment is that "after a number of seemingly unrelated scenes unfold the viewer discovers for himself that computers are not mysterious." Through these discontinuities—

FIG. 2.17__*Think* multi-channel projection film by Charles and Ray Eames for IBM Pavilion (1965). First image shows diagrams for the installation projection piece demonstrating the flow chart approach. The Work of Charles and Ray Eames, lot 13186–5, no. 31, Photography and Print Division, Library of Congress. © 2013 Eames Office, LLC. The following images, from the movie *IBM at the Fair* (1965), trace the presentation that made analogies between computers, modeling and life, family trees and corporate organization charts, all representable through flow charts and programs, to illustrate the prime point of the exhibition—"think"—with humor. Video transfer, from "IBM at the Fair," The Films of Charles and Ray Eames, Vol. 5, Image Entertainment (2005).

"seemingly unrelated scenes" in the words of IBM, that replicate thought — the viewer will come to understand, or perhaps "think," like a machine.[87]

Eames and IBM took this very literally. Large portions of the presentation were dedicated to a hostess's planning of a dinner party. She needs to make many strategic decisions about whom to invite, and where they should sit so as to ensure social balance and harmony. The implication of this movie was that these decisions made by a hostess were just like those made by a machine. In fact, the Eameses created algorithmic diagrams of the hostess's decisions, demonstrating how her thinking and the machine's logic were the same.[88]

But, more significant, this machine would operate through random, "seemingly unrelated" sequences. From the hostess in the living room the film zoomed to vast infrastructural images of railroads and urban space, to numbers and equations, to football games and the strategic plans of a sports coach. There were no historical narrative structures for the presentation. With the users incorporated into a feedback loop, each being trained similarly to create a personalized pattern from the data, there was no predetermined order to the pattern. There was, however, an established method, process, or approach that was amenable to any linking points in any data set . . . even if "unrelated."

The image of the wall (fig. 2.17) demonstrates the kind of idiosyncratic connections the Eameses were making between numbers, people, diagrams, patterns in nature and society. Organizational charts and family trees are jointly displayed to demonstrate the idea of a "model" and abstraction. In the middle image, the screens are collectively organized around the mathematical number 2; this segment of the installation is timed to give the spectator an understanding of an idea of two-ness, not a description of any specific individual two things. Mobilizing this aesthetic strategy, the Eameses hoped to force the spectator to understand generalizable concepts, not to remember any specific data point.

The Eameses were thinking about connections between data, not individual screens. In their films often multiple types of data would be shown simultaneously to express a mathematical concept, or they would use repetitive patterns of scaling — going from micro to macro images of a phenomenon or environment — to make this point. Charles said of his filmic installations that "you get a feeling about relationships you didn't have previously. In thinking about decision making in the future — whether it be recombinant DNA or what have you — we believe some attention should be given to honing the techniques of showing critical things simultaneously. We had hoped that by now this might be a rather general procedure."[89] This language already evokes a future, generated out of a structure, like DNA, emerging from the relationships between images. Eames wants this to become a "procedure." He desired a new technology for seeing. He viewed attention as related to DNA: a perception offered autonomy to self-reproduce. As the language of DNA and the discourse of "decision making in the future" suggests, one can also assume that this is an aesthetics of truth and knowledge on which reasonable, perhaps rational, spectators might act.

Affective Rationality

These analogies between objectivity and sensation, modulation, vision, and memory should also force us to consider commonly articulated beliefs often ingrained (even if through dispute) in our theories and histories of media that separate objectivity, science, and rationality from embodiment, sensory affect, and emotion. If we have long correlated rationality with objectivity and modernity in the sciences, the work of these designers challenges us to define what we speak of. For here is a sense configured to be both affective and logical. As interior and exterior were reconceived in terms of communication, the sensual, perceptual, and cognitive became part of a single order, a rational and algorithmic set of processes or logical patterns that could be studied, built, modulated. This point takes significance when viewed as the very infrastructure for interactivity. It is the reformulation of abstraction as material, of perception as cognition, of sense as logic that makes possible our contemporary interactive environments. Deferring concern with documentation, and refocusing on process, facilitated a vertiginous reformulation of the relation between mind, body, and machine that made seeing and thinking both material for a communication architecture.

This procedure, a new DNA of vision, would rest on an endless genealogy of infinitely recombinable data, using the past to generate the future. This archival discourse redirected previous discourses of indexicality and objectivity into a technical drive to accumulate more data in order to encourage more analysis. The very concept of information was in transition, balancing between older histories of photographic and image capture, that still correlated information with the past and the documentation of an external and "real" world, and an emerging redefinition of information from the *Mathematical Theory of Communication*, which was always in the future tense and based on mathematics of chance and probability. Never resolving this friction, these designers utilized the imbalance between the past and the future to produce another pedagogy, at once familiar and entirely novel, that shifted the discourse away from organizing data to producing organizations.[90] In deferring the problems of recording, design shifted to a focus on process that could make thinking and seeing part of one channel.

Control

For Eames this was, indeed, a scalable form of nonconscious reason. Eames always spoke of himself as an "architect." In a letter addressed to Ian McCallum, the editor of *Architectural Review* in 1954, Eames wrote of his initial foray into the realm of feedback and communication theory, the film "Communication Primer":

> one of the reasons for our interest in the subject is our strong suspicion that the development and application of these related theories will be the greatest tool ever to have fallen into the hands of the architects or planners. . . . If ever an art was based on handling and relating of an impossible number of factors, this art is architecture. One of the things that makes an architect is the ability to include in a concept the effect of and affect on many simultaneous factors—and a precious tool has been his ability to fall back on his own experiences, . . . If however, a tool should be developed which could make possible the inclusion of more factors—and could make calculable the possible results of relationships between combinations of factors.—then it would become the responsibility of the architect and planner to use such a tool . . . We may have the possibility of such a tool in the "Theory of Games."[91]

Relating intuition (see also chapter 3 on reason) to computation, Charles Eames believed that linear programming and game theory might be tools to direct architecture. These were in his opinion "pure mathematical systems that can be used in relation to human problems."[92] For Charles, who adored the idea that complexity could be translated into ones and zeroes, these principles were concepts for producing architecture and planning. He argued that "it is unfortunate that in this time much of the really creative thinking . . . is shrouded with the panic of secrecy."[93] But, he wrote to McCallum who was also editor of a section called "Townscapes" involving urban planning, the application of such concepts will change the townscape, and must be included in architectural practice.

Eames regularly discussed the management of urban space in terms of control, communication, and the logistics of game theory (fig. 2.18). His installations of perceptual manipulation were also imagined as encoding mobile tactics for the management of territories:

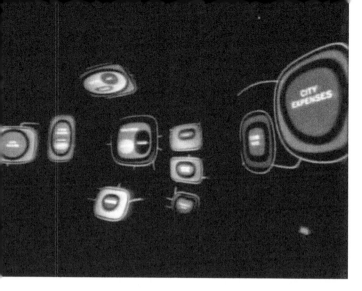

FIG. 2.18__City flow chart. Screen shot from *Think*. Video transfer, from "IBM at the Fair," The Films of Charles and Ray Eames, Vol. 5, Image Entertainment (2005).

in the management of a city, linear discourse certainly can't cope. We imagine a City Room or a World Health Room (rather like a War Room) where all the information from satellite monitors and other sources could be monitored. . . . The city problem involves conflicting interests and points of view. So the place where information is correlated also has to be a place where each group can try out plans for its own changing needs.[94]

Eames's concepts were never developed in urban planning, but were continually reaffirmed in his many movies and multimedia installations. The idea of visually displaying information and attenuating attention was his concept of "architecture." The very idea of architecture here becomes despatialized, and instead involves the production of complexity. The "art" of architecture is about the ability to handle an "impossible number of factors," and good design is about the "inclusion" of more factors in the interest of making outcomes "calculatable."[95]

For Eames, unlike Lynch, control is about predicting future outcomes, and the city is reconceived as a calculating machine for managing and resolving multiple conflicts. Urban planning and design should be a simulation exercise where different outcomes and goals can be tested in relationship to one another. Counter to Lynch, Eames appropriated a notion of changing needs but systemic homeostasis. Implicit in his discourse is an assumption that minimax solutions between highly complex interests are the goal of urban management.

At the time Eames was not alone in seeking to apply the logistics of the war room and game theory to urban space. The Rand Corporation by 1969 had a specific division labeled "urban center," focused on examining problems of New York City's police, hospitals, and fire. In a 1970 report, Rand determined that it was necessary to convert military-oriented research for civilian use and that urban planning might provide a major site of interest. In earlier reports from 1963, Rand had proposed system analysis for urban planning.[96] The city was increasingly discursively rendered as being in a state of insecurity and crisis analogous to a state of war. But if analysts at Rand created endless statistical documents, the Eameses proposed the training of attention as a fundamental element to being able to administer these networked territories. The trained engineer, architect, or planner was one whose attentive capacities were enabled to cull patterns from vast data.

In this moment, where the future of territories, populations, and identities were being negotiated through the increased application of computational methods and the automation, if not autonomy, of sensory experience, what are we to make of these varied understandings of communication, cybernetics, and feedback? Kepes, Lynch, and Eames all contributed in complicated ways to producing perception as a channel to scale between individuated bodies and vast networks, and a pedagogy and epistemology that reworked space through a new territory of algorithmic logics and cognitive processes. As urban space was being reformulated in a context of massive changes in economy and race relations, designers and social scientists increasingly turned toward new strategies of modulating sense.

Disorientation

There is nothing, however, inevitable or obvious about these contemporary forms of interactivity and ubiquitous computing; a great deal of work, and imagination, went into convincing us that vision could be designed and machines and minds both "thought" while organizations were reformulated as organisms with unclear boundaries or interiorities.

This smooth, self-referential world built of multiscreens and "glass" psychologies was not, therefore, without its own self-induced conflicts. At some point the logic of data inundation appeared to fold back onto a history of meaning and representation. Not all spectators found these tactics successful, and it may be a marker of historical change that what we now take as given was once potentially disorienting. One reviewer, Mina Hamilton, on seeing the Eames installation remarked in *Industrial Design*, "visually the show is a sensation." But, she added, sometimes the show was a "deluge," implying that there might be a threshold beyond which too much information became threatening. While the writer was astute in noting that: "Eames did not expect everybody to see everything. Quite the contrary, he expected each person to come away with separate sets of information and experience." Ultimately, "the pace of the show, however, is so fast that a person does not have enough time . . . to weed out what he wants to see or not. . . . The Kaleidoscope-like result is overwhelming and 'spectacular' but too fragmented to be entirely successful." This reviewer immediately identified the individuating aspects of the show, the element of giving a sufficient data field, allowing each user to come out with "separate sets" of information and experience while deploying the same tactics as everyone else of pattern seeking and connection making.[97]

What are we to make of this sudden deluge, where the user cannot reduce the data field, fails to find the pattern, and is "overwhelmed"? The reviewer implied, even if in critique, that this mode of viewing was contingent on a capacity within the viewer to assimilate certain quantities of data, denoting that the Eameses were producing an observer understood as possessing thresholds for information processing that could be enhanced or overwhelmed. Hamilton also noted that the show was "sensational" and provoked affect but without linear narrative or space for identification. ("The person does not have enough time . . . to see what he wants to see or not.")

It might be concluded after reading such an account and surveying the data-dense, and temporally and spatially disorienting, displays of these classrooms, labs, and installations, that the implied observer of this environment must be an interdisciplinary observer, capable of relating to the significance of a statement by a famous author and a scientific image simultaneously, but no Renaissance man in the colloquial understanding of the term. This was an observer who might be able to recognize, but not necessarily identify, or authoritatively define, what he or she was looking at, immersed as he or she was in pedagogical and spectacular environments offering no establishing shots of stable perspectives on the data field. In keeping, perhaps, with the idea of camouflage, these designers opted for the mimetic perspective, the perspective from within the machine, not the perspective of or the perspective external to the information field. It might even be suspected that the observer here might be psychotic in his or her inability to find stable points of reference, lacking a clear boundary between stimulus, perception, and cognition, and without any clear emotional or affective relationship to objects in the world. If so, and this observer was quite logically and algorithmically psychotic, the pattern was stable, even if the subject was not.

Kepes made an interesting statement to this effect. He wrote that the purpose of good design and art "is to bring the outer and inner world in correspondence," implying that the psychic space was now flattened into a perceptual process that was also cognitive and structural, a "vision of felt order."[98] Order, which was implicitly defined here as stable, rational, patterned, and logical, was equivalent to sense, while the subject was envisioned as a smooth space for the transfer of information between the inner and the outer worlds, between the registers of analysis and stimulus. Kepes's intimated observer was like the user of the Memex or the envisioned future user of the computer, privately scanning data in the interest of participating in an innovative process without any clear-cut end. This subject was both rational and orderly and was affective, touchable, seeable, visible—a subject whose consciousness, cognition, perception, and memory were now envisioned as part of one interactive process. A flat space, without inside or outside, only feelings and order; a subject who is, perhaps, also a channel, a conduit for the ongoing circulation and reorganization of data.

Conclusion

Remaking the boundaries of subject, the definition of rationality, and reconceptualizing the screen, these older midcentury environments have complex relations to the present. The critical reviewer still perched between different histories of visuality gives us insight into a moment of potentiality and vacillation. She apprehended the display not as a given and desired form of life but as a choice, and not one defined by a mathematical communication theory. This choice would be contingent on the relationship between the interface and the older histories of storage and representation on which it rested. It is still not clear, even in our present, how this deluge that "overwhelms" the subject will be organized—do these communicative forms of interaction offer the possibility to remake the body, self, and other? Or to descend into a self-referential vortex of consumption without any imaginary of futurity or the possibility of encounter with difference? At this moment in history when a new hyperindividuated yet networked observer emerged accompanied by a transformation of discourse from one of planning and images to one of environment and cognition, we are forced to ask about the possibilities of such a condition. These architectures of labor, pleasure, and education found themselves superimposed on older histories or representation, politics, and structure; the relationship between these strata would become a matter of concern for both ethics and politics.

The designers of these spectacles also had reservations. In a moment reflecting on the events of the war that had precipitated his revision of vision,

Kepes ended his first book, and most famous design text, *The Language of Vision*, on a very strange note. After a book dedicated to advertising for the nascent information economy, he turned in one of the final moments to the ideal of art. Kepes ends one of the final chapters with a discussion of Picasso's *Guernica*. His language is affective. "*Guernica*," he writes, "evokes an optical fury" and "the shrieks of a danger siren." This opticality that shrieks warning—what, might we ask, does it warn us of?[99]

Guernica itself, of course, was a reference to the Fascist bombing of civilians in the Spanish Civil War. It bears witness to an event that Picasso himself never saw, an event perhaps that was never in the realm of the visible, ignored by much of the Western world—an event that speaks to the limits of the infinite archive that is so readily assumed to be always/already available for recombination in the postwar period.

In looking at this image, Kepes argues that it succeeds because it creates contact between a plastic, modular structure and the trace of history—"social events" in his language. Kepes writes: "Picasso, stirred to a fury of indignation by a human drama caused by the regressive social forces and their significance today, in a visual projection of the discrepancy between life as it is and life as it should be, represents human figures in a distortion of pain and suffering. . . . Tears are in action like a bursting bomb. The plastic interconnection of the lines, planes, and texture surfaces acts as do suffering individuals. . . . Two contradicting systems, plastic organization—the message of order—and the organization of a meaningful whole—the messages of chaos are wielded in an indivisible whole."[100] This image works in a manner different from advertising, in his argument, because of the disjunction between what has happened ("life as it is") and what should happen ("life as it should be").

Kepes returns here to interrogate a theme that has largely gone undiscussed in this chapter but that preoccupied him throughout his work: signification. In his expression, concerns about time, language, and translation are returned to the smooth world of communication theories. Kepes perhaps hopes to utilize the very substrate of this communicative objectivity—the assumed recordability of the world—to produce speech. This speech emerges at the moment when each medium has reached its capacity and encounters another form of action. The visual and the aural pressure each other in the inability to render either complete.[101] Rather than seek to suture the time between the infinity of recording and the impossibility of representation, Kepes hopes to explore it. He seeks to ask which histories will inform the future. Kepes wants to produce images of temporal disjunction.

His concept of temporality infused his logic of action. The work of art, for

Kepes, was not to make engineering more playful or creative, in his parlance, but to challenge our imagination of life and technology.[102] He desired, and demanded, that differences and translations must exist between the arena of artistic practice, which must not serve functional purposes, and the work of technologists. In his development of an integrated art and science curriculum and center at MIT, he took an alternative path to other art and science collaborations of the time, such as E.A.T. His organization did not valorize corporate structures and technology but rather more timid and minute actions, adherence to older institutions, more sustainable structures. Kepes appeared to literally fantasize a space between life as it is and should be by asserting a distance between practices of nonfunctional fantasy and those of documentation and authority. Kepes relied on and supported artists who were less oriented to engineering and more oriented to environment, nonfunctionality, and other practices that rethought the boundaries of the sense and bodies.[103]

Simultaneously conservative and institutional, while ethical and concerned, his organization, CAVS, while in service of a classic industrial military space, MIT, highlights the variety and internal complexities that emerge from inside institutions and from within the structure of our contemporary information networks.[104] Kepes offered sense autonomy and produced a new form of scanning seeing and communicative objectivity, but he sought to recuperate this now embodied and mobile eye for many purposes. In his work he separated himself always from the more essential and often pure search for form following function that preoccupied Gestalt psychology and the original Bauhaus.[105] This "abandonment" opened him to conceive of art and science as intimate lovers, but never unitary subjects. Kepes insisted on a translation between the discourse of truth and that of imagination.

This conservative and progressive logic reveals to us something about history. The discourses of cybernetics and communication were enacted by individuals who were often conservative. Their own innovations were attacking their own pasts. In their efforts to recuperate their memories of previous organizations of sense, art, and vision, however, we can trace the binding and unbinding of meaning to action; and recognize the subtle possibilities that exist between the total amnesia of the present, often radically embraced in 1960s countercultural responses to art and technology, or the absolute reactionary return of nostalgia, identity, and nation that marked the rise of the conservative politics of the time. In place of these extremes, we find more subtle configurations and recombinations of past and future.

Kepes was not alone in seeking to excavate the subjective, interactive, and temporal nature of perception for ethical purposes—the not yet familiar future. Lynch in 1972 released another landmark book on planning, summarizing much of his research, titled *What Time Is This Place?* Covering examples from London's Great Fire in 1666 to the place of revolution in Havana in 1959, Lynch sought to introduce the dimension of time into place in order to match the internal state of the human biology and psychology with the external space of the city. The desirable image of the city, he wrote, "is one that celebrates and enlarges the present while making connections with the past and the future."[106] Lynch studied multiple sites, from Ciudad Guyana to London to Boston, in the attempt to refine his method of mapping perception onto space with time as a variable.

It is, he argued, the subjective nature of time that offers planners an opportunity to avoid either historical preservation or creative destruction as the only options for urban planning. He fantasized a sense optic that, like Picasso's painting, would bring the future and past into contact, allowing new forms of life to emerge. "But surely," he implored, "we can envision public devices— films, photos, signs, diagrams—that could bring those invisible processes [of change] within everyone's grasp. There might then be 'mutoscopes' on the streets, which speed up past and future changes or slow down present vibrations so that we can see them, just as public microscopes and telescopes would extend our perceptual reach."[107] This perceptual reach that would map the environment to our interior states and bodies would be the infrastructure for the future city. "Our earthly environment is a very special and perhaps unique setting for life. It should be conserved; it cannot be preserved." Writing against preservation but in support of ecological conservation, Lynch sought to produce change and create the possibility, and space, for many forms of life— human and otherwise.

This mandate could only emerge, he thought, by modulating the time of spatial existence. His fantasies of a new optic, an extended perception, eerily invoke our present of ubiquitous sense and computing devices, our fantasies of sentient cities; but he also sought to produce a world of "dynamics" where human beings might encounter each other in new ways, and the urban space might be, in his language, made "humane." "Our real task," Lynch wrote, "is not to prevent the world from changing but to cause it to change in a growth conducive and life-enhancing direction."[108] He embraced change if it could be diverse and nonhomogeneous. While turning away from classic discourses of class, race, or structure, Lynch evokes environment, sensation, and time as the frontier on which the future of urban life is to be developed. He offers an ideal of a fuller temporality for development, a speed that might, perhaps, not be as homogeneous as the speculative time of development in such spaces as Songdo.

These efforts to recuperate certain pasts of history, sentiment, and affiliation transformed the engagement with planning and the vision of the city. While Eames reenvisioned urban space, and attention, as a control center, with game theory as the central model for planners, Lynch deliberately misconstrued communication theories, Fourier transforms, and other computational tricks to smooth over the space between the binary logic and the continuum of space and instead insisted on making the disjuncture between these models, rather than their convergence, a tool for social science. For Lynch the world could not be perfectly broken down into ones and zeroes, but ideas of communication and computation could enliven our understandings of complexity. Whether one views Lynch's work as more ethical than RAND's reports to New York City and Los Angeles throughout the 1960s is a matter of perspective. I would argue that they were—that Lynch's models opened to the world, and produced a series of mobile methods that could still be reattached innovatively to history. These models work on multiple scales, both temporal and spatial; allowing planners to both consider global conditions, while creating subjective and contextualized understandings of the environment.

This statement is made in full consciousness that Lynch's work corroborates our contemporary discourses of local action and antagonism to large, governmentally sponsored programs. Such planning strategies would be defined as neoliberal by many geographers. Perhaps ethnographic observation and urban gardening are not the ideal tools for city improvement. But at the time when systems theories and social science were arguing against any forms of planning intervention, either on the grounds that the city was too complex for planning intervention, and that urban plans never took into account social forms, or on the basis of approaching the city in absolutely computable models, as in the logic of IBM as smooth space for the extension of consumption and computation, Lynch provided an alternative model. In his work, planning is simultaneously conceived of as cognitive, process oriented, and data driven, but capable of attaching data sets to context in tying individuals to environments to create new sites of intervention without recourse to overwhelming structures.

History troubles our present. These designers and planners, without question, demonstrate historical transformation in the organization of power, knowledge, and vision. They contribute to our contemporary forms of governmentality and territory. But as Foucault notes, governmentality is productive of new spaces, technologies, and subjects. As Deleuze argues, the purpose of philosophy is always to excavate the "homogeneity" but also the "irreducible differences" between the images that induce thought within any historical strata.[109] One of the most regularly repeated arguments about information design at this period is its relationship to an amorphous concept of "control" and "organization," usually implied negatively. Rheinhold Martin, for example, identifies an "organizational complex," which he views as an extension of the "industrial-military complex" and as derived from cybernetics: "networked, systems-based, feedback driven—this organicism, and the circuits of power that it serves, sustains myths of dynamic deregulation, corporate benevolence, and dispersed, de-hierarchicized interactivity."[110] For Martin, the "pattern-seeking" endemic to Kepes's practice was part of creating the architecture for this space.

I wish to extend and complicate Martin's discussion to attend not to architectural space but to epistemology and to biopolitics. What needs to be elucidated in the study of organizational spaces is how systems complicate and self-differentiate. What types of organizations and networks are emerging—are different assemblages coming into being? And finally, how do these assemblages interact with older forms of territory, subjectivity, power, and social hierarchy? These questions are absent in the architectural discourse on Kepes or Eames.

These stumbling efforts to articulate a discourse of temporality can be said to constitute an effort to give voice to what might constitute politics and ethics under such conditions. Kepes and Lynch both adhered to a pragmatic ethics of the image—of perception and self—seeking to build accumulation through time, rather than just encourage circulation. They highlight that numerous options were produced from within these logics, and different practitioners mobilized different histories to reformulate and understand the technicization of vision and the nature of communication.

This historical interface between design, aesthetics, and communication sciences and discourses reveals an important fact. Our contemporary media environments were not technically determined; aesthetic practices were central to the production of the condition of possibility for the acceptance not only of computing but also of our contemporary ideas about psychology, attention, and space.[111]

The critical work of scholarship now is to make this network between vision and cognition haunt our contemporary present. How do we activate the internal possibilities, collapses, and fissures—between the archive and the interface, between the assumption of an already recorded world and the memory of capture, between cognition, perception, and action—all simultaneously encoded into our lived environments and pedagogical spaces?

We are left to ask what other dreams for our perceptual future emerge from our interfaces . . . where we believe sense, perception, and cognition have been compressed into process, program, and algorithm to be regularly fed-back to us at every click and choice made at the screen. Every day we negotiate this question in front of our many screens and interfaces, through our communication networks and channels, in our massive spatial product "smart" cities, in the very architectures of our contemporary perceptual field.

3_RATIONALIZING

Cognition, Time, and Logic in the Social and Behavioral Sciences

"What we thought we were doing (and I think we succeeded fairly well) was treating the brain as a Turing machine; that is, as a device which could perform the kind of functions which a brain must perform if it is only to go wrong and **have a psychosis**. The important thing was, for us, that we had to take a logic and sub-script it for the time of the occurrence of a signal (which is, if you will, no more than a proposition on the move)."—**Warren McCulloch**, "The General and Logical Theory of Automata" (1948)

N 1948 AT A CONFERENCE ON CIRCUITS AND BRAINS, IN PASA-dena California, the prominent cybernetician and neural net pio-neer Warren McCulloch introduced the idea that rationality could not only be both physiological and logical, but also *unreasonable*. Addressing a room of the most prominent mathematicians, psychologists, and physiologists of the day, discussing the nature of mechanisms in both human brains and logical machines, McCulloch sought to titillate his respectable audience by offering them a seemingly unintuitive analogy. Finite state au-tomata, those models of calculative and computational reason, the templates for programming, the very seats of repetition, reliability, mechanical, logical and anticipatable, behavior, were "psychotic" but brain-like.

These statements should not, however, be thought in terms of human sub-jectivity or psychology. McCulloch, while trained as a psychiatrist, *was not* discussing psychosis in relation to patients in mental clinics. Rather, he was

responding to a famous paper delivered by the mathematician John von Neumann on logical automata.[1]

McCulloch might not have been an expert in Turing machines, but the invocation of "psychosis" was deliberate. McCulloch explicitly called on the medical definition of psychosis, particularly of the schizophrenic type.[2] Since the late nineteenth century, with the advent of standardized psychiatric taxonomies for such ailments, schizophrenia had been the imagined territory of repetition automism, mechanical speech, speculative confusion, and improbable visions. With its name literally denoting a "splitting of the mind," the disease had an etiology classically defined by symptoms such as the inability to organize and translate thought into coherent historical narrative, the suffering of hallucinations, the perception of false sensations and voices, and trouble recognizing and relating affectively to others.[3] Reductively stated, schizophrenia has long been considered a pathology of temporal and spatial organization.

According to the medical definitions McCulloch was playfully citing, to be schizophrenic is not to be reasonable, but it is to be logical and mechanical. It is also a potentially violent condition. But McCulloch was not speaking as a clinician in this discussion, and neither reason nor violence was explicitly on the table, although from a historical perspective they haunted the room. What McCulloch *was* doing was responding to a famous paper that the mathematician John von Neumann had delivered at the same conference on the ability to build logical automata. McCulloch had no intention in arguing about the essential characteristics, the ontology, of machines or minds. He recognized that computers were not yet the same as organic brains. This question of equivalence was not at stake. What *was* at stake was the set of methodologies and practices that might build new machines—whether organic or mechanical. And the answer, both McCulloch and von Neumann suggested, was to develop a new form of logic, both psychotic *and* rational, that might make processes usually assigned to analytic functions of the brain, perhaps associated with consciousness and psychology, amenable to technical replication.

McCulloch gave voice to an aspiration to turn a world framed in terms of consciousness and liberal reason, into one of control, communication, and rationality. And he did not dream alone. At this conference where many of the foremost architects of Cold War computing, psychology, economics, and life sciences sat alongside each other, we hear a multitude of similar statements arguing for a new world, now comprised of "psychotic" but logical and rational agents.

I want to take this turn to a new discourse of cognition and rationality as a starting point to consider the relationship between memory, reason, and temporality in cybernetic discourse. If in chapter 2, I discussed the transformations in attitudes to perception, here I examine the corollary of the algorithmic eye—the algorithmic mind. Just as designers and planners produced a sensory infrastructure supporting a faith in communication and interactivity, social scientists and human scientists produced its complement—an epistemology of rationality that also substantiated a faith in information, and ultimately data visualization and mining, as the source of wealth, autonomy, will, and action.

But if McCulloch was calling into being another type of machinery, he was only doing it through the language of an older psychoanalytic psychology. While explicitly antagonistic to psychoanalysis, his definitions of psychosis largely conformed to the definitions of the day in psychoanalysis. His new model was troubled by questions of memory and time, features he assigned to human psychology:

> Two difficulties appeared [when making neurons and logic gates equivalent]. The first concerns facilitation and extinction, in which antecedent activity temporarily alters responsiveness to subsequent stimulation . . . The second concerns learning, in which activities concurrent at some previous time have altered the net permanently, so that a stimulus which would previously have been inadequate is now adequate.[4]

Stated otherwise, for McCulloch, historical time presented challenges to making thought and logic equivalent. Having managed to build a new machine-mind, McCulloch continued to be troubled by problems of temporal organization, change, learning, and perhaps, consciousness.

For cyberneticians, cognitive scientists, and neuroscientists seeking to represent thought logically, the literal mechanisms of thinking always haunted computational models. Older histories of science, psychology, and philosophy, invested in the surplus and unrepresentable elements of human thought, troubled the new machinery of social and behavioral science. Despite, therefore, proclaiming that "mind no longer goes ghostly as a ghost" and psychoanalytic concepts could now be disavowed in favor of neurophysiology, McCulloch continued to be haunted, and animated, by these ongoing problems inside networks.[5]

It is my contention that this troubled and troubling relationship between logic, history, and memory continues to animate our machines and digital networks; driving a dual imaginary of instantaneous analytics and collective intelligence, while encouraging the relentless penetration of media technologies into life through a frenzy to record and store information.[6] This chapter is therefore not a theory of "psychosis" but rather a historical investigation into the cybernetic relationship to temporality that suggests that cybernetic concerns with how time would be organized in circuits was fundamental to the reformulation of intelligence as rational, and produced a new epistemology of pragmatic behavioralism, embodied and affective logic, and nonliberal agents that continue to inform contemporary practices in fields ranging from neuroscience to machine learning to finance.[7]

To do so, this chapter examines with specificity the forms of knowledge and the types of truth claims being made around rationality as related to time, as evoked through a discourse of "psychosis" or incomplete reason, in cybernetics. The chapter will map the relationship between time and reason in the work of Warren McCulloch and many of his compatriots in the human and behavioral sciences in relationship to previous efforts, in sciences such as psychoanalysis, to account for both the mechanistic and unreasonable nature of human minds.

My argument is that the cybernetic reformulation of reason produced new forms of measurement and methods in the social and behavioral sciences, encouraging a shift toward "data-driven" research adjoined to a valorization of visualization as the benchmark of truth, and as a moral and democratic virtue. In the course of this epistemological shift in values, techniques developed that transferred older questions of political economy, human desire, and social structure into discourses of communication, personalization, and visualization. This is, therefore, not a comprehensive history of reason but a particular investigation into cybernetic attitudes to psychoanalysis, history, and time, and how the cybernetic idea of mind—inculcated in neural nets—impacted a series of other fields, with implications for our present and the forms of speculation, temporality, and futurity architected inside our digital networks, "smart" cities, and interactive displays.

Historicizing Rationality

My purpose in this chapter, therefore, is not to offer a specific definition for "psychosis"—rather to take this language as used by cyberneticians and affiliated operations researchers, social scientists, political scientists, and economists as a vehicle for historicizing reason. The question is not what *is* "psychosis," but what work does it do in the cybernetic and communication science milieu, and how does this differ from earlier utilization of the term either as a medical or social diagnostic?

Just as psychoanalysts struggled to discover and document the truth and laws of the unconscious, logicians and mathematicians also concerned themselves with overcoming the subjective and mediated nature of human reason. Throughout the nineteenth and earlier twentieth centuries, mathematicians and philosophers struggled to define the limits of logical representation. Could the world be subsumed and represented by human logic, representable in orderly and clearly defined steps, translated into questions with clearly defined endpoints, and definitively answered as true or false? Could, in short, all forms of calculation be made knowable and verifiable? Gottlob Frege, Kurt Gödel, David Hilbert, Bertrand Russell, Alfred North Whitehead, and Alan Turing were among a (literal) army of mathematicians and logicians struggling to find a structural and theoretical objectivity a priori of human sense perception. The answer to their search, however, was definitively no. It was as a negative proof demonstrating the impossibility of fully representing all statements in first-order logic that the Turing machine was conceived.[8]

But McCulloch did not have a normative or mathematical agenda. Accepting that there were many things that could not be known or computed, he operationalized the question Turing had posed and chose to ask instead: what if mental functioning could now be demonstrated to emanate from the physiological actions of multitudes of logic gates? What could be built? — not: what could be proven? If instead of seeking an absolute reasonable foundation for mathematical thought, as Bertrand Russell did, or the never-achieved objectivity of Sigmund Freud, perhaps, McCulloch implied, we should turn instead to accepting our partial and incomplete perspectives, our inability to know ourselves, and make of this "psychosis," in his words, an "experimental epistemology"?[9]

At this conference on cerebral mechanisms, sponsored by the Macy Foundation, where some of the foremost architects of Cold War rationality in machines, economies, and governments met, there was, therefore, much at stake. McCulloch's comments demonstrate that postwar comprehensions of logic and reason were far from those of the liberal tradition or even the nineteenth and earlier twentieth centuries' aspirations for mechanical objectivity and statistical regularity that had preoccupied the social and behavioral sciences. His statements must also be positioned against the search in Gestalt psychology and even psychoanalysis for a "structural objectivity," the hunt for the invariant structures and theoretical laws that condition subjectivity and sense.[10] In their place, cyberneticians, cognitive scientists, neuroscientists, and social and behavioral scientists reframed their epistemological goals and their definitions of logic to focus on a new site of study—algorithmic processes as cognition.

What has been erased from the historical record in our present is the explicit recognition in the aftermath of World War II and the start of the Cold War that rationality was not reasonable, had no relationship to consciousness, and demanded different concepts of systems, markets, and agents. It was not only in critical and poststructural theory that critiques of modernity and Enlightenment reason emerged. These concepts were also explicitly articulated from within the human, computational, and social sciences. The historian of science Lorraine Daston reminds us that we would do well to recall that those things that today are considered virtuous and intelligent, such as speed, logic, and definitiveness in action, were not always so. She is explicit in pointing out that rationality in its Cold War formulation, despite the insistence of technocrats, policy-makers, and free-market-advocating economists, is not reason as it has been understood by Enlightenment thinkers, liberals, or even modern logicians. Quantitative and calculative capacities, for example, from the eighteenth century and throughout the nineteenth century were considered debased mechanical functions unrelated to invention, genius, or consciousness—hence the regular use of women, for example, as computers feeding data to Hollerith and other calculating machines in the late nineteenth and earlier in the twentieth century. Mathematicians and logicians, including Alan Turing and Bertrand Russell, had proved, conclusively, in struggling with the *entschiedungsprobleme*, that many analytic functions *could not* be logically represented or mechanically executed, and therefore machines were not human minds.[11]

What moves, therefore, had to occur to allow this commensurability between logic and thinking that would turn mental processes into cognition, and transform reasoning into algorithmic rationality? What instruments and methods accompanied this transition? And finally, we may ask a pressing historical, political, and ethical question: why today have we forgotten this history and do we still regularly equate reason (perhaps consciousness) with rationality? It was only in the 1970s that rationality became clearly attached to liberal concepts of reason in political discourse and in economic mathematical models of finance.[12] Even today, as economists take up neuroeconomics, and other highly financial and behavioral models, the insistence on the individual as the unit of analysis persists; a seemingly amnesic action neglecting a history of alternative models from within the social and behavioral sciences.[13] It was not only disenchanted philosophers who labeled modernity's reason psychotic, for twenty years after the war, in the behavioral, computational, and social sciences discussions of psychosis were not supplementary, but actually necessary to reformulate rationality and produce the forms of measurement and computation that underpin so many of our contemporary systems.

The ethical corollary is that if we wish reason to be rational then what type of agency preoccupies our imaginaries and dictates our values? In the course of this chapter and the next I will draw a preliminary topography from the emergence of a new model of mind to the reorganization of models of government and economy to make a case for the emergence of a new methodology conveyed in terms of aesthetic diagrams and data visualization and framed in a language of psychosis and logic.

I open, therefore, with this redefinition of logic as psychotic because this was not an isolated concept relegated to the corridors of the Institute of Advanced Study, or contained within the laboratories of MIT. Rather, this materialization of psychic processes, previously labeled "psychotic," as algorithms produced the substrate for a new methodology across the social and behavioral sciences. As Joseph Dumit has shown, neurotic, psychotic, and schizoid circuits proliferated in the diagrams of cyberneticians and information theorists from cognitive science to political science.[14] From the workings of brains to the structure of governments and the operations of financial markets, subjects previously considered abstract, analytic, or organic might now, if scientists but reframed their standards of truth, be demonstrated as material, logically representable, and amenable to reproduction, circulation, quantitative analysis, and data visualization.

The Path to the Neural Net

The cybernetic model of mind already posited a complex relationship to representation, temporality, and knowledge that emerged from the convoluted biographies of its practitioners. McCulloch himself had a complex history that contributed to his imaginary of the mind. He had been raised as a devout Catholic and extreme patriot. As a youth he studied first at Haverford College, leaving due to his pro-war sentiments during World War I and attending Yale University for officer's training. He studied philosophy and psychology at both places. Immanuel Kant, Gottfried Wilhelm Leibniz, Alan Turing, and Bertrand Russell were his inspirations. He then enlisted in the navy during World War I. His (albeit limited) experience inspired him to develop concepts of redundancy of command and decentralized control that were later influential in his conception of the nervous system and mind.[15]

McCulloch proceeded to be trained as a psychiatrist, a vocation that was not to be his calling. He spent the late 1920s at Bellevue Hospital and Rockland State Hospital in New York, during his residencies after completing a medical degree at Columbia University College of Physicians and Surgeons. He spent little time in actual practice but acquired, in his words, a "life time" of experience, that he would always call upon in the future to legitimate his speculative models of minds and machines. Most of his future work would commence at a desk at MIT far away from clinics and patients. But what psychiatry had seemingly offered McCulloch was an archive of cases and a collection of pathologies to mobilize as instruments in his later research.

As a psychiatrist, he was constantly obsessed with pathologies as routes to explanation and often thought about psychotic and neurotic patients, the truly very nervous. In his experiments he often theorized ideas about circuits using medical analogies about the function and working of neurons in making patients "healthy" and operable within their environments. He often spoke of traumatized causalgics, paranoid soldiers, neurotic housewives, and delusional teenagers, all in the name of demonstrating how circuits make the mind both mad and organized, with the steadfast interest of rewiring (as in literally rewiring the circuits of neurons) patients in the interest of health.[16] Even in the cases that were clearly unrelated to physical injury to the brain or the body, he wanted to find a way to link the nervous system to the psychology, a causal mechanism between the two. Perhaps, he postulated, such disorders were still physiological, in being related to some sort of misfiring or reverberation in the nervous nets—a confusion from within the system of where signals were coming from and when, a problematic feedback loop that could, perhaps, be interrupted. McCulloch was steadfastly physiological in approach and had an almost pathological dislike for psychoanalysis.

He bemoaned, for example, in a BBC interview in early 1953, the plague of psychoanalysis. Addressing an entire nation, Britain, that had suffered no small amount of wartime trauma, McCulloch hoped to perform an exorcism, perhaps of memory. He opened his account by recounting a scene of a patient, also a "friend for forty years." This patient had been blown out of a shell hole in World War I and suffered from what was then diagnosed as "shell shock." Shipped back to America, the patient suffered the jitters, his hands sweated and trembled, and his sleep was tormented by dreams of battle. If woken in the midst of these traumas, the patient lashed out violently. For thirty-five years, McCulloch lamented, this poor individual was forced into psychotherapy, fifteen of those years wasted in psychoanalysis. McCulloch argued: "his analyst is sure he will recover when he can recall the trauma of being expelled from his mother's womb." It was not Oedipus, he argued, that really impacted anything here; "what he really cannot remember is what happened between the time he jumped into the shell hole and the time his friends gave him a drink and a cigarette. His hands still tremble and he still dreams of battle."

McCulloch continued the saga of psychological therapy. "In World War II such men were more fortunate. They were promptly and effectively relieved of their shock and terror by the use of chemicals of a kind which we know reduces the turbulence of an upset nervous system." Advocating the use of sedatives and then stimulants (preferably barbiturates followed by cocaine) in what does sound like a very effective treatment to reorganize the nervous system, if not wipe out a few neurons entirely, McCulloch calmly propagated the rising support of psychopharmaceuticals, of which the first classes were only formally introduced after the war.[17] This psychiatrist did not seem to think that renarrating memories would help anyone. Rather, he urged the simple rewiring of the circuit. If these memories of battles did not keep feeding back and circulating in the system, the former soldier would no longer suffer so. This attitude—of returning the nonfunctioning to functioning—was normative and disciplinary. McCulloch had a medical doctor's normative conception of the healthy and the sick and an interest in making the diseased behaviorally functional.

On the other hand he also marveled at the autonomous capacity and prodigious liveliness of the very elements of the nervous systems—the neurons—as somewhat wily and unpredictable creatures—creatures of logic and probability seeming unrelated to normal functioning in people. He talked about neurons as "ghosts,"[18] "little plants,"[19] "telegraphic relays," and "machines that think and want."[20] The transformation, which McCulloch pioneered, of pathologies into circuits that could be moved from within people to the construction of other machines departed from any normative or absolute concept of how bodies or nets should work and instead focused on what types of machines, and circuits, could be built, modulated, and enhanced.

This was a radically different concept of perception and mind, one no longer attached to human bodies, and concerned not with limits, disabilities, and pathologies but with capacities, circuits, and channels—in many different types of machines. As McCulloch never stopped reminding his adherents and reviewers, the question that guided him in all his research, whether on the logic of neural nets or the cognitive capacity of animal optic nerves, was a basic question: "what is a number, that a man may know it, and a man, that he may know a number?" Put otherwise, if other cyberneticians like Norbert Wiener were guided by ideas of truth and norm, of finding diagrams still linked to truths in nature, then McCulloch asked more philosophical questions about finding the logical building blocks that precede even "man," and that could produce a computational logic to describe, and perhaps make, thinking beings. More significant, rather than answering his own question, as earlier logicians had tried, McCulloch proceeded to displace problems of truth—"what is man or number?"—with problems of agency: what do men or numbers do? What machines might be built? These problems, however, would always be animated by McCulloch's own nervous past with psychology and psychoanalysis.

The Nervous Nets

In 1943 McCulloch and the logician Walter Pitts first introduced an account of the brain directly linking computational logic with neurons, in answer, perhaps, to the problem of thinking about the nature of men and numbers. Their article "A Logical Calculus of Ideas Immanent in Nervous Activity," which appeared in the *Bulletin of Mathematical Biophysics*, has now come to be one of the most commonly referenced pieces in cognitive science, philosophy, and computer science. The model presented in the piece would forever impact future concepts of minds and machines and lay at the heart of models ranging from the treatment of schizophrenia to the architecture of the stored program computer.

While much has been made of the development of game theory and computers as emerging from military imperatives, it is important to note that unlike Wiener's antiaircraft predictor, the neural net was not the side product of any direct effort to build a medical or military technology. Rather, this model was the result of a philosophical inquiry. Pitts and McCulloch sought Kantian a prioris to thought, using Bertrand Russell's *Principia Mathematica* as a guide.

But while their imaginations might have been moved by histories of mathematics and logic, McCulloch and Pitts also still worked within a more pragmatic context in which both battlefield trauma and signal processing were emerging as central concerns—concerns abetted by McCulloch's own history as a psychiatrist and naval officer. Whether by intention or by coincidence, McCulloch and Pitts provided a model to unify psychiatry with communication, and link logical reason with the mechanisms of biology.

It took upward of two years to conceive of this calculus of the net. At the time of their introduction, in 1941, McCulloch was conducting work on a range of topics, including the impact of different drugs and chemicals on neurotransmitters and the nervous system at the Neuropsychiatric Institute of the

University of Illinois, Chicago.[21] This work may have also included research on chemical agents related to warfare, supposedly in the interest of treatment in case of exposure, and extensive research on psychosis and the effects of battle on the nervous system. The laboratory was steadfastly physiological in focus and had few (if any) psychoanalysts.[22]

McCulloch had long been interested in merging psychiatry with psychology to seek physical infrastructures for cognitive and conscious thought. It was, therefore, in the interest of inquiring about the mechanical capacities of the nervous system that McCulloch came to attend seminars run by the Russian founder of mathematical biology, Nicolas Rashevsky, as a forum for exposure to alternative approaches to the study of biological systems. It was at these seminars, held at the University of Chicago, that McCulloch was introduced to the young prodigy Pitts, who had trained under Rudolf Carnap in logic.[23] Pitts had run away from his family and was purportedly always attempting to complete one degree or another, to no avail. He was basically adopted by McCulloch, living with him for two years.

The decision to move the research in the direction of neural nets was also influenced by the proliferation of research on control and communication happening at the time. McCulloch had been introduced to Norbert Wiener through a mutual friend at the time, and McCulloch and Pitts were particularly influenced by Wiener and Rosenblueth's piece "Behavior, Purpose, and Teleology" (discussed in chapter 1) and proceeded to apply ideas of black boxes, probability, and feedback to the brain. McCulloch went on to chair the Macy Conferences on Cybernetics, and both McCulloch and Pitts later went to MIT in 1952 at Norbert Wiener's behest.[24]

The Logical Calculus of the Nervous Net

By reformulating the problem of what is a mind as the problem of what it can do, cyberneticians and affiliated communication scientists hoped to produce new forms of machines—both biological and mechanical. There are a series of moves by which neurons could be made equivalent to logic gates, and therefore "thought" and higher cognitive functioning could be represented logically and proven as realizable from the material physiological actions of the brain. These moves reformulated psychology, but they also demonstrated a broader transformation in the constitution of evidence and truth in science.

The model of the neural net put forth in *A Logical Calculus of Ideas* has two characteristics of note that are critical in rethinking cybernetic conceptions of representation and temporality.[25] The first claim is that every neuron

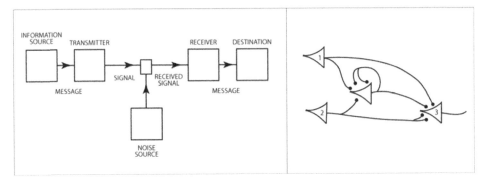

FIG. 3.1__Two types of channels. Diagram (left) based on Claude Shannon and Warren Weaver, from *Mathematical Theory of Communication*, 1963 ed., 7; Neural net diagram (right) by Warren McCulloch and Walter Pitts (1943), from McCulloch, *Embodiments of Mind*, 36.

firing has a "semiotic character"; that is, it may be mathematically rendered as a proposition.[26] To support this claim, Pitts and McCulloch imagined each neuron as operating on an "all or nothing" principle when firing electrical impulses over synaptic separations; that is, the pair interpreted the fact that neurons possess action potentials and delays, as equivalent to the ability to effect a discrete decision. This effect affirms or denies a fact (or activation). From this follows the claim that neurons can be thought of as signs (true/false), and nets as semiotic situations, or communication structures (just like the structured scenarios of communication theory, fig. 3.1).[27] This discrete decision (true or false, activate or not) also made neurons equivalent to logical propositions and Turing machines.

The second element of the model is a strictly probabilistic and predictive time. Neuronal nets are determinate in terms of the future (they are predictive) but indeterminate in terms of the past. In the model, given a net in a particular time state (T), one can predict the future action of the net (T + 1) but not the past action. From within the net, one cannot determine which neuron fired to excite the current situation. Put another way, from within a net (or network) the boundary between perception and cognition, the separation between interiority and exteriority, and the organization of causal time are all indifferentiable.

McCulloch offered as an example the model of a circular memory neuron activating itself with its own electrical impulses (fig. 3.1). At every moment, what results as a conscious experience of memory is not the recollection of the activation of the neuron, but merely an awareness that it *was* activated in the past, at an indeterminant time. The firing of a signal, or the suppression of firing, can only be known as declarations of "true" or "false"—true there was an impulse, or false, there was no firing—not an interpretative statement of context or meaning that might motivate such firing.

Within neural nets, at any moment, one cannot know *which* neuron sent the message, *when* the message was sent, or *whether* the message is the result of a new stimulus or merely a misfire. In this model the net cannot determine with any certitude whether a stimulus comes from without or from within the circuit; whether it is a fresh input or simply a recycled "memory." Put another way, from within a net (or network) the boundary between perception and cognition, the separation between interiority and exteriority, and the organization of causal time are indifferentiable. But rather than being a disadvantage for the capacity of a neural net, McCulloch and Pitt's brilliance was to see this as an advantage.

They end on a triumphant note, announcing an aspiration for a subjective science. "Thus our knowledge," they wrote, "of the world, including ourselves, is incomplete as to space and indefinite as to time. This ignorance, implicit in all our brains, is the counterpart of the abstraction which renders our knowledge useful."[28] If subjectivity had long been the site of inquiry for the social and behavioral sciences, now, perhaps, it might—in its very lack of transparency to itself, its incompleteness—become an explicit technology.

Inventing Cognition and Redefining Memory

These elements of the model—neurons as semiotic propositions, and time as indeterminate and predictive—reconfigured both the definition of cognition and memory. By deduction, this could not be a Cartesian mind, capable of clear separation from the field of perception. The psychiatrist made clear that the model presumed a "psychosis" due to the inability to differentiate the origin or location from which signals arrived, and that this psychosis was necessary for the model to work.[29]

As stated at the start of this chapter, McCulloch and Pitts declared that in order to make neurons logically representable, two problems had to be dealt with: the issue that neurons in real brains can alter responsiveness according to previous activity (the problem of facilitation and extinction that makes the synaptic firing an undefined temporal event) and the problem of learning, and history, in which previous events alter the structure of the net. These are both problems of memory, the past, and accounting for the historicity of signals. Eliminating these two "problems" by envisioning the brain as a "psychotic" machine, in McCulloch's terms, allows each event to be discrete, definitive, and conclusive.

Neural nets were thus made equivalent to Markov chains and also compliant with the Russian mathematician A. A. Markov's famous definition of an algorithm as possessing three variables: definiteness, generality, and conclusiveness.[30] They are definite in having only a set number of states that can be described and are always predictive (T + 1, not T − 1); they are general in being logic gates capable of true/false statements and not specific descriptions of the content of the signal; and they are conclusive—either the net fires or it does not, the statement is thus absolutely true or it is false, there are no other interpretations of the situation. I can make a final extrapolation and argue that for McCulloch and Pitts processes of reasoning can be directly equated and represented as algorithms and derived from the physiological mechanism of the neuron's actions.

If the time of the net and its logics are always predictive, then memory must also serve new functions. The ideal of memory was rethought as a physiological and mechanical process. For memory to be logically represented as Boolean operators in this model, then it must also be malleable and erasable, in order for new impulses to enter and circulate through the network and to facilitate the processing of orders in an algorithmic pattern. As McCulloch argued, "in looking into circuits composed of closed paths of neurons wherein signals could reverberate, we had set up a theory of memory—to which every other form of memory is but a surrogate requiring reactivation of a trace. Now a memory is a *temporal invariant*. Given an event at one time, and its regeneration at later dates, one knows that there was an event that was of a given kind."[31] By implication, memory was amnesic and neurotic, it carried data, and it affirmed the truth or falsity of something having happened (events exists and leave marks), but only at an indeterminate time. Memory neurons could not assign with fixity when that time was or what had incited the impulse. Memory, therefore, was not dedicated to establishing temporal organization, only to executing behaviors (in which conscious thought and analysis are included). The time of the net, like that of all cybernetic and communication theories, was teleological, and all reactions were irreversible (nonlinear).

Labeling logical minds amnesic, psychotic, neurotic, and speculative all at once, McCulloch and Pitts end on a triumphant note, announcing an aspiration for a subjective science. "Thus our knowledge," they wrote, "of the world, including ourselves, is incomplete as to space and indefinite as to time. This ignorance, implicit in all our brains, is the counterpart of the abstraction which renders our knowledge useful."[32] If subjectivity had long been the site of inquiry for the behavioral sciences, now, perhaps, it might—in its very lack of transparency to itself, its incompleteness—become an explicit technology.

McCulloch and Pitts were explicit that their work was a *Gedankenexperiment*, a thought experiment that produces a way of doing things, a methodological machine. Cheerily, McCulloch admitted that this was an enormous "reduction" of the actual operations of the neurons.[33] "But one point must be made clear: neither of us conceives the formal equivalence to be a factual explanation. *Per contra!*" At no point should anyone assume that neural nets were an exact description of a "real" brain.[34] In fact, nets are not representations, they are methodological models, and processes. McCulloch and Pitts discussed this logical reasoning as an experiment, a machine perhaps like the ones described by Deleuze and Guattari in *Thousand Plateaus*, that does not describe a reality, but rather helps scientists and engineers envision new types of brains and machines, and challenge what scientists thought they knew about how mental processes work. These circuits are labeled by McCulloch as "psychotic" because they challenged scientific perspective, and reformulated the boundaries of interiority and exteriority, between knowledge and practice.

Reductive or not, the pair had established that a capacity for logic and very sophisticated problem solving might emerge from small physiological units such as neurons linked up in circuits. In doing so, and by way of exploiting the amnesia of these circuits, McCulloch and Pitts were able to make neural nets analogous to communication channels, and shift the dominant terms for dealing with human psychology and consciousness to communication, cognition and capacities. Their conception of the neural net informs a change in attitudes to psychological processes that makes visible an epistemological transformation in what constituted truth, reason, and evidence in science.

Let me outline this new epistemology briefly here. It rests on three important points that are seemingly unimportant alone, but are significant when recognized as joining a history of logic, engineering practices, and the social and behavioral sciences in a new assemblage. The first is that logic is now both material and behavioral, and agents can be an- or non-ontologically defined or "black-boxed." The second is that cybernetic attitudes to mind rest upon a repression of all questions of documentation, indexicality, archiving, learning, and historical temporality. And third, the temporality of the net is preemptive, it always operates in the future perfect tense, but without necessarily defined endpoints or contexts.[35] Nets are about T+1, the past is indeterminate: McCulloch regularly argued against caring about the actual context, or specific stimulus that incited trauma in patients, or systems.[36] Together these points meant that rationality could be redefined as both embodied and affective and good science was not the production of certitude but rather the account of chance and indeterminacy.

Control and Computing

The historical redefinition of rationality demanded therefore a reconsideration of what "control" might mean, and what forms of futurity and memory exist in nets and machines. In most contemporary scholarship control was correlated with prediction, knowing the future, and command (often military).[37] The most prominent impact of the nervous net came in the form of the architecture for the stored program computer, arriving in 1945. The concept of a memory capable of erasure (the neuron must fire, and then be prepared to receive further signals) but still able to recycle past impulses was highly amenable to the demands of a computational machine and fulfilled the logic set forth by Turing. This ideal of memory is still encapsulated within our contemporary computers whose architecture continues to be inspired by the von Neumann architecture for the first stored program computer (EDVAC). The ongoing compartmentalization of the arithmetic unit, memory, and control unit embodies these principles.[38]

In their 1948 book, *Planning and Coding*, von Neumann and Herman Goldstine introduced flow charts and circuits for stored program computers. However, in describing their circuits they wrote, "[w]e propose to indicate these portions of the flow diagram of C by a symbolism of lines oriented by arrows. [fig. 3.2] ... Second, it is clear that this notation is incomplete and unsatisfactory ..."[39] In other words, control is not definable, its operable imagining and its explicit definition are incommensurate. But rather than treat this failure in representation as a problem, this threshold became a technological opportunity; this emergent space between the definable and the infinite provided the contours of the engineering problem—an opportunity to turn from logic to technology.

This model of the cycling memory neuron, and an indeterminate control in the stored program computer, directly refracts an earlier concept of control in the Turing machine. Control in the Turing machine is the head that "reads" the program from memory and then begins the process of executing it according to the directions in the memory. On the one hand control directs the next operation of the machine. On the other hand control is directed by the program. The control unit, or the reading head in a Turing machine, is directed by the tape it is reading from memory, not reverse. Control is that function that will read and act on this retrieved data, inserting the retrieved program

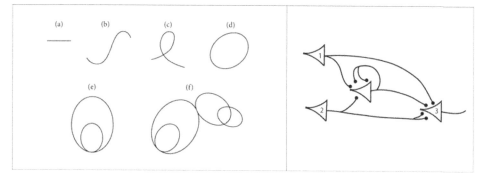

FIG. 3.2__**Control**: early flow diagrams depicting "C." Goldstine and von Neumann, *Planning and Coding*, 157; **Memory**: circular nets in minds, neural net diagram based on Warren McCulloch and Walter Pitts (1943), from McCulloch, *Embodiments of Mind*, 36.

or data into the run of the machine. Such machines do not operate top down but in feedback loops between storage, processing segments, and the interface for input and output. In his 1946 report on building a computing machine, the ACE *Report*, Turing reiterated that only the possession of memory "give[s] the machine the possibility of constructing its own orders; i.e. there is always the possibility of taking a particular minor cycle out of storage and treating it as an order to be carried out. This can be very powerful."[40]

If there is a feature that allows minds and machines to act in uncanny and unexpected ways, it is this surprising capacity to change the pattern of action by way of insertion of a program from storage. Control in computers is like reverberating circuits in brains, and both are classically defined in previous psychologies, such as psychoanalysis, as psychotic in receiving memories without history (fig. 3.2).

Even today our machines cannot truly differentiate between inputs and programs; the two are the same from the perspective of the machine, although programmers and engineers do much to signal their separation. We have all experienced, for example, "pathological" breakdown, when programs crash, and we are forced to witness "error" messages. This is usually a control problem in one's machine—too many conflicting programs making the machine incapable of acting. Control makes systems operate and gives them their emergent properties—an older memory can always interrupt the run of the machine and change the course of operation. But the best control is also neurotic, perhaps even psychotic, in cycling memories into programs without knowledge of causality or history. If today we regularly assume we know what control is, and deploy the term critically at will, from within the machine it is far less clear, and far more dynamic.

Significantly, for us, McCulloch and Pitts, and following them von Neumann, can be said to have operationalized the problem posed by the original negative proof of the *entscheidungsproblem* that is the Turing machine. The problem could be inverted from seeking the limits of calculation to examining the possibilities for logical nets. What had been an absolute limit to mathematical logic became an extendable threshold for engineering.

This reformulation made reason, cognitive functioning, those things labeled "mental," algorithmically derivable from the seemingly basic and mechanical actions of the neurons. Rationality was redefined as both embodied and affective, and good science was not the production of certitude but rather an account of chance and indeterminacy. As Joseph Dumit has argued, for neural net researchers the inquiry then turned to determining not whether minds are the same or different as machines but what difference it makes to be in one network or another. The performative, not the ontological question.[41]

This logic of a rational network found itself embedded within numerous fields. Reformulated concepts of reason and intelligence proliferated in human minds through cognitive science and in the flow charts of operations research and organizational management (fig. 3.3).

Remaking the Psychological Subject

Perhaps the most infamous revision of psychoanalytic concepts and psychology in cybernetics, inspired by these models of mind, came in the form of the 1956 introduction of the model of the "double bind," by Gregory Bateson, into both the human and social sciences. Bateson was to become the guru of an emergent West Coast communal culture and an icon for many of the ideals of a new information society being birthed in the nascent Silicon Valley and elsewhere. Most prominently displayed in venues like *The Whole Earth Catalogue*, Bateson's ideas and concepts were widely part of a new techno-utopic culture based in communication theories, cybernetics, and environmentalism. Bateson's name, and his reconceptualization of mind, ecology, and schizophrenia, were regularly cited by Stewart Brand and many in the counterculture.[42] Today we are also familiar with Bateson through his impact on theorists like Deleuze and Félix Guattari, whose ideas of "plateaus" come from Bateson's work.[43]

At the time Bateson developed the "double bind," he was working as an ethnographer under the influence of cybernetic ideas at a veteran's hospital in Palo Alto. His job was to provide additional data to supplement the work of a team of psychiatrists. Working with patients coming from the Korean War and World War II, he recognized that many of the pathologies of addiction and battlefield trauma might have emerged from the result of conflicts in signals emerging at the same time without hierarchy between sets—in this case the signals being "Thou shalt not kill" and "Thou should kill." His innovation was to recognize that the environment that patients were in might be part of their psychological system, and that if two contradictory messages arrived at the same time this presented problems for the subject. Applying communication theories to minds, Bateson predicted that perfectly reasonable people might become dangerously paranoid if faced with incommensurable messages.[44]

Bateson was one of the most active participants in the Macy Conferences on Circular Causal Feedback Mechanisms in Biological and Social Systems.[45] He corresponded regularly with McCulloch and with Norbert Wiener and on the basis of that correspondence he developed his ideas of applying logic to embody the mind. Irrespective of the medical veracity of his theories (which may or may not have worked, and Bateson was not the most focused of researchers), they impacted ideas of psychology and society in many fields.

Inspired by cybernetics, Bateson was steadfast in applying theories of mathematics, logic, and communication to his ethnographic observations. Already in 1954, Bateson wrote to McCulloch wanting his opinion on the validity of his theory that "the syndromata of schizophrenia are describable as pathologies in the use of those signals which should identify the 'language games' in Wittgenstein's sense; and that the etiology of schizophrenia goes back to infantile learning in which the relevant traumata have patterns reminiscent of type confusion."[46] At stake in this discussion was the relationship between logic and memory. The disease reenacts "patterns" from earlier in life, like a child, and does so in the manner of a logical "game"; for Bateson language had become a matter of patterns and redundancies as existing in neural nets and communication theories. Bateson, therefore, chose to focus on the structure of the interactions rather than their content in contemplating psychological trauma.

Witnessing the internal divides of patients, he began to think about schizophrenia as a problem with typing, or finding "sets," as defined by Bertrand Russell, by which to hierarchize and organize information. Schizophrenia, according to Bateson, was the result of closed systems for whom there are two self-negating terms. This dilemma is classically explained in the following terms: person must do X, X negates Y, person must also do Y, therefore neither can be executed. The person cannot ignore X or Y but also cannot comment on the absurdity of the situation.[47]

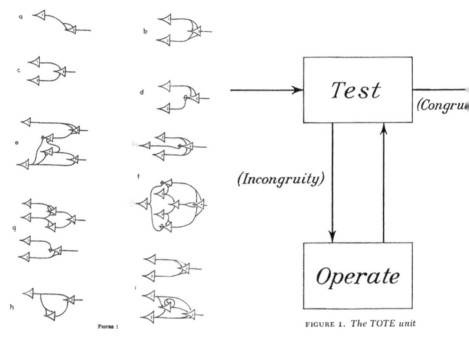

FIGURE 1

(Incongruity)

Test

(Congru

Operate

FIGURE 1. *The TOTE unit*

FIG. 3.3__**Network**: nervous nets, from McCulloch and Pitts, *Logical Calculus* (1943). **Cognition**: tote units organizing cognitive processes, from Miller, Galanter, and Pribram, *Plans and the Structure of Behavior*, 26; **System**: flow diagram of distributor sector, by Jay Forrester. From Forrester, *Industrial Dynamics*, 163.

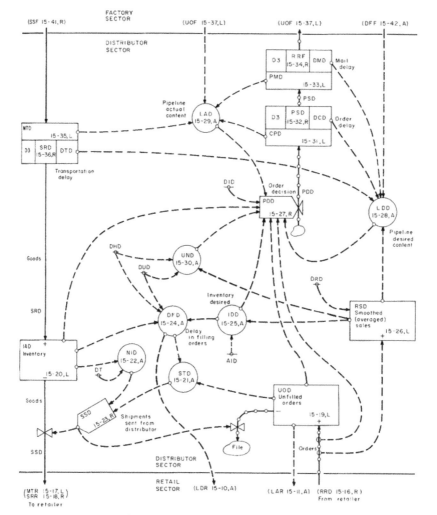

Figure 15-15 Flow diagram of distributor sector.

For example, a child is commanded to "be spontaneous" or "creative." To obey the command is to fail to be either. To disobey is to break a rule. Either way, the subject is caught, unless he or she is able to suggest an external option out of a different framework, for example, retorting: "that is not possible" or "I don't understand the rules." These are responses that operate at a different logical level, with points of reference in a different communicative situation.

Bateson argued that schizophrenia was not an automatically pathological state. But it was a communication structure that was modelable, like theories of communication, as a system and a channel between and within the body, different actors, and the environment. For Bateson this situation made the mind an "ecology."[48]

Bateson's conclusion was that the pathological form of schizophrenia was an inability to comprehend the structure of the communicating situation. Schizophrenics receive signals but cannot respond in a purposeful way (i.e., with a deferred known goal) because they have no other, different, communicative situations to compare, contrast, or use outside of the immediate message. Schizophrenia was a problem of failing to circulate information and accumulating too much in one place, too many signals interfering with each other—a circuit problem. Paranoid symptoms, therefore, emerged from dysfunctions in storage and memory. If an intervening program cannot enter from memory, the repetitive actions cannot be disrupted. The system cannot learn, and it becomes stuck or jammed—the irony being that the more rationally and logically an animal, machine, or human acts under particular conditions, the more self-destructive it can become.

Bateson went so far as to suggest that schizophrenia is one of the standard states we all live in within contemporary information societies. It was, in his estimation, quite common for individuals to be embedded within contradictory communication structures without a Cartesian or external perspective on the situation.

Bateson's model facilitated new attitudes to treating not only individual patients but, more important, patients in relationship to other individuals and the environment. Under Bateson's influence a number of psychiatrists began working on family therapy at the Mental Research Institute (MRI) in Palo Alto.[49]

Family therapy reframed disease away from individual subjects to interactions *between* subjects and objects. This was not only a frame shift in what constitutes the fundamental unit of psychological intervention but also a transformation in documentary practices. A number of groups at MRI adopted methods inspired by Bateson's work and began applying film, then videotape, to therapy for replay and analysis. The introduction of film through two-way mirrors and by having cameras present in sessions refocused psychiatric assessment and therapeutic concern on new details. For example, therapists focused on experimenting and reflexively evaluating the timing of the lengths of the sessions. The tempo and cadencing of conversation in the session, its form, became as much an object of study as the specific content of discussion between patients and doctors. (Not incidentally Lacan, who was also revising the ego in psychoanalysis at this period, was also highly invested in the cadencing and timing of analysis.) The very feedback interaction, both in gesture and language, between the analyst and analysand was explicitly studied as much as the specific etiologies of any pathology.

One of the lead analysts, Don Jackson, explicitly analyzed the possible dependencies developing in the session between the analyst and the analysand. Reflexively, the communication paradigm made the therapy session itself, in his words, a "double bind" where the two figures of the patient and the therapist were taught to remain pathological so as to continue the interaction. This bind, arguably impossible for psychoanalysis to fully recognize, could, Jackson argued, be made visible and therefore reprogrammable through the self-reflexive externality of the recording systems.[50] Analysis itself became both the experimental setup and the self-referential object of intervention and study.

In this work, and that of many of Bateson's compatriots in the social, communication, cognitive, and computational sciences, we hear similar statements transforming a world of stable identity, conscious subjects, and discrete objects into one of interaction, patterns, and networks. If a half century earlier Sigmund Freud had diagnosed schizophrenia as the result of one communicative structure—the Oedipal complex—Bateson allows these structures to take many forms, to be unmoored from a normative situation. By substituting communication for the language of pathology and psychology, and replacing performance with the representative function of language, both the work of representation and the function of illness are transformed. It is the inversion of psychosis from a pathology to an instrument for social research that makes a new entity take precedent—methodology and measure. This reformulation of the site of psychic action in interactions, rather than within subjects, permitted the formulation and scalability of cognitive models, from clinical setting to sociological organizations.

What had once been a vexing problem to be surmounted by science was now an opportunity for technical development for methods of self-reflexive analysis. If the actual experimental setup was always the one thing modern scientists wanted to repudiate as a site of knowledge production, here the collapse between analyst and analysand is embraced. Like the "incompleteness" of McCulloch's knowledge, these therapeutic techniques rescripted what constituted evidence (interactions between subjects, the structure of the situation, the therapeutic interaction itself) and what constituted knowing.

By 1957 "cognitive dissonance," in the landmark study of it by Leon Festinger,[51] emerged as one dominant model for human behavior. Festinger worked at MIT in 1945, as part of the newly formed Research Center for Group Dynamics, before going to Michigan, Minnesota, then Stanford University, and finally the New School for Social Research. He was quick to adopt ideas of cognition and communication emerging at the time in these Universities. The head of the MIT Center for Group Dynamics, Kurt Lewin, who initially recruited Festinger, had himself attended the Macy Conferences on Circular and Causal Mechanisms.[52]

Festinger's theory posited all subjects as split, and "dissonant." Coherent worldviews are, he noted in his studies, rare. Instead he proposed that individuals reconciled seemingly incommensurable possibilities, of which he names things like being against racism but not wishing to have African Americans in the neighborhood, for example, through the careful construction of information environments that supported and abetted the cognitive knowledge even if it was not commensurable with "reality." "The major point to be made is that *the reality which impinges on a person will exert pressures in the direction of bringing the appropriate cognitive elements into correspondence with that reality.*" Festinger's theory envisioned a split creature living in a self-produced mediated environment, rather than a reasonable subject being deceived sometimes but fundamentally possessing a capacity for objectivity. "Knowledge," he announced, was a "belief," an "opinion"—a term that applied to any form of psychological faith that facilitated activity, cognitive or otherwise.[53] For Festinger, human beings were integrated into communication channels from which they carved resonances by attempting to avoid, channel, and divert information that did not cognitively map to a self-referential and deeply subjective worldview. Disjuncture between belief and action was the very fuel for change. Rather than assume that our beliefs guide our actions, for Festinger, the human being might act first and think as a result. Knowledge, for Festinger, was a performative entity, not a stable truth. And as a social psychologist, he, along with his colleagues in many of the social sciences of the time, began to view systems in manners simultaneously psychic and rational, but never reasonable in the sense of possessing full information about an external environment. Festinger's innovation in cognitive science refracted another movement toward producing an "ecological" but also "algorithmic" mind. If minds worked like nets, with series of steps leading to circuits, then these circuits could be discretized, modeled, rebuilt, reenacted. More important, if social scientists reframed their truth claims in terms of operability and pragmatism rather than making causal or fundamental claims about the nature of people or circuits, then, rather than focusing on the nature of the subject, the scientist could focus on the structure of the experiment.[54]

Psychoanalysis in the Machine

We might, then, seek to further historically situate the relationship between these cybernetic and computational discourses and older modernist discourses of psychology and drives by asking: what is at stake in this movement from consciousness to cognition? Immersed in efforts to build computational machines, revise psychiatric treatments, and transform scientific practice in the behavioral and social sciences, cyberneticians often negotiated explicitly and implicitly the residues of earlier nineteenth- and twentieth-century concepts of thought, logic, consciousness, and reason.

If cybernetics, mathematics, and neuroscience had provided the direct inspirations for the model, however, there was another science that lurked in this machine. One of the single most regularly repeated themes in cybernetics, and its affiliated sciences, such as game theory, was a discourse of exorcism. This discourse focused with ferocious intensity on concepts of subjectivity, desire, and the unconscious, whether defined in terms of Oedipal structures, affective behaviors, or utility. So regular a repetition of a theme might imply proximity, rather than distance, between these postwar discourses of rationality and cognition and prewar concepts of desiring drives and consciousness. McCulloch and Pitts themselves appeal to the divine and the spiritual in their effort to expunge the past and affirm the validity of their present. The neural net paper ends with an exorcism. "Mind no longer 'goes more ghostly than a ghost,'" McCulloch and Pitts write; "Instead, it is now 'deducible from neurophysiology.'"[55]

And the one science, in McCulloch's estimation, that devilishly possessed psychiatry and psychology was psychoanalysis. Psychoanalysis, with its insistence on narrative and talk therapy, appeared to McCulloch to offer a religious and occult understanding of the mind as entirely spiritual and in the realm of representation, narrative, and human language. Not an animated, vital, mechanical mind as he envisioned.[56]

However, despite McCulloch's linking psychoanalysis with the occult, he also appeared to struggle with ghosts. If minds were material, this was not to say that they were not still supernatural in their ability to trouble the present and evade historical and causal explanations. "On more than one occasion I have said, not quite in jest, that in their essence gremlins and neuroses are demons, sentient and purposive beings, exercising in their own right properties more mental and more physical than any psychoanalyst has yet had the courage and clarity to claim for them, as much when they haunt machines as men." These demons resulted from reverberating circuits, operating "autonomously," according to McCulloch. Offering pathology both subjectivity and purpose, McCulloch gave mental phenomena material properties of being and sovereignty as subjects. That psychosis and neurosis might now possess a material and embodied capacity to act at will makes visible a historical moment where psychic phenomena were rendered functional as technologies to be extracted, materialized, and perhaps resynthesized in separation from the locality or anatomy of the human body.[57]

Arguably, this was already the aspiration of the psychological sciences at the start of the twentieth century, and it is not coincidental that it was by way of this earlier science of the mind—psychoanalysis—that McCulloch endeavored to legitimate his research. The psychiatrist regularly appropriated a previous era's concerns with the autonomy of drives, desire, and the unconscious to propagate a new technology—not of subjectivity but of computation. McCulloch would later write:

> [the psychoanalyst] makes a sharp distinction between processes which are conscious, and others which were or might have been conscious, but are now shut out of consciousness. In the latter he puts the core of the neurosis. Now the psycho-analytical use of the term unconscious, if somewhat indefinite, is quite a special one. In essence, I believe, it is nothing but the suppression of information as we have just described it. The memories and learned reactions subserved by circuits connected with the regenerative circuit can no longer be evoked ... simply because all the stimuli which once evoked them ... now produce only the invariable neurotic response which displaces all the others. .... We may say the neurosis is autonomous, almost completely so; it is not isolated.[58]

In this return to psychoanalysis and the pathologies of neurosis McCulloch offered neurosis an autonomy, a will and a capacity to act as isolatable and material reverberating circuits. This autonomy of the neurotic cathexis is in keeping with many of Freud's observations, particularly on the compulsion to repeat and the death drive.

On the other hand McCulloch was clear that these circuits were not "repressed"; there could be no such operation in the science of cognition. For repression demands the separation of the conscious from the unconscious and, by derivation of thought, from the body or brain. Rather, McCulloch envisions a series of information channels feeding back and amplifying or disturbing other signals coming from parallel processes occurring at the same time in the nervous system. These disturbances create "displacement" and can no longer be managed by the usual control mechanisms of the mind's circuitry. Cyberneticians thus relocated long-standing concerns in the modern sciences of the psyche with the management of information and the circulation of nervous impulses in the psychiatric network, deferring questions about how the mental apparatus produces legible and conscious comprehension of the environment in order to maintain a focus, instead, on managing the circulation of information in the system.

McCulloch was often quick to assert the possibility for reformulating subjectivity and even quicker to dispense with ideals of an objectivity untempered by subjective experience or embodiment. In a series of later essays labeled "Of, I and It," he asserted that the decentralized nature of human cognition made the clear delineation of otherness murky. He wrote: "yet, from the use of *I*, *me, mine* in the communications of daily life in health and disease, we are entitled to infer that the vagrant solid which the speaker labels *I* in the moment of the experience consists only of events that occur as he intends. The rest he calls *it*." McCulloch went on to argue that concepts of resistances and thresholds should replace any "explanations" that psychoanalysis might provide.[59] For McCulloch, resistances, which may occur from lovers who do not return our affections, or even from within the body, such as a causalgic arm or a nervous tick, produce a fluctuating threshold of differentiation instead of a clear, set boundary between analyst and analysand or between subject and object.

If there was a critical pivot on which cybernetics would separate from previous histories of science, therefore, it involved a reformulation of the standards of authority and truth. For the psychoanalysis of the earlier twentieth century, struggling for credibility under the terms of objectivity offered at the time, psychosis, and schizophrenia especially, linked as it was to both the media technologies of the day and the transgression of the subject, presented a special problem. These pathologies might appear strikingly proximate to the practice of transference in psychoanalysis.[60] Psychosis made visible the threat to the clear-cut separation between the analyst and the analysand. These proximities and intimacies violated the forms of objectivity that Freud aspired to.

For McCulloch, Pitts, and their many interlocutors in the emerging cognitive and social sciences of the time, psychoanalytic concerns with pathology, normalcy, desire, and in fact consciousness were repressed. Gregory Bateson made this reorganization of the terms of desire into algorithm explicit:

> Classically Freudian theory assumed that dreams were a secondary product, created by "dream work." Material unacceptable to the conscious thought was supposedly translated in to the metaphoric idiom of primary process to avoid waking the dreamer. And this may be true of those items of information which are held in the unconscious by the process of repression. As we have seen, however, may other sorts of information are inaccessible to conscious inspection, including most of the premises of mammalian interaction. It would seem then to me sensible to think of these items as existing *primarily* in the idiom of primary process, only with difficulty to be translated into *"Rational"* terms. In other words, I believe that much of early *Freudian theory was upside down*. At that time many thinkers regarded conscious reason as normal and self-explanatory while the unconscious was regarded as mystical, needing proof, and needing explanation. Repression was the explanation, and the unconscious was filled with thoughts which could have been conscious but which repression and dream work had distorted. Today we think of consciousness as the mysterious, and of the computational methods of the unconscious, *e.g.* primary process, as continually active, necessary, and all-embracing.[61]

Bateson radically externalizes the unconscious. His statement implies that science now has a new technique—"computational methods" of the unconscious—that not only accounts for the behavior of individuals but is "all embracing," extendable to systems, ecologies, and organizations. He went so far as to label these unconscious and computational methods "algorithms of the heart, or, as they say the unconscious."[62] Bateson's statements suggest a transformation of psychological inquiry and concern from the conscious to the unconscious and the displacement of what had once been a source of vexing scientific concern and a limit to knowledge (mainly the recognition of the subjective nature of human perception and consciousness) into a "method" and an "algorithm."

At stake in the emergence of psychotic logic, therefore, was the stability of older histories of objectivity, truth, and documentation. But also up for negotiation were the terms of encounter between bodies and subjects. This question of both authority and intimacy, however, had been displaced; it was no longer the center of debates in cybernetic and computational research on the psyche or behavior. Within twenty years of the war, the centrality of reason as a tool to model human behavior, subjectivity, and society had been replaced with a new set of discourses and methods that made "algorithm" and "love" speakable in the same sentence and that explicitly correlated psychotic perspective with analytic logic. Cyberneticians sought to make the very space between rationality and reason, or the unconscious and conscious, amenable to logical, perhaps, mathematical, representation. The impossibility of visualizing this process, an impossibility already faced by Freud in his turn to dream work, now attached to a language of channels, circuits, and communication. This distance between reason and rationality was reformulated into calculative and statistical technologies. At the same time, the absolute inability to visualize the unconscious and represent time, as discussed in chapter 1, was transferred into an extendable threshold driving a frenzy of data visualization.

The Measurement of Rationality

What the cybernetic reformulation of logic as psychotic permitted, therefore, was an abandonment of ontological concern with the present, or a desire to access the "real," in the interest of focusing on future predictive interactions. These models measured not what is happening but what will happen as a result of finding patterns of past data, which ironically is devoid of historical temporalities. It is the circuit, the net, the structure of the game or the organization that matters. Process was materialized, made visible, transformed into interfaces, and became both the object of and the tool for new forms of measurement. The transformation in truth claims and epistemology opened a new frontier for study—subjective interactions in environments without complete information. Just as Kevin Lynch could reformulate the study of urban planning, as discussed in chapter 2, many fields followed the same logic of making subjectivity a new form of objectivity through a focus on postcollection analysis, pattern seeking, and prediction.

These literally nervous networks and logical rationalities proliferated in the methods and ideas of the social and behavioral sciences. Cybernetic concepts of mind were part of a broader shift at the time in concepts of reason, psychology, and consciousness. But nets, in particular, influenced certain figures such as John von Neumann. Premised on the same behavioral approaches, probabilistic frameworks, and logical apparatus as McCulloch and Pitts, *The Mathematical Theory of Games*, introduced by von Neumann and Oskar Morgenstern in 1944, does not start with a subject imbued with consciousness, or, perhaps, even with a subject. The first feature of the model is that there are rules, and that out of these rules one can define winning or losing, and in fact "rationality." von Neumann and Morgenstern explicitly define "rationality" as definable and rule-bound, implicitly making rationality equivalent to algorithms: "we described in [section] 4.1.2 what we expect a solution—i.e. a characterization of 'rational behavior'—to consist of. This amounted to a complete set of rules of behavior in all conceivable situations. This holds equivalently for a social economy and for games."[63] Like those who conceived of the neural nets, von Neumann and Morgenstern insisted on a world of zero-sum or absolute decisions that could be likened to logic. In order to make rationality equivalent to algorithms, and to extend the logic of the algorithm out of the simple world of arithmetic to that of choice and behavior, they made a move against both political economy and liberal economists. In a move that arguably underpins the popularity of their model in supporting ideas of deregulation and "free" markets, von Neumann and Morgenstern dispensed with both the labor theory of value and marginal utility. "Utility," they wrote, is "difficult," and preference is "impossible" to logically represent. To get past issues of usefulness, taste, sensibility, or desire, all unrepresentable by mathematics and the logics available in the early 1940s, they opted to be, in their words, "opportunistic" and to argue that the aim of all participants in an economic system was money. This decision was a technical trick to make different types of interactions comparable and, more important, quantitatively representable with finite values that could be assessed behaviorally. Money is not an ambivalent or ill-defined number. Money, when only considered in relationship to other monetary values, does not suffer the problem of translation. There cannot be an accounting for an action not taken (such as utility loss or gain). Another way of framing it is that there is no language or representational system for desire in game theory. One can speculate, but that speculation can only be defined as happening or not happening, not as the diversion of one set of wants toward other objects, nor can there be libidinal costs related to fetishes or nonrealizations.

Where von Neumann and Morgenstern even introduce the term "rationality," it is in scare quotes, and the authors evince a comprehension that it is an ill-defined entity. They argued both tautologically and self-referentially that an individual playing by the rules of the game and getting the most money can be said to be acting "rationally." They explicitly state that being rational depends on your assumption of what the rules are and what the best results are.[64]

Rationality was divorced from any particular immanent qualities of the subject (such as sovereignty, humanity, reason, or property), and this ontological instability was what made rationality material for computation. While the original theory perhaps maintained certain vestiges of subjective agency in the minimax solution as optimization, von Neumann was never pleased and went on to apply himself to theories of logical automata in an interest to create better logics that might represent more complex and less reasoned situations.[65]

The postwar social sciences were repositories of these techniques that transformed what had once been a question of political economy, value production, and the organization of human desire and social relations to problems of circulation and communication by way of a new approach to measurement. Liberated from direct causality and object-oriented analysis, social scientists started assiduously examining the space between an organism and its environment.

The stake of this transformation from the occult to the erotic to the machinic, therefore, has everything to do with contemporary structures of networks and capital. Desire, it appears, was one discourse that had a troublesome place in the epistemology of psychotic logic that underpinned emergent concepts of cognition, processing, and rationality. Nowhere is this more evident than at the very heart of rational choice decision-making theories and in the revision of "utility" and "choice" in these theorems.[66]

Speculative Nets

But it was particularly in the emergent fields of finance, as exemplified by the work of individuals like Herbert Simon, that these forms of measurement, parading under the label of rationality, took hold. Simon, who worked primarily in the business school at Carnegie Mellon, was an innovator in applying psychology, communication theories, computing, and cybernetics to the study of organizations and business. He was also a central figure in developing artificial intelligence throughout the 1950s and 1960s. In 1955 he was temporally working at the Air Force–sponsored think tank the RAND Corporation in Santa Monica. This corporation, whose name has become synonymous with Cold War game theory, security analysis, and military strategic studies, was dedicated to applying statistical and quantitative instruments to assessing human behavior, strategic policy, and markets. Initially created to assess the efficacy of the Air Force's tactics, RAND very quickly became the premier site of simulating and strategizing the unthinkable scenario of nuclear war. However, in recruiting the statistical and operational knowledge to simulate superpower interactions, RAND built the expertise to expand to other types of social sciences and problems. As a result RAND found itself temporary home to many of the top human and social scientists of the 1950s and 1960s, often even in fields seemingly extremely distant from anything involving nuclear deterrence or containment.[67] By the late 1950s, recognizing the limits of modeling nuclear deterrence alone as a business model, RAND expanded and redefined national security to include many regions of expertise, including, by the 1960s, urban planning, economics, the environment, and health.

Simon, then working on administrative behavior, came upon the idea, under the influence of von Neumann's games and computers, Wiener's cybernetics, and McCulloch's neural nets, that rationality had to be redefined, even in game theory. His first objection was that "rationality" usually assumed a separation between the organism and its environment, and thus a subject that could process information from its system without being of it, or inside it. He later explained his theory as emerging from the "acute schizophrenia" suffered by the social sciences that dually posited an omniscient rational actor with full information or an affective and stupid beast guided only by Oedipal complexes and pleasure principles. Perhaps, Simon suggested, by way of cybernetics there is a compromise. He imagined a new subject capable of cognitive capacity for making systematic and apparently "optimizing" decisions according to preset rules but who simultaneously is no longer reasonable and rational in an Enlightenment sense of possessing a perceptual field external to the environment. Simon imagined a subject incapable of objectivity. "We must," he wrote, "be prepared to accept the possibility that what we call 'the environment' may lie, in part within the skin of the biological organism." Organisms, he argued, are bounded by virtue of their physiology, biology, computational capacities, and access to information. That said, organisms are still often "rational" in the sense of being capable of making systematic, discrete decisions, in logical order, with set endpoints. Even nonreasonable subjects can act algorithmically, often because they lack an outside perspective on their situation.[68]

In the 1960s while beginning to work on problems of artificial intelligence, Simon elaborated on this vision of an agent by recourse to the figure of the ant. The ant, he argued, is only as intelligent as its environment; it is coupled with the exterior world intimately, its choices decided as much by the outside as the internal workings of the nervous system. The ant "deals with each obstacle as he comes to it; he probes for ways around or over it, without much thought for future obstacles."[69] For Simon, revising the agent in this way refocuses the search for cognitive intelligence on the production of situations and patterns for actions rather than on the effort to understand language or consciousness. Perception and cognition become the same as the act of "probing" and the act of decision-making is made equivalent, and the idea of intelligence can be linked to small idiotic decisions rather than the comprehension of concepts.

This rationality is also sensible—a situation that revises many of our dominant attitudes toward digital and computational mediums as distancing, disembodied, or abstract. Simon's definition implies a rationality guided by the data gathered through embodied sensory perception, not by a reason that is separate from sensation and perception. Von Neumann, Morgenstern, and Simon are widely presumed within the histories of economics and finance to have introduced the mathematics, statistics, logics, and models to begin quantifying economics and producing modern technologies of finance like derivative pricing equations, and other futures technologies. If this is so, then our markets, like our governments, are sentimental, affective, psychotic, and rational. These circuits networking our minds to our governments possess uncanny and wily capacities and qualities.[70]

If it is one of the dominant assumptions in the study of modern history and governance to assume that liberal subjectivity and economic agency are defined as a logic guided by a reason separate from sense, then these discourses mark a clear separation.[71] In making the subject slightly psychotic, incapable of assigning itself set coordinates in space and time, Simon proposed a new notion that would facilitate the modeling of ever larger behaviors, and even organizations, in algorithmic and logical terms.[72] Simon's vision, or perhaps, to use the word of the economic sociologist Donald Mackenzie, "jibe," seemed to find fruition in the later development of option markets and the Black-Scholes equations in the 1960s and 1970s. As Mackenzie puts it, "Herbert Simon . . . would no doubt have felt that option theory in the 1960s was a perfect manifestation of his point [concerning behavior in economics]. The theoretical formulas for the values of options and warrants always involved unobservable parameters, in particular the parameter on whose unobservability Simon had focused, a stock's expected rate of return." This is another way of saying that if complete information and observability are removed, one can still create markets, even if the mathematics itself does not conform to the logical parameters and rigor of a previous era's calculative standards. Even if proof was not definitive in the sense that Turing or other mathematicians sought, or desires and intuitions remained outside of logical calculation, behavioral actions could still be calculated and programmed. More important, the disjunction between the micro and macro levels of behavior, the discrepancy between what individuals do and what networks manifest, was actually an opportunity to import the very equations developed by mathematicians like Norbert Wiener, of Gibbsian physics, into finance.[73] As the Swedish economist Axel Leijonhufvud has put it, "the Economic Man has given way to the Algorithmic Man." Even in economics.[74]

Governance, Calculation, Territory

These theories also entered models of government, often directly by way of cybernetics. However, in the course of these many translations the language of rationality was often reformulated and contested. For political theorists, psychologists, and financial theorists inspired by neural nets, concepts of sovereignty, agency, information, and control often took very different valences than they did for the rational choice decision-makers who were inspired mostly by game theory without major contact with cybernetic ideas of feedback or unconcerned with problems of learning, change, and evolution in systems.

For example, one of the preeminent political scientists of the time, the MIT, Yale, and Harvard professor Karl Deutsch, offers a complex example of this territory where economics and politics became equivalent. While Deutsch had a biography common at that time in elite American universities for an expert in central and eastern European politics, his methods deviated from many of his compatriots. Having participated in anti-Fascist politics, he was forced to flee the former Czechoslovakia in 1939 for the United States. Unlike some of his fellow émigrés, such as Hannah Arendt,[75] Deutsch did not share in total skepticism toward the role of technology in democracy. His investment in quantitative methodology cannot, however, be understood as a dismissal of qualitative interests in human life and society. Rather, he turned to technologies and instruments from mathematics, statistics, and cybernetics after the war as a way to reinvigorate, in his terms to "humanize," politics.

This interest in quantification, communication theory, and probability emerged through his encounter with cybernetics while at MIT during and immediately after the war and his subsequent collaboration with Norbert Wiener on topics ranging from world peace to urban planning (fig. 3.4). As early as 1949 Deutsch was already applying concepts of communication, probability, feedback, and physics so as to ground political science in an empirical methodology. But he was no reductive bureaucrat or functionalist; rather, Deutsch turned to these forms of thinking because they redefined the relationship between empiricism and theory ("quantity" and "quality," in his words) in a manner that was amenable to producing new methods that he believed would make the human and progressive nature of systems emerge out of the recombination of data.[76]

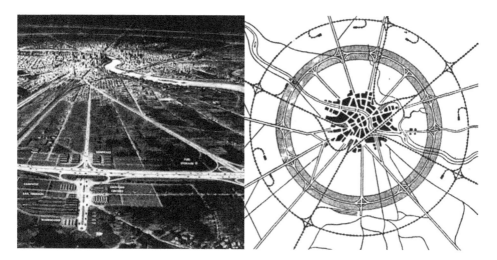

FIG. 3.4—Images of city as cybernetic network. "The city is primarily a *communications center*, serving the same purpose as a *nerve center in the body*. . . . All this applies to the nature of cities in times of peace as much as in times of war . . . the traffic jams . . . the jams on the telephone circuits—jams which fray our nerves and squander our time—are evidence that our cities as they now exist are not nearly in their best possible form." Karl Deutsch, Georgio de Santillana, and Norbert Wiener, "How US Cities can Prepare for Atomic War—Cities vs. A-Bombs," *Life*, December 18, 1950, Norbert Wiener Papers, box 29A, folder 638, MIT Institute Archives and Special Collections. Courtesy of MIT Libraries, Institute Archives and Special Collections, Cambridge, MA.

He elaborated on the theme of communication in a speech on cybernetics, delivered to a conference of prominent social scientists in the early 1950s at Yale, in which he asserted that the study of society would be the study of

information processing systems and their capabilities. We are interested in information channels that are capable of carrying information and in the configuration of these channels, comparable to the telephone wires in a telephone system of the information storage registers in a computer; and *we distinguish between a channel and the capacity of an information channel on the one side and the actual flow of symbols and information through the channel.* In this manner we distinguish between the channel that can carry the information and the information carried in the channel, distinguishing between the twelve inch Vinyl record and the music which it carries and which could be a symphony by Beethoven or a jam session of the Beatles. We have thus a way of attacking what the philosophers have long treated as the body-mind problem. We may think of a mind as the flow of information through these channels and as information stored or recorded in these channels. Human beings from this viewpoint are then indeed what Shakespeare called "the stuff dreams are made on."[77]

Deutsch intimated that any system, whether a market or a political structure (and he often slid between the two), can be thought of as a telephone system, or a computer, and the separation is not between material and abstraction but rather between capacities and actions. In his words, "we distinguish between a channel and the capacity of an information channel on the one side and the actual flow of symbols and information through the channel." Measurement moves to a focus on self-referential assessment of a system's performance in relationship to its expected outcomes.

If political economy was a language and a critical discourse for considering how relations of capital and relations of power were related earlier in the century, this discourse was now to be subsumed by "communication." For a political scientist to abandon the language of power for communication is itself a historical marker of a transformation in method.

These older problems of mind and body, and perhaps also of the unconscious with its desiring drives and the conscious with its repression, were now replaced in social theory by a focus on form and communication. The separations between material labor and production and abstract value and money were seemingly eliminated. Politics and economics appeared to disappear or act interchangeably in terms of their representation, measurement, and modeling. Both political and economic systems were rendered equivalent through the same method.[78] What cybernetics does, Deutsch wrote, is redefine quality as "the matching of two structures. . . . *Quantity* in this view would appear to be really a more complicated notion than quality."[79] Communications theories make interactions, formerly qualitative forms of action, visible and logically representable.

Deutsch was zealous in the pursuit of these patterns. His archive is full of research reports and questionnaires. He had a data-collecting obsession, keeping thousands of reports, running hundreds of studies, seemingly inventorying every last piece of information that crossed his desk. The result of this data obsession, however, was not a declaration about the nature of particular countries, or deep studies about any one territory, ethnicity, or group. Deutsch insisted almost excessively on comparison. Instead of assuming stable and ontological categories, identity could now be modeled and measured as the result of interactions between groups, the teleological output of nondeterministic disturbances, resistances, and limits to the transfer and translation of language, signs, or behaviors.[80] Race, for example, he redefined in a lecture in 1969 following the race riots in American cities not as stable category but as the result of communication differentials. Approaching the problem of ongoing segregation and inequality in the face of federal legislative changes after the civil rights movement, Deutsch turned to a new approach to identity that might both account for history while refusing essentialism.[81]

Deutsch implied that race, ethnicity, and nation were archival artifacts. Identity, he realized, emerged from the assemblage of characteristics that finally produced signals or blocked circulation of messages or interactions. For example, Deutsch describes a fictive situation where there are blue and green skinned individuals equally arrayed in a population and having no other differences. If it happens that more blue-green people are poor, or speak a different language or eat different food, and so forth, these education and cultural characteristics might be attached to a biological feature of color and therefore agglomerated into a "race" that would finally become, in his words, "visual." For Deutsch vision is thus the site of resistance inside the circuit. The visibility of race emerges as a result of a resistance in a channel to smooth information flow that imposes a limit to recognition or translation between groups, producing arbitrary linguistic, visual, or behavioral signs identifiable as "difference." Race, he continued, could thus be disaggregated into proxies like wealth, geographical allocations, linguistic structures, perceptions and attitudes toward similar events, and shifting linguistic patterns. Social scientists studying sociological phenomena like race, class, and gender should therefore not focus on identity but should commence to run an infinity of studies on psychological perception, newspaper coverage, behavioral differences between groups in their responses to similar messages, comparative national assessments of attitudes to media, and so forth. What was important for Deutsch was the application of the same algorithm and method to many situations.[82] Race riots in the United States and postcolonial conflicts could all be compared, and regularly were. The lessons of one revolution or conflict seemed immediatcly transferable to another. Nations for Deutsch were no longer stable geographical territories but interactive spaces, emerging through comparison with other nations and states.[83]

Inside these diagrams (figs. 3.5 and 3.6) older categories of territory, nation, and population were redefined. At once providing new types of relations between subjects, populations, and environments, these wish-images of political

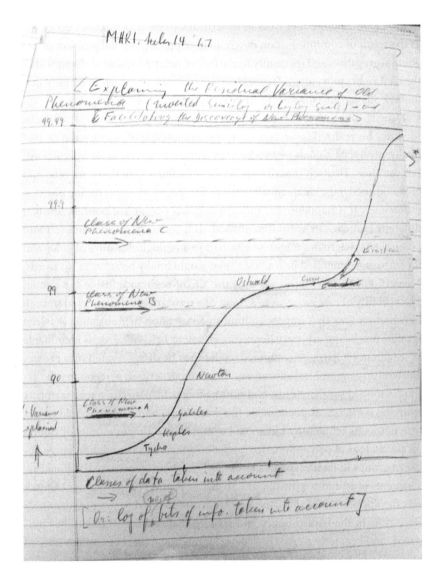

FIG. 3.5__Notes on "Connectivity and Systems Levels," by Karl Deutsch (1963), Papers of Karl Wolfgang Deutsch, HUGFP 141.50, box 1963, Harvard University Archives. Courtesy of Harvard University Archives.

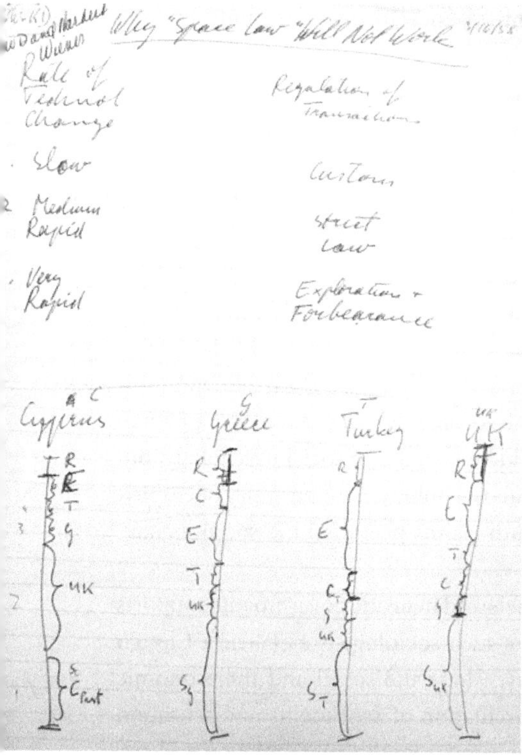

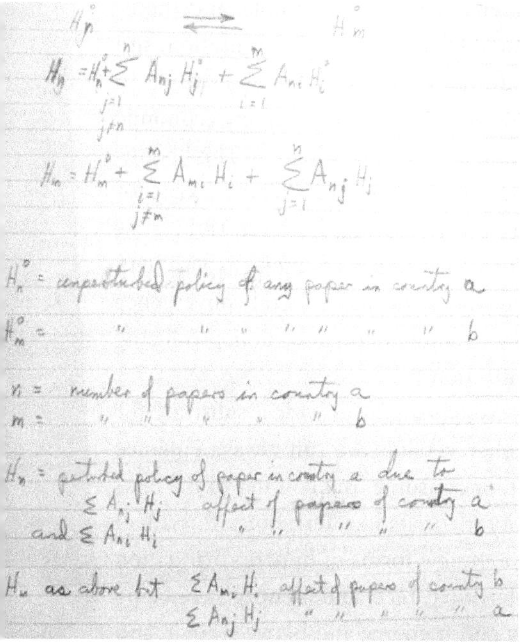

FIG. 3.6__Lecture notes, by Karl Deutsch, from course taught with Norbert Wiener in 1958 at MIT. Deutsch and Wiener present various methods by which to measure communication and rate of change differentials. They argue that preliminary ideas concerning, for example, the administration of outer space will not work to account for the relative and comparative rates of exchange and change among different countries. In the lower sheet they lay out equations for contemplating the rate of openness and communication in different nations' media systems as proxy measures for democratic openness. Papers of Karl Wolfgang Deutsch, HUGFP 141.77, box 1, folder "Periodicity and Interacting Systems" w/ Norbert Wiener, March 16, 1958, Harvard University Archives. Courtesy of Harvard University Archives.

organization speak to both the possibilities of new forms of encounter and the dangers of self-reference.

If other political scientists and theorists of the time, such as Hannah Arendt, whom Deutsch fought with vociferously, lamented the loss of individual sovereignty and freedom to the mass and insisted on the qualitative and impossible-to-represent element of human reason, Deutsch stood behind the virtues of data inundation and diagraming as a route to freedom.[84]

Deutsch in many ways pioneered methods that were spreading throughout the social sciences of the time. There are many other examples. In sociology, similar phenomena, equally influenced by new notions of knowledge, rationality, neural nets, and action, proliferated, making the study of communication central to the theorization of the social.[85] The social took on a mental aspect, and individual psychology could be scaled into social psychology, but in a break from earlier theories of social psychology, a new armory of empirical methods and different types of truth claims pervaded the social, behavioral, and human sciences.[86]

This networking between nervous nets and neurotic governments supports the contemporary arguments made by such sociologists as Patricia Clough that we live in an "affective" economy, where the social and the economic emerge from the manipulation and circulation of sentiments and sensation before consciousness, and without representation; this manipulation and circulation are based in forms of measurement that make visible, and therefore manipulatable, the interactive circuit between perception, sensation, and cognition. The media theorist Tiziana Terranova extends this observation to argue that today we live in "a whole new political rationality where economic and vital processes are from the beginning deeply intertwined."[87] This "political rationality" she labels, following Foucault, "biopolitics." What makes economic and vital processes equivalent in this new sense, however, is this new definition of "rationality" that created practices of measurement, experiment, and visualization. These speculative networks allowed a literal convergence between the interior of the mind and the structure of organization, producing new spaces for measurement and visualization.[88]

This unbinding of desire from objects appears to provide a substrate for constant growth. There is always another version; we can always enhance our capacity to be productive, efficient, or consume. But counter to the relatively disciplinary concerns put forth by theorists like Terranova, this affective economy is also an emergent space. If new forms of hyperindividuation were emerging, older formulations of territory, race, and identity were being remade. In this space between history and memory and between logic and reason arguably lies an architecture of our contemporary forms of governance, and perhaps, economy.[89]

Memory as a Cyclical Machine

Having, however, displaced older terms of consciousness, reason, and desire from the algorithmic rationality of the network, these terms would return in cybernetics in the guise of visualization, time, and memory. Behavioral and social scientists arguably found a new basis on which to reintroduce consciousness into the machine. Counter to expectation, the new dream of consciousness emerged through a correlation between time, memory, and vision. Rather than dispense with the tools of computation and communication, social scientists sought to intensify their effects through recourse to a valorization and aestheticization of data; feedback appeared as a route to reintroduce reflexivity and perhaps self-awareness into systems. Cyberneticians and their affiliated disciplines became obsessed with data visualization as *the* mechanism for consciousness.

At the sixth Macy Conference in 1949, memory was increasingly problematized between its dynamic and stable elements and problems in storage. McCulloch opened the conference with a warning from John von Neumann. The warning was that even the entire number of neurons in the brain, according to von Neumann, could not account for the complexity of human behavior and ability. McCulloch went so far as to discuss "lower forms . . . such as the army ant where you have some 300 neurons that are not strictly speaking sensory or strictly speaking motor items, and that the performance of the army ant . . . is far more complicated than can be computed by 300 yes or no devices."[90] McCulloch, however, went on to say that there was no way that these capacities can be understood as illogical or analog. Rather, he turned to another model that might retain the logical nature of the neurons, but still account for the capacity to learn, and behave at scales beyond the comprehension of computation.

The answer, coming through a range of discussions about protein structure and memory within cells, emerged as one of refreshing in time. Wiener argued: "this variability in time here postulated will do in fact the sort of thing that von Neumann wants, that is, the variability need not be fixed variability in space, but may actually be a variability in time." The psychologist John Stroud offered the example of a "very large macro-organism called a destroyer," a "ship" that had endless "metabolic" changes of small chores throughout the day but still retained the function of a destroyer. This systemic stability, but with internal differentiation and cycling, became the ideal of agency and action in memory.[91]

Memory and mind came to be seen as mirroring split systems operating between the real-time present of reception and circulating data from within and outside the organization, and memory in time, a cyclical "refreshing," as in a television screen system, where change, and differentiation—between the organism and the environment, between networks—became possible through the delay and reorganization of circuits from "within" an organism, now defined through temporal delays.

McCulloch and Stroud presented models of memory as bifurcated between perfect retention of all information with retroactive selection or memory as a constant active site of processing of information for further action, based on internal "reflectors," or "internal eyes." Stroud argued: "we may need only very tiny little reflectors which somehow or other can become a stimulus pattern which is available for this particular mode of operation of our very ordinary thinking, seeing, and hearing machinery. This particular pattern of reflectors is what I see as it were with my internal eyes just as what I see when I look at a store window, is a pattern on the retinal mosaic." In other words, all mental functions could be seen as a matter of processing information into particular self-referential "patterns"; functional patterns that might be learned . . . or programmed.[92] These internal "eyes" from within the psychic apparatus allowed a self-reflexive mechanism for processing and deferring decisions to become the site of agency.

Here fissures appeared between different sciences of communication, as feedback gained value as a moral buffer against repetition. "[Game theory] answered a psychological need," Karl Deutsch wrote, "since it embraced a relatively comforting picture of a static world, and of largely unchanging political habits, with a tacit reassurance that superior stamina and courage of one's own government, ethnic group and national culture, in comparison to those of the governments and peoples of the west's adversaries." Cybernetics, he continued, on the other hand introduced elements of probability and, most important, a "learning net" that could be "studied step by step, and recombined into more efficient patterns."[93]

In this formulation, game theory took the psychological position, and in this state it was obsolete and even damaging. Counter to this psychological mode, where there was only a single direction of stimulus and response, cybernetics in this account produced the possibility for self-reflexivity, facilitating "autonomy," "self-awareness," and "freedom." Psychology here was seemingly opposed to consciousness. Deutsch made a subtle distinction here between systems that were self-referential, composed of their own rules, and systems that were self-reflexive, capable of learning from their own behavior.

Deutsch appeared to imply that it was cybernetics that might return, or perhaps for the first time realize, the humanity of the system by way of reflexivity, now labeled feedback. "Freedom," Deutsch quotes Rousseau as saying, "is obedience to the law which we prescribe to ourselves." This freedom, never realized in the Enlightenment, Deutsch argued, might now be returned to us through our communication channels. "Implicit in Rousseau's words is memory," he wrote, "for without it law cannot be formulated, but without openness to new information from the environment, self-steering organizations are apt to act like projectiles [missiles] entirely ruled and driven by their pasts." The ability to channel the past without linearity or determinism, and the ability to be open to the environment, could become a rare site of opportunity to achieve what had never been realized—freedom.[94]

As if to doubly assert this possibility of finally realizing our humanity by way of the communication channel, it was vision, above all, that dominated Deutsch's account. Game theory, he argued is "hard to imagine visually or as a flow of processes." This impossibility or resistance to imagination and to flow made games, in his estimation, always insufficient and reductive as models. What cybernetic feedback offered was a tool to visualize process. Deutsch's words belie a residual concern with representation and consciousness, now always understood in terms of lateness, or deferral. If systems could see themselves in time, they could adjust their own operations and be both open to their present and nonreactive due to the positive repressive force of the past. Vision equates with freedom but is no longer a vision of omniscience; it is now a vision of self-reflexivity.[95]

Deutsch returned to Kant and Rousseau, those very thinkers expunged by rational choice decision theory, to lay claim to a potential for reason. But this reason only emerged through self-reflexive performances of data, through a valorization of memory, diagrams, and performance, and a turn toward self-reflexivity as an ethic. Perhaps it can be understood as a folding of Kant's subjective crisis into the logic of communication, but there is no skepticism here of the sensory apparatus, perhaps because it is subjectivity that has become the technology.[96]

The prominent textbook in political science in the early 1960s *Nerves of Government*, makes this clear. The diagrams (fig. 3.7), informed by computer programming, cybernetics, and Deutsch's engagement with the circuits of neural nets, define the behavior of organizations. Deutsch's methodological imperative came through a new form of diagram that, like those of McCulloch

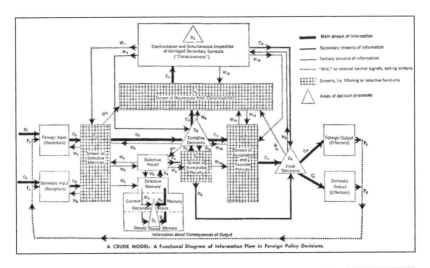

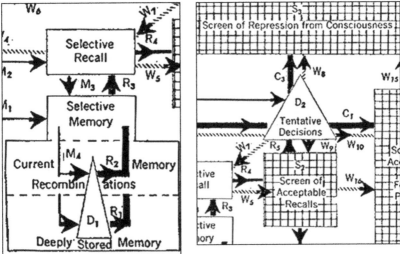

FIG. 3.7__**Nervous System**: **Memory**: Deep memory and selective memory in the nerves of government. **Neurosis**: Repression and deferral in memory. All images from diagram of foreign policy decision-making process, by Karl Deutsch, from Deutsch, *Nerves of Government*, appendix.

and Pitts, did not represent any one system or particular institution but served as a conceptual experiment and tool in visualizing process. These nonmimetic diagrams, producing their own conception of political organization, began appearing, containing circuits inspired by cybernetic ideas of neural nets and feedback loops, along with massive displays of quantitative data.

Organizing the text was a diagram, first displayed in the introduction, designed as a flow chart influenced by computer programming and neural nets (fig. 3.7). These diagrams were the images of generic processes, of process as an object and material in and of itself. They were not representations of a particular institution or organization. In these diagrams there are no consolidated entities, only inputs, outputs, and "screens" that act to obstruct, or divert, incoming data.

Political will for Deutsch was the deployment of an action based on past data that can temporarily ignore incoming environmental data. "Will" collapsed with the concept of machine control. As we may note, the little arrows of will are "internal control signals"; they set screens. This is a machine that demands a little repression to work; delaying the arrival of stimuli can be a good thing. The diagrams depict a series of screens that operate like RAM in a computer getting data (screened) from areas of deeper memory.

Throughout Deutsch's models there were screens that acted to repress incoming data, to separate signal from noise, but also, and perhaps most important, to allow the system to delay and defer immediate reaction, and to facilitate the recombination of memory, as with the memory circuits of the neural net, in order to generate multiple potentialities for action. A sound government, he wrote, was one that makes decisions without immediately responding to outside stimuli while simultaneously still recording and being able to respond to changing environmental considerations. By implication, the soundest government was neurotic—capable of deferring impulses. A little repression might be a good thing.

Deferral equates with logical governance—a defense, he argued, against fascist and totalitarian tendencies. The government that responds too quickly is amenable to the same dangers afflicting cybernetic mechanisms—overoscillation, leading to increasing, not reducing, the error between the desired outcome and the action.[97] Memory in the cybernetic formulation, therefore, takes on a messianic function, capable of rescuing systems from their own self-generated imperatives for speed and allowing for sound predictions.

Moving beyond a dialectic opposing close observation to theoretical abstractions Deutsch's image world was simultaneously empirical and abstract. For Deutsch the visual tactics of "scanning" and pattern seeking might create a bridge between the national and international. His diagrams producing a new "scale" of observation, turning discrete and nonsensical data points into coherent, and meaningful, patterned flows. In another text he murmured in awe at this world of informational abundance. Writing a review of Gyorgy Kepes's book *The New Landscape of Art and Science*, Deutsch marvels:

> these are pictures that could not be seen by human eyes until quite recently, pictures that communicate both a sense of awe and beauty to all of us, as well as conveying a sense of aesthetic and intellectual excitement to scientists. . . . [These images] offer an unusual and important statement of the relations between these two realms of human experience, the experience of science and the experience of art.[98]

Assuming an always recorded world, Deutsch found the new realm of possibility to lie in this aesthetic terrain, where human experience, perhaps imaginary, and "awe and beauty" might be returned through the virtues of data presenting itself. The new challenge is to find ways to show that world to itself, in its shininess . . . and beauty.

Zealously collecting data, producing endless charts and visualizations, these diagrams enacted an imaginary of social science as a self-reflexive interface, a fantasy that perhaps continues to speak through our discourses, ranging from open source activism to movements for open knowledge and knowledge access, to an obsession with data visualization.[99] Feedback would disrupt programmatic behavior, as in the memory of a computer, and would provide systems with emergent capacities. Self-reflexivity through data analysis and visualization became a democratic virtue and obligation. This virtue and obligation found itself being demonstrated in the sociological turn to structured interviews, the compulsive gathering and keeping of notes by Norbert Wiener, the courses dedicated to pure data analytics provided by Gyorgy Kepes, George Nelson, and Charles Eames, and the list goes on. We might think back to Bush's fantasy of an always recorded world in the Memex or Lynch's fabulous "images" of the city, or we might contemplate other machines, such as the proposal by the Columbia sociologist Robert Merton of an "inspectrometer," a magical machine for sociologists, never built, that would allow the full display and constant replay of interview data from focus groups, or the constant replay of film and video in family therapy. These magical, and sometimes built, machines fueled a growing concern for replaying and reviewing data analysis as an ethical and truth-producing practice.[100]

The media theorist Wendy Hui Kyong Chun has posited the paranoia of control and the fantasy of total sovereignty or freedom as a fundamental feature of contemporary digital media. Between fear of surveillance and the dream of information making us free, the drive to increase the use of the medium of the Internet is enhanced. To her account I have added here a historical grounding that demonstrates how paranoia and psychosis are not part of a dialectic but are the constitutive discourse by which ideas of sovereignty and governance support networks and drive a frenzy of visualization.[101] What needs to be recognized, however, is that these pathologies were not diagnostic categories for cyberneticians but were part of a discursive shift in truth claims that described ideas of thought as algorithmically patterned. However, Chun's description of a social condition organized around two poles: control or freedom leaves untouched the liberal agent as an ideological cultural norm whose definition is stable over time. Today's attitudes to networks put Chun's analysis in question.

Deutsch's models do not support an omniscient Enlightenment or liberal vision, but a partial perspective whose only truths are a residual consciousness of a network whose entirety cannot be comprehended, whose ontology is unavailable. Deutsch's models whisper of a faint fantasy of a messianic capacity embedded in these models; bereft of stable ontologies, identity, territory, race, ethnicity become malleable—for good and for bad—and different forms of action are available through a reimagining of the data patterns.[102] It appears that many cyberneticians and social scientists embraced proximities and bodily disintegrations while simultaneously hoping to achieve consciousness and truth through temporal displacement and feedback. This frenzy of visualization in the social sciences and design (chapter 2) was coming to be embraced as the very key to democracy and "freedom." This contradiction between the desire to produce knowledge and the demand to circulate information, otherwise understood as a space between older histories of knowledge and the emerging paradigm of bounded rationality, was driving a constant demand for calculation and visualization—a logic that perhaps continues to underpin our contemporary cultural attitudes to the screen, the image, and data.

Fissures in the Machine

In a 1969 conference at the National Institute for Mental Health on the "double bind," Gregory Bateson returned to his work in the 1950s to explore the theme of a dynamic and internally differentiating memory. Standing before an audience of prominent psychiatrists and psychologist, he proceeded to offer the example of a porpoise.

This porpoise had been trained at a navy research facility to perform tricks and other acts in return for fish. One day, he recounted, her trainers started a new regimen. They deprived her of food unless she produced a new trick. Starved if she repeated the same act, but also if she did not perform, the porpoise was trapped. This experiment was repeated with numerous porpoises, usually culminating in extreme aggression and a descent into what from an anthropomorphic perspective might be labeled disaffection, confusion, and antisocial and violent behavior. Bateson, with his usual lack of reservation, was ready to label these dolphins as suffering the paranoid form of schizophrenia. The anthropologist was at pains to remind his audience, however, that before rushing to conclusions about genetic predeterminacy, or innate typologies, the good doctors should recall that these psychotic porpoises were acting very reasonably and rationally. In fact, they were but doing exactly what their training as animals in a navy laboratory would lead them to do. Their problem was that they had two conflicting signals. They had been taught to obey and be rewarded. But now obedience bought punishment and so did disobedience. The poor animals, having no perspective on their situation as laboratory experiments, were naturally breaking apart, fissuring their personalities (and Bateson thought they had them) in efforts to be both rebellious and compliant, but above all to act as they had been taught.

One porpoise, however, appeared to possess a good memory. She was capable of other things. Bateson related how, between the fourteenth and fifteenth time of demonstration, the porpoise "appeared much excited," and for her final performance she gave an "elaborate" display, including multiple pieces of behavior of which four where "entirely new—never before observed in this species of animal."[103] These were not solely genetically endowed abilities; they were learned, the result of an experiment in time. This process in which the subject, whether a patient or a dolphin, uses the memories of other interactions and other situations to transform his or her actions within the immediate scenario, can become the very seat of innovation. The dolphin's ego (insofar as we decide she had one) weakened sufficiently to be reformed, developing new attachments to objects in its environment and to memories in its past. This rewired network of relations could lead to emergence through the recontextualization of the situation in which the confused and conflicted animal finds itself. Schizophrenia, therefore, could be the very seat of creativity.[104]

Bateson ended in triumph, having now successfully both made the psyche intersubjective and simultaneously amenable to technical appropriation in family therapy:

This story [of the porpoise and its trainer] illustrates, I believe, two aspects of the genesis of a transcontextual syndrome [double-bind schizophrenia]:

First, that *severe pain and maladjustment* can be induced by putting a mammal in the wrong regarding its rules of making sense of an important *relationship* with another mammal.

And second, that if this pathology can be warded off or resisted, the total experience may promote *creativity*.[105]

For social scientists, the capacity to perform these actions, to simulate and see the disasters and potentials of systems in self-reflexive feedback loops, was the key to futurity. The poor tortured porpoise was a visible exemplar of the dangers and creative possibilities of control in a networked system. Instead of analyzing data, Bateson performed it, made the situation visible, as spectacle, not as conclusion. The porpoise's pain was a testimony to the virtues of visualizing scenarios in the interest of speculating on different outcomes; taken to an extreme, the porpoise serves as a vital exemplar of a new aesthetics of visualization. It is terrible to cause such suffering, but it is virtuous to learn from such self-inflicted pain. Paraphrasing Bateson, torture and creativity appear related; a little pain might be a good thing. Bateson ended in triumph, having now successfully made the psyche inter-subjective and simultaneously amenable to technical appropriation via family therapy.[106]

There is, however, a third term, "creativity," in the matrix. Bateson's performance redefined object relations in terms of probabilities and communication. The productivity of a schizoid situation rested for Bateson on the discovery made by both communication theory and physics that different times could not communicate directly to one another. Only temporal differences resist circulation from within the definition of communication that was being put forward here. Bateson applied this understanding liberally to animals. In cybernetic models the ability of a subject to differentiate itself from its environment and make autonomous choices is contingent on its ability to simultaneously engage in dangerous proximities spatially with other objects and its ability to achieve distance through memory or time. For individuals like Bateson, memory ceased to be a stable place and instead came to understood as a communication structure, distantiated in time from behavior in the present.

At stake in the negotiation over the nature of networks and the time-scale of analysis was nothing less than how to encounter difference—whether between individuals, value in markets, or between vast states during the Cold War. A question that perhaps started in psychoanalytic concerns over psychosis, found technical realization in cybernetics. For cyberneticians the problem of analogue or digital, otherwise understood as the limits between discrete logic and infinity, the separation between the calculable and the incalculable, the representable and the non-representable, and the differences between subjects and objects, was transformed into a reconfiguration of memory and storage; a transformation that continues to inform our multiplying fantasies of real-time analytics while massive data storage infrastructures are erected to insure the permanence, and recyclability, of data.

While the time of neural nets and communication theories is always preemptive, the shadow archive haunting the speculative network is one of an endless data repository whose arrangement and visualization might return imagination and agency to subjects. The relationship between rationality and control drives the ongoing penetration and application of media technologies as the result of an imperative to seek consciousness through better visualization and collective intelligence through the collaboration of many logical, but hardly reasonable, agents. Architecturally these dual desires incarnate themselves in a proliferation of interfaces and a fetish for visualization and interactivity, merged with an obsession to amass and store data in huge systems of data centers and server farms. What had first been articulated as a problem of memory and time has now become a compulsion for analytics.

Bateson also performed another act, however, one that resonates twenty years later in critical thought. In his story of the porpoise and theory of the "double bind," he splits paranoid forms of psychosis from schizophrenia. In their first text, *Anti-Oedipus*, Deleuze and Guattari set out to "analyze" capital and schizophrenia from the standpoint of Marx and Freud. *Anti-Oedipus* arrived in 1972, perhaps not coincidentally at the moment of the dissolution of the Bretton Woods convention and the transformation in finance that saw many of the aforementioned instruments finally massively formalized as commodities and technologies. This text works somewhat dialectically, affiliating paranoid "tendencies" with an overcoding of power that relates to structural social sciences (Marx, Freud) and certain forms of philosophy that territorialize, or impose constraints on, subjects and societies. Opposing this paranoid style is a schizophrenic style that releases desire and the social modes of production (and reproduction) from stable relations and into a state of "permanent revolution." The paranoid style seeks to create and stabilize meaning through a violence of self-reference, not unlike McCulloch's paranoics in tales of wounded soldiers and circling nets, or the stuck porpoise of Bateson's example. Schizophrenia, however, designates surrealism and word play from which conclusions are impossible to draw, perhaps the more unclear world of affiliations and connections that is made possible through the flexible bodily and mental contours of the networked and embodied mind of cybernetics, or the possible creativity of the porpoise.[107]

Eight years later, in 1980, *A Thousand Plateaus* appeared. Despite the English title still containing the subtitle "Capitalism and Schizophrenia," for all intents and purposes the language of psychosis and schizophrenia has disappeared, as has the dialectical deconstructive structure of the initial text. Instead, Deleuze and Guattari take it upon themselves to call on Gregory Bateson's notion of a plateau: "a plateau is always in the middle, not at the beginning or the end. A rhizome is made of plateaus. Gregory Bateson uses the word 'plateau' to designate something very special: a continuous, self-vibrating region of intensities whose development avoids any orientation towards a culmination point or external end."[108] Whereas *Anti-Oedipus* focused clearly on a move between the paranoic and schizophrenic, *A Thousand Plateaus* multiplies the findings of *Anti-Oedipus* in an arboreal structure that resists any conventional teleology. Vastly increasing the points of reference and the movements of concepts, *A Thousand Plateaus* is an ecology of rapidly multiplying terms, such as bodies without organs, that then become multiple other kinds. The clear-cut work of de- and reterritorialization, or signification and designification, that accompanies the paranoid and schizophrenic in the first text become a much more convoluted set of movements. The literary critic Eugene Holland, to whom this reading is indebted, notes that the second text in fact recognizes a new deterritorializing force to capital, arguing almost conservatively that "small supplies of signification" might even be a recommended antidote to the smooth nervous networks of capitalism.[109]

Another way to read this transformation is historical. Deleuze and Guattari appear to recognize that paranoia and schizophrenia can no longer serve as productive machines for thought, or as critical intervention into social forms. Perhaps that is the success and failure of these histories: the automation of schizophrenia as a perceptual condition to the point of making it only a technology, whether in medicine or computing. It is perhaps not a side note that the next *Diagnostic and Statistical Manual* for psychiatry, the fifth edition (DSM-V), no longer has a classification for the paranoid forms of schizophrenia at all (all subtypes have been eliminated). Analysis of paranoia, as a named and separate category, may no longer serve a political, ethical, or even medical diagnostic function.[110]

History for the Future

I opened this chapter arguing that cybernetics and its affiliated communication and human sciences aspired to the elimination of difference in the name of rationality, a dream of self-organizing systems and autopoietic intelligences produced from the minute actions of small, stupid, logic gates, a dream of a world of networks without limit, focused eternally on an indefinite, and extendable, future state.

What Bateson and Deutsch articulated in their critique of game theory was the worry that in the real-time obsession to entangle life with calculative logics, learning, and by extension thought would itself be automated in such a way as lead to violent harm, and perhaps, the destruction of the world. This condition only becomes inevitable, however, if we ourselves descend into the logic of immediate and real-time analytics. We must avoid this conclusion, and this condition. Like Bateson's porpoise, torn between reactionary return, and self-referentiality, we are forced to ask about the other possibilities that still lie inside our machines and our histories. The cycles of the porpoise reenact the telling of cybernetic history where ideas of control and rationality are often over-determined in their negative valence, and the inevitability of the past to determine the future is regularly assumed.

But here I would also like to note that we might take heart and think historiographically about Deleuze and Guattari's acts of multiplication. Bateson, Deleuze, and Guattari all participate in an act of redefining the site of language. In their scenarios, particularly for Bateson, language, if understood as the act of translation (as first articulated in chapter 1), takes place between different times and communication structures. These translations are never complete and, rather, make evident the impossibility of directly converting or communicating between one temporal state and another; as a result the possibilities for action multiply, rather than converge. Hence, despite Deleuze and Guattari's explicit antagonism to any discourse of representation, we can still witness the multiplications in *A Thousand Plateaus*.

In response to this vision of an inevitable future, rationality must be understood as a contested interface. The discourses that made mental and organizational processes both methodologically commensurate produced new concepts of subjectivity, truth, and cognition. While the time of neural nets, Markov chains, and communication theories is always predictive and speculative, the shadow archive haunting the speculative network is one of an endless data repository whose arrangement and visualization might return imagination and agency to subjects. These wavering interactions—between the networked individual and the fetish of data—preoccupies us in the present, speaking through our contemporary concerns with data mining, search engines, and connectivity.

It may be good to remind ourselves that the minds and machines envisioned in these sciences were unruly, emergent, and never completely known, networks that encompassed our selves and renegotiated our relationship to the world. Policy-makers and social scientists struggled over time-scale and the unit of analysis. What should be regulated? Where should interventions be introduced, whether in economy or international relations? At the level of the individual or of the system? What was a system? How should history be used, and what institutional structures might encourage emergent rather than reactive behavior?

For the twenty years after World War II, we see multiple contesting understandings of what rationality might mean and how memory might be organized. These imagined networks and machines were affective and material, multivalent and decentered. Terms like "control" were unclear, were subjective, and were sites of emergence. Cybernetics did not dispense with subjects or consciousness but aspired to achieve these forms through a self-reflexive capacity to reenact and make visible failures and processes.

If this cybernetic turn to reframing the world as a computational medium is often considered reductive, abstract, disembodying, McCulloch's experimental epistemology demonstrates otherwise. Rather than assume the opposition between a modern depth psychology and conception of time and storage and a more contemporary computational model of prediction, speculation, and speed, it might be worth asking how the feedback between these two epistemologies produced so many of the instruments that measure and thereby help govern our contemporary world.

For psychoanalysis, psychosis, and schizophrenia most of all, was a question of encounter, love, and difference. For cybernetics these problems were never erased. The question for the present is whether our current discourses of paranoia, security, and risk serve an amnesic or reactionary function in relocating our desires elsewhere. We might, then, contemplate the divide between self-reference and self-reflexivity so well articulated by Karl Deutsch. In the 1950s and early 1960s the language of psychosis still had a role in reflexively revealing the dangers and possibilities of control and communication, whether in the example of the porpoise, the neurotic government, or the reverberating neurons. But today, while there is no lack of discourses concerning psychosis and paranoia, we no longer appear to take pleasure, or suffer, at witnessing the dangerous cycles and marvelous interactions made possible by our nervous networks.

4_GOVERNING

Designing Information and Reconfiguring Population circa 1959

The eye speaks to the brain in a language already highly organized and interpreted.
—**J. Y. Lettvin, H. R. Maturana**, and **W. S. McCulloch**, "What the Frog's Eye Tells the Frog's Brain" (1959) reprinted in *Embodiments of the Mind*, 231

My problem is that I have been persecuted by an integer. . . . The persistence with which this number plagues me is far more than a random accident. There is, to quote a famous senator, a design behind it, some pattern governing its appearances. Either there really is something unusual about the number or else I am suffering from delusions of persecution.—**George Miller**, "The Magical Number Seven" (1956), 81

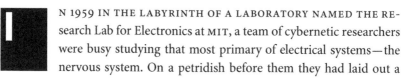 N 1959 IN THE LABYRINTH OF A LABORATORY NAMED THE RE-search Lab for Electronics at MIT, a team of cybernetic researchers were busy studying that most primary of electrical systems—the nervous system. On a petridish before them they had laid out a poor *Rana pipiens* frog. The flap of skin behind the frog's eye had been cut open, and the frog lay pinned, unable to move for pain, on the dish. A number of phototubes and cameras finely tuned to detect and image the electrical movements of the optic nerve were arrayed around the poor animal. The endeavor at hand was to isolate and analyze the actions of the nerve separate from the brain. The researchers sought to extract the process of vision itself. Beautiful (to the researchers) images of "frog-like" environments were displayed to the frog's eye, with no reaction. But when small rotating disks and small black objects were twitched, a volley of electrical impulses was released and transmitted through the nerve fibers. Having recorded the actions of the

FIG. 4.1__*Glimpses of the USA*, installation by Charles and Ray Eames, Moscow (1959). The Works of Charles and Ray Eames, lot 13393, no. 14, Photography and Print Division, Library of Congress. © 2013 Eames Office, LLC.

nerve isolated from the brain, the researchers came to the conclusion that "a frog hunts on land by vision. He escapes enemies mainly by seeing them. . . . The frog does not seem to see or, at any rate, is not concerned with the detail of stationary parts of the world around him. He will starve to death surrounded by food if it is not moving." The eye, without the brain, could recognize moving targets.

These words should immediately call our attention to two features: the mobility of vision and the capacity of vision to act (or hunt). The very title of the piece, "What the Frog's Eye Tells the Frog's Brain," suggests an eye autonomously speaking to the brain; an eye capable of cognition. Vision, the research group argued, can act—it operates on algorithmic principles, making decisions such as identifying "prey" or an "enemy."

Since when, we might ask, could eyes talk and think? I have opened this chapter with the discovery of the minute isolated optic nerve because the implications of this revision of perception did not end at the boundaries of the Research Laboratory for Electronics at MIT. This emergent discourse of vision as a channel endowed with capacities to act linked the nascent neurosciences of the period to broader changes in governmentality relating to how perception, cognition, and power were organized. In the last two chapters, I began to intimate a relationship between changing forms and practices of planning and design, an algorithmic optic, and the emergence of rationality and data visualization as democratic virtues and economic values. In this chapter I want to return to these themes by continuing to interrogate the historical and conceptual relationship between cybernetics, design, vision, and cognition. How do circuits link from within the eye to the structure of cognition and to the governance of attention? I want to run a cumulative experiment surveying the practices and objects introduced throughout this book—linking the transformations in design to the cybernetic reformulation of memory and cognition to demonstrate the affective, and aesthetic, infrastructure of Cold War politics, and perhaps to begin asking about the forms of power and media that operate in our present.

To do so, let me offer two further examples, culled from the archives of the Cold War that reflect this transformation of our ideas about how we think and how we see. If the mind has long been considered the site of the *cogito* and human autonomy, and if optic nerves were capable of cognition—then what might a cybernetic account of the mind look like when applied to seeing and remembering information? A few years earlier in 1956, one of the most famous articles in psychology (to this day, if assessed by number of citations in professional literature and downloads) was published—"The Magical Number Seven, Plus or Minus Two: Some Limits on Our Capacity for Processing Information." Reframing memory and decision-making in terms of information, recoding, and data compression, the psychologist George Miller, in keeping with the account of "rationality" discussed in chapter 3, created a new account of psychology, arguing that cognition was an algorithmic process that could be manipulated. In cognitive science, as in neuroscience, cognition and perception were rendered equivalent—both treatable as communication channels, and subsequently both subject to new forms of intervention. Just as communication channels in telephones and computer systems can be manipulated and constructed in different ways to increase capacities, thresholds, and action potentials, the same, Miller implicitly argued, could be done with the mind. This historical change in the accounts of the physiology of perception and psychology laid the groundwork for a novel science of the mind, and arguably, brain.[1]

In the same year that the frog's eye learned to speak, 1959, a new political structure of spectacle also emerged. In Sokolniki Park in Moscow, the United

States and the Soviet Union embarked on the first of a series of programs in "cultural exchange." Here, at this site, many new forms of presentation were paraded. As audiences were shown multi-screen-channel exhibits (fig. 4.1) based on psychological theories of feedback and communication theories of information, Khrushchev and Nixon debated the variable merits of American and Soviet kitchen design and technology in front of the new television cameras. Five hundred American corporations displayed their wares, along with models of the "splitnik," the Levitt-style suburban tract homes into which these commodities were to be placed. Fashion shows paraded the ideal nuclear American family, selected to appear normal, without distinguishing features or abilities. Beneath the images of commodities and gigantic screens of American landscapes was exhibited the pioneering photo-essay *The Family of Man*, which documented a biologically diverse but still singular and universal human species.[2] The pavilion was an architecture of perception and data that revealed both new forms of spectacle . . . and politics. Or perhaps politics as spectacle.[3]

The chief administrator of this design spectacle, George Nelson, spoke of this exhibition as an "enlargement of vision." Whereas his predecessor designer and influence on display structure, the Bauhaus member Herbert Bayer, spoke of stretching vision along horizontal and vertical axes, the postwar American designers discussed their work not as prosthesis but as an autonomous visual terrain capable of expanding infinitely and moving through many territories. Nelson critically argued that people view the world "atomistically"—"everything is seen as a separate, static object or idea."[4] For Nelson such a form of vision could not serve the newly integrated and sociotechnically dynamic postwar world.

His attitude to building the U.S. pavilions directly mirrored, in his terms, the disappearance of the author in modern art. He wrote: "it is of no coincidence [of late] that . . . the individual dissolved into an almost incomprehensibly abstract network of relationship and that the same thing happened to his concept of inanimate matter. Both developments, you will note, tended to substitute transparency, in a sense, for solidity, relationships for dissociated entities, and tension or energy for mass."[5] It is possible to deduce from this statement that Nelson was arguing that subjectivity in its stable and egocentric form could no longer be the basic substrate for either design or vision. We might also take a moment to notice that the dissolution of the individual artist was related to a transformation from matter and representation to "tension," "abstract network," and "transparency." Nelson forcefully insisted on substituting "relationships" for "entities." In design, as in the anatomies of neuroscience, visuality gained a new logic to be encoded into the architecture of USIA Cold War propaganda and the nervous system of the spectator.

If, as Jacques Rancière argues, politics is the organization of the sensible, then I can only presume that this moment in the 1950s marks a profound reorganization of affect, with logistical implications for governance, autonomy, and subjectivity.[6] Furthermore, in that vision and power have long been linked in Western philosophy and scholarship, the mention of autonomy, survival, capacities, and channels as related to sight might also inspire us to reconsider our contemporary concerns with security, information, and biopolitics.[7] In the course of this chapter I will draw a map linking the three aforementioned examples to show how circuits and networks travel between nervous networks and eyes to overstimulated spectators and to the logic of government. I want to ask how the act of surviving through the identification of the prey and enemy becomes an autonomous and self-referential technology embedded in our machines, media environments, and medicine. In tracing this topography, I argue that we can begin a historical excavation into how vision and power were reconfigured through discourses of control and communication in the immediate postwar period and pose some preliminary questions about the implications of this condition for our present. Most critically, these practices demonstrated internal foldings between older concepts of territory, population, and subjectivity and emerging computational models of perception and cognition. These internal multiplications are the infrastructures to our present, and they complicate any simple understanding of what it means, in the words of the art historian Beatriz Colomina, to be "enclosed by images."[8] While so much has been written about the place of aesthetics in supporting politics, and the place of the Eameses, and design more broadly, and the military in establishing our present conditions of attention,[9] much less attention has been paid to defining what "vision," "politics," "attention," or even an "image" constitute at this historical threshold. More centrally, my intention is to extend the important work already done on Cold War propaganda, and the work done by this exhibition in particular, to complicate our understanding of today's biopolitics. To do so, it is not sufficient only to examine the design practices. I argue we must also link them back to the cognitive and emerging neuroscience of the time, so as to examine more completely how vision was being reformulated and scaled from within the optic nerve to the massive global demonstrations of superpower ambition.

In this final moment, I want to accumulate artifacts from the book to make visible these biopolitical processes, by which life itself is now governed; and to ask about how historical concepts of identity and space relate to emerging forms of sensorial territory and globalization. I seek to ask the one question that architectural and art historians have not asked: how was population transformed through the integration of emerging cybernetic models of perception and cognition into design?

Cybernetic Vision

While today it may appear self-evident that vision is a material process that can be performed by machines, and rebuilt as medical prosthesis, it was not always so. The nascent neurosciences and cognitive sciences took the concepts of perception as an autonomous process as outlined in chapter 2 quite literally.

That vision and the identification of "prey" and enemy (even in frogs) would be of prime interest to cybernetically informed researchers should come of little surprise. Cybernetics, as I mentioned earlier and is by now well documented, emerged, after all, within the context of antiaircraft defense and radar research in World War II. Cyberneticians who were focused on shooting down planes came to treat the relationship between the gun and the plane behaviorally and statistically; their analysis thus shifted from documenting the present to predicting the future on the basis of the extraction of patterns from past data on system action.[10]

Vision became an algorithmic process for pattern extraction. Vision, as already demonstrated at the time and in this book by the work of Béla Julesz, Gyorgy Kepes, and Charles Eames, came to be compressed with cognition as a channel capable of autonomously analyzing data and patterns out of information-rich fields. Warren McCulloch—neural net pioneer, cofounder of cybernetics, and psychiatrist—and his colleagues at the Research Laboratory for Electronics at MIT, with their autonomously cognating optic nerves offered another example of this process by which perception was reconfigured.

In their article, Lettvin, Maturana, and McCulloch opened with a seemingly simple question: assuming a world of informatic overload, how can we assume that all processing occurs in the brain? Their answer was revolutionary from the vantage point of history: it does not. Cognition, they argued, *does not* happen in a centralized location (the brain). They argued that the management of data emerged through the networked organization of the sensation-perception-cognition system.

Their initial logic was critical. They hypothesized that the optic nerve does not transmit every piece of data (light) it contacts. Such an assumption reconfigured their experimental practice. Rather than test discrete stimuli, they exposed an optic nerve to *variations* in light. They created a test environment

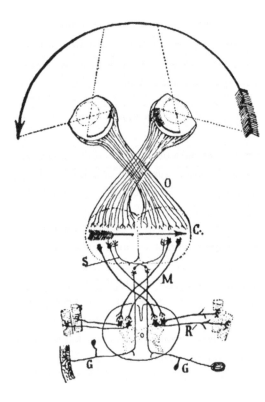

FIG. 4.2__Frog optical system. From Lettvin et al., "What the Frog's Eye Tells the Frog's Brain" (1959). Courtesy of MIT Press.

where a series of myelinated and unmyelinated fibers in the intact optic nerve were exposed to variations in light stimuli. Working on these moving edge detectors, the team discovered a fiber that "responds best when a dark object, smaller than a receptive field enters that field, stops, and moves about intermittently thereafter." From the measurement of subsequent electronic impulse activity, they wished "to discover what common features are abstracted by whatever groups of fibers we could find in the optic nerve."[11] What they discovered was that when the eye was exposed to stimuli simulating a moving insect or an enemy (stimuli that moved or changed from light to dark) the electrical impulse given off changed *before* ever arriving at the brain, demonstrating that the eye—and they were studying only the actions of the optic nerve—was capable of making decisions between such binaries as prey or enemy and nonprey and nonenemy.

They concluded that "the eye speaks to the brain in a language already highly organized and interpreted, instead of transmitting some more or less accurate copy of the distribution of light on the receptors."[12] Their colleague Michael Arbib summarized this finding as proof that the frog's eye could deal with universals like "prey" and "enemy."[13] In summary, eyes were found to be Turing machines. Perception, therefore, became the same as cognition, as autonomous entities, like eyes, began the process of abstracting and processing information. This analysis opened up the possibility that perception as an autonomous process could be technologically replicated—a conclusion further substantiated by the fact that this research continues to underpin much computer science work on vision.

The emerging postwar neurosciences did not understand the image as a representation being transmitted and then translated on arrival in the brain but redefined vision as encompassing the entire relationship structuring the act of observation: a communication channel.

These neurosciences thus produced a flexible barrier between the realms of stimulus, the form of the data, the organs of reception, and the site of processing. While such subjective perception had been found in nineteenth-century physiology and psychology, it was now no longer a problem for scientific objectivity and knowledge, and was positively embraced for technological potential in neural nets.[14] The very nerves, extracted from any particular body, are capable of processing and analyzing data. I would even argue that ontology and epistemology were both collapsed into another approach, which focused on method, process, and feedback. The act of processing information and the act of analyzing it became the same, and the possibility emerged that this decontextualized seeing process could be rebuilt in other locations.

This is not an insignificant experiment in the histories of visuality.[15] The cybernetic model of perception desired a purely technical and autonomous eye. If one wished to see an insect, then one built a frog's eye; if one wants to see a missile silo, perhaps one builds a different form. Vision circulated. There was no single norm for vision. The ideal of a singular, or objective, form of vision was replaced by a fantasy of effectiveness serving particular functions. Historically I wish to focus on the critical function that the lack of concern for static ontologies played in facilitating a shift in the conception of sense perception as an interactive process and a material technology in design, cognitive science, and cybernetics. This was an eye extended into the body and out into the world, a vision that was material and could now act on its own—flies eaten and airplanes blown up, for example—a networked cognition beyond the brain and a new way to understand the differences between subjects and objects. There was no ontological stability in cybernetic visuality; there were no stable enemies or preys.[16]

There was, however, a curious indexical and temporal nature to this ability to materialize vision. To focus on how eyes "speak" to the brain demanded a lack of regard for, or perhaps an automation of, recording and an assumption of an informatically dense world. The impossibility of ever accessing and processing all this data was no longer the problem. Instead the question became how to manage and utilize the unknown. This subtle but important revision of attitudes to knowledge and objectivity was first articulated in McCulloch's classic piece, written with Pitts in 1943, establishing the equivalence between neurons and Turing machines and conceiving of a "neural net." As I explained in chapter 3, McCulloch ends the piece with an astonishing statement concerning scientific claims: "thus [this research proves] that our knowledge of the world, including ourselves, is incomplete as to space and indefinite as to time. This ignorance, implicit in all our brains, is the counterpart of the abstraction which renders our knowledge useful."[17] This "ignorance" or subjective quality of all cognition was now the "abstraction" that produced "use." Subjective perception was equated with technological potential without concern for mediation, and efficacy replaced the concept of an absolute reality as the measure of truth. McCulloch not only took a non-Cartesian perspective but also resolutely declared any split between the mind and the body, or reality and cognition, both undesirable and impossible.[18]

Cybernetic Cognition

Cyberneticians thus existed in an environment of infinite potentiality, a chaos of informational excess, out of which process, order, and meaning were carved. Theirs was a world of infinitely available data from which patterns, techniques, and potentials for actions—psychological, technological, behavioral—could emerge.

Structuring the frog's eye was a broader claim that perception could be modeled as a communication channel, and subsequently enhanced, mobilized, reconstructed, and modulated. Underpinning the equivalence between the senses and communication channels was a revision of the relationship of perception to cognition. If eyes could think, then minds and now brains were also part of the communication structure. An entire science—cognitive science, a field that continues to dominate contemporary psychology and particularly administrative and organizational psychology—emerged whose concern was to model thought processes in algorithmic manner. Like the autonomous speech of eyes, the ability to study cognition separate from psychology or physiology identified an important change in the constitution of knowledge and the experimental practice of psychology.

A few years earlier, the purported founder of cognitive science, George

Miller, had written the aforementioned paper that continues to be the single most cited piece in contemporary psychology: "The Magical Number Seven, Plus or Minus Two: Some Limits on Our Capacity for Processing Information." A young faculty member at Harvard, Miller was working at the high temple of behavioral psychology. Perhaps as a matter of proximity (or rebellion), Miller was explicit throughout his writing in the 1950s that his prime concern was deploying cybernetics and information theory to challenge behavioralist psychologies and psychoanalysis simultaneously.

His explicit use of cybernetics was not coincidental. As a soldier and a graduate student, Miller had worked during the war on signal processing and speech perception, and later on radar engineering. As a result of this crossover between studying signal processing and psychological processes, Miller was very familiar with dominant theories in computing, cybernetics, and communication.[19]

It is somewhat telling, then, that this most famous of all psychology articles was an act of archival recombination. The text is an analysis of other researchers' work on memory, recall, and identification. No experiments were done by Miller to write the piece. The scientists' practice was already seemingly grounded in an epistemology of informational surfeit.

"The Magical Number Seven" opens with a political reference that serves as a gateway to rethinking psychology, politics, and aesthetics. Miller appeared haunted, perhaps deluded, by the very histories of psychology he was attempting to overturn. "My problem," wrote Miller, "is that I have been persecuted by an integer. For seven years this number has followed me around, has intruded in my most private data. . . . The persistence with which this number plagues me is far more than a random accident. There is, to quote a famous senator, a design behind it, some pattern governing its appearances. Either there really is something unusual about the number or else I am suffering from delusions of persecution." That senator was, of course, McCarthy, and the delusional form of persecution Miller described was none other than a faith that seven defines a natural and normative limit, structurally and mechanically ingrained into our minds; a pattern produced by nature or a higher power, a pattern not amenable to change. However, just as organizations like governments could assume paranoid formations, scientists could be deceived by their own minds; a pathway was thus opened by which the single psyche and the networked organization could collapse and be similarly treated by science through the notion of "information."[20]

The article then proceeded to survey an entire history of research on judgment and memory recall, first covering articles on the ability of individuals to distinguish tone, then pitch, then visual data points. The article treated any form of stimulus the same. But Miller repeatedly argued that cognitive limit reappeared. Test subjects in psychological studies, given certain amounts of data, can only seem to remember or differentiate between seven data points.

Before, however, being deluded into believing that there was only one normal standard for human memory and information processing (in relation to which individuals may either be normal, subpar, or geniuses), Miller argued that thinking of psychology in terms of information might transform a set and stable limit to a permeable threshold.[21]

What does Miller mean by information? Here he called on the mathematical theory of communication and the communication sciences to revise the current idea of psychology. Miller equated information with the amount of variance in a study. "The equations," he wrote, "are different, but if we hold tight to the idea that anything that increases the variance also increases the amount of information we cannot go far astray."[22] Another way to understand this equivalence is to think that if a study demonstrates a very wide number of different cognitive responses, then finding the single unitary pattern is more difficult, or wields more information about psychology than a study in which all test subjects perform the same. One study (the more variant) reveals there are many more options or possibilities for action than another. This increase in possible outcomes is equivalent to the probabilistic nature of cybernetic and communication sciences.

The advantages of this new way of talking about variance were, he continued, "simple" enough. Variance was always stated in terms of the unit of measurement. By rethinking variance as equal to information and "amount of transmitted information" as equivalent to "covariance" or "correlation," Miller opened the possibility of a stable scale or curve becoming mobile and relational. Instead of thinking in terms of set units, we could begin to think, in his words, in terms of "channel capacities." Here the explicit introduction of "capacity" mirrors the engineering imperative of finding out how much information *can* be compressed into a particular channel or structure. By removing any set or stable scale or unit of measurement, it is made immediately evident that cognition can now be thought about at different scales (organizational to individual) and extracted as a material process to be modeled and enhanced.[23]

Having established both the apparent repetitive nature of the "magical number seven" and the equivalence between psychological responses and communication theory, Miller then proceeded to explain the implications of this finding. He argued that if we consider an observer in terms of channel capacities and information, we can also begin to think about the limitations to recall and judgment in terms of compression and recoding—computational terms. Miller claimed that the way to increase the amount of information an observer could listen to was by changing either the number of items per input (chunks) or the bits of information (number of decisions or relations per input). Chunks were related to immediate memory and recall, and bits were related to immediate decisions or judgment. Miller's analysis thus splits psychic processes into space and time—immediate assessments of difference between visual and aural stimuli versus durational recollections of data points. This separation between the location of immediate decision-making and the transfer of past data into the present mirrors the structure of computers, allowing Miller to consider how to store data separately from how to operate on it.[24]

So, for example, in memory, Miller noted that we can remember and differentiate many faces (more than seven) and that test subjects had easier times remembering sentences or full grammars than discrete letters. For visual data, subjects judged differences between multidimensional inputs, for example changes in color, when given a comparison color field, better than when shown one input at a time separately. These were all examples of information compression and, in Miller's language, "recoding"; each of these experiments dealt with relational data points rather than discrete stimuli. Recoding for Miller is putting more bits into each chunk. So for example, when giving subjects decimal numbers to remember, if the numbers are grouped together in some pattern then subjects can remember and recall many more numbers then if just shown a series of discrete numbers with no "chunking" (this finding is supposedly the reason we have seven digit phone numbers in the United States). The studies he mentioned also noted that speed was a factor in cognitive performance. The velocity by which data was given as well as the tempo of data delivery, for example how long each input stimulus lasted, or how long one had to watch or listen before given a break, all impacted the specific ability to recall or discriminate between inputs. If information could be recoded, ever more of it could be processed and remembered by the individual, thus the psyche for Miller was elastic.[25]

The article thus paralleled the relationship developed by McCulloch and his colleagues in their work on the eye. It should come as little surprise that Miller's article was written at the same laboratory that McCulloch, Lettvin, and Miller were all members of (the Research Laboratory on Electronics at MIT) or that there was a close correspondence between all these researchers concerning the use of cybernetic principles in their research.[26] More significant, it was this deferral of interest in static ontologies (what we might label "content"), and the shift to examining interactions made possible by using the frameworks of communication theory, that transformed the nature of psychological inquiry, making cognitive processes visible, modelable, and technically replicable.

Miller's work created a new way to approach perception and cognition. Like the neuroscientists, he was not concerned with separating the senses; the process of perception was interchangeable between acoustic and visual stimulus. More important, Miller treated the perception and cognition of stimuli as a relational feedback interaction. He wanted to model the *process* of perception and memory, not delineate the divide between an external stimulus and an internal response. In rethinking the "observer" in terms of "channel capacities" instead of as stimulus-response or a conscious-unconscious subject, Miller opened the path to the augmentation, and perhaps automation, of both memory and decision-making. No longer concerned with normative performance, or a single locus on which the subject was to focus his or her attention, cognitive science turned to understanding how channel capacities could be modulated and directly acted on. The primary concern was enhancing the subject's ability to consume information. Psychologies and machines became epistemologically equivalent, and the intent was to figure out how to model the process of perception as a channel capable of operating in informationally dense environments.

The final lingering problem that Miller returned to, at the end of his article, was whether seven was arbitrary or necessary, a reality or a construction. He answered that even posing such a question would induce pathology. Perhaps, if instead of obsessing about the number seven, we noticed a pattern and stopped looking for causal reasons, we could avoid sinking into any delusional or paranoid states at all. Miller implied that a brain reconfigured through communication theory would be less susceptible to paranoia and delusions. Or perhaps, in keeping with the extended networked intelligence of the frog's eye, a psychotic structure would become the normal model for cognition. Asking such paranoid questions, Miller intimated, such as whether seven was really a natural, immutable, and eternal truth or whether it was a "construction," or contemplating whether the number seven came from nature or G-d or from environment and society to persecute our minds, would only make us ill. Since we are all now provably subjective in our perception and cognition, and unable to tell if input is from inside or outside our own nervous systems, there could be little productivity in attempting to reassert those boundaries in scientific research.

The scientific question, Miller implied, was what work does seven do? What can it teach you about how to recode information? Miller never assumed a genetic stability architecting the mind's cognitive processes. In a moment before either minds or machines were as well known as they are today, a mind-machine was conceived that was both biological and emerging, computational but not mechanistically reductive. This ontologically unstable mind-machine emerged logically from psychologists' embrace of communication, since communication was where the threshold between the exterior and the interior and the stability of reference or scale were both abandoned. The observer configured as a communication channel stood prepared to serve as a conduit for ever more data.

Political Spectacle

But why stop with the reconfiguration of vision and psychology? In the same year the frog's eye learned to speak to the brain (fig. 4.1), in the midst of the Cold War, another architecture of perception emerged at Sokolniki Park: a vast cavern filled with seven screens, built by Buckminster Fuller and designed by Charles and Ray Eames. I want to return here to chapter 3's intimation of a relationship between politics and aesthetics to ask about the implications of this emergent technical and autonomous form of vision imagined into being through cognitive science, cybernetics, and design.

A "totally new type of presentation," in the words of Charles and Ray, the seven channel presentation was envisioned as a "letter," perhaps of love, between two nations in a world where writing would no longer suffice; a "glimpse" of the United States, a day in this foreign country's "life" that by the end would cease to be foreign. In the face of this imagined textual and lingual collapse, the designers believed visual images might serve as a new mode of human interaction. The Eameses believed in spectatorship as choice. The idea of spectatorship invoked in their work was to force users to interactively choose the images they wanted to watch and to find their own patterns in the vast data field of images. Twenty-two hundred images were shown on seven screens for thirteen minutes—data inundation as design and pedagogy principle. Charles Eames has been quoted (chapter 2), in discussing his design and pedagogy for engineering, architecture, and business students, as arguing that vision was a new "language" and that the function of his multimedia displays was to test "how much information *could* be given" to a spectator in an allotted time.[27] Hundreds of thousands, if not millions, of people saw the installation.[28] Everyone, according to the architecture curator Peter Blake, "had tears in their eyes as they came out" of the opening show, rumor had it especially Khrushchev. American officials were alleged to have called it one of the most successful acts of psychological warfare ever conducted. Ray Eames called it an "affective" experience.[29]

Despite the universal acclaim of the success of the American Pavilion, it might be noted that it was also deeply contentious. Rather than emerging from a clearly defined and authoritative government plan, the pavilion emerged from a series of accidents and assembled interests. While officials at USIA claimed that the exhibition was about promoting "improved understanding" between the two nations, this was hardly the only agenda. William Benton, a former assistant secretary of state, remarked that the State Department "was in the propaganda not art business."[30] The art exhibition was embroiled in attacks by congressmen against the supposed Communist associations of participating artists. George Nelson personally had to guarantee that the Eames exhibit could be produced without government oversight or prescreening, because the designers were concerned about excessive interference.

More broadly, the United States struggled with its "soft" tactics. On the one hand U.S. government officials realized the salience and popularity of American products and entertainments; on the other hand officials desired to educate viewers in the evils of Communism. These two goals often seemingly warred with each other as pedagogical propaganda demonstrating the moral virtues of capitalism and the supremacy of American science and technology warred with the (always seemingly more popular) pleasures of Hollywood film entertainments that offered a spectacle of the United States as one of fast women and criminal men obsessed by material wealth and bereft of soul.[31] Public furor and congressional anxiety aside, the show went on.

Scales of the Image

The movie opens with the view from above, the aerial flight. In the night sky, the voiceover intones, Soviet and American cities look the same. The scene from the plane is ominous in the midst of concerns over thermonuclear war. But this view from above rapidly condenses into the network on the ground, such that the recollection of aerial bombing is not memorialized but rather becomes an aesthetic device to enter the circuit. Central are images of highways, modular housing, speeding cars and transport, infrastructures of power and industrial plants, conspicuous consumption, seven screens of Marilyn Monroe winking from *Some Like it Hot* (the Soviets clapped every time) and perhaps most critically, signs of a perfectly racially integrated society (fig. 4.3).[32] The irony that it was the very infrastructure for an emerging American spatial apartheid, highways and modular tract housing, that was the substrate for a

FIG. 4.3__Stills and production shots for *Glimpses of the USA*, installation by Charles and Ray Eames, Moscow (1959). Aerial views; images of cities and infrastructure; scenes of racial integration on the playground. The Work of Charles and Ray Eames, lot 13234–1, no. 24, 42, Photography and Print Collection, Library of Congress. © 2013 Eames Office, LLC.

collective vision of a single humanity should not be lost, but the movie's major statement was that flow, and communication, would overcome difference—between nations, between people.[33]

The democracy of viewership was affirmed by the computer display outside the pavilion. Immediately at the entrance was a RAMAC computer by IBM, programmed to interact and poll entering viewers about their attitudes to the United States. The computer asked questions about what the Soviets thought about the United States and allowed them to "vote" for presidents and policy decisions, while simultaneously responding to questions about the culture, people, and politics of the United States. From the very opening of the pavilion, it would appear, an aesthetic of assimilation, interaction, and consumption operated (fig. 4.4).[34]

As part of the complex in a smaller pavilion that served as a corridor leading to this enormous installation, viewers experienced another pathbreaking exhibition in the history of visuality—*The Family of Man*. Curated by Edward Steichen, the chief photography curator of the Museum of Modern Art in New York, based on 1930s Farm Security Administration documentary project aesthetics, the show was a photo-essay depicting the "human" condition on earth. Drawn from Magnum and other press photo archives as well as work solicited from many of the most famous documentary photographers of the time, the display consisted of hundreds of photographs with titles documenting a single species; an ode to biological diversity in humanity, framed by the narrative

of a standard heteronormative life cycle with the nuclear family at its center. Compared to the multimedia spectacle accompanying it, the tempo of *The Family of Man* was slower and mapped the photograph to a text to mobilize melodramatic sentiment or empathetic relationality (depending on your standpoint toward the show). The dominant reading of Steichen's work by such critics as Roland Barthes, Susan Sontag, and Allan Sekula was as a sentimental representation of the human condition in the name of propagating American imperial ambitions, in fact a pivotal part of the aesthetics of empire by which the United States presented the face of a new global consumer species.[35] *The Family of Man* arguably still operated on a model of legibility produced through the aesthetic conventions of documentary realism and the temporal narrative of a linear life pattern. As an exhibition, and in contrast to the Eames work, it maintained fidelity to textuality as necessary for, and abetting, visuality.[36]

Algorithmic Cinema

But this story of an aesthetic of imperial ambition operating through still modern forms of sentiment, identification, and ideology may be too simple. The installation accompanying Steichen's show had a different logic. The Eameses were not interested in life as a linear progress through stages. While children and mothers appeared, and faith, certainly, in *Glimpses*, overwhelmingly it was infrastructure, roads, electric bridges, material pleasures such as food, nightclubs, winking women, flying kites, and other moments of pure gesture such as the cycles and rhythms of mobility and labor as people dashed into cars for work. If Steichen's exhibit has been understood as presenting a single idea for how human beings share a world and a definitive historical and life cycle/biological time, the Eameses had a different idea—one of choice, patterning, reverberations, and redundancies.

FIG. 4.4_The floor of the Moscow Exhibit (1959), with multiple types of display operating; integrated children play at a fountain in front of the *Glimpses of the USA* installation. *The Family of Man* exhibit was in a pavilion, to the side of the central oval that led into this space. The Work of Charles and Ray Eames, lot 13234–2, no. 27, Photography and Print Collection, Library of Congress. © 2013 Eames Office, LLC.

Anticipating our arguments today that that moment in history heralded both new forms of perception and novel forms of economy, Charles Eames insisted on the historical specificity of this form of visuality and media. He was explicit that this form of exhibition was an architecture of multimedia and *not* cinema.[37]

The installation was, indeed, a case study in communication theory—a critical experiment in information management. The seven-channel installation was carefully timed. The flow charts made by Charles Eames and his editor, miming those in computer science (with which they were familiar through their major client, IBM, see fig. 4.5), cadenced the presentation; breaking editing velocity and acoustic flow into clear cut steps. The Eameses viewed communication theories as central to their design principles and regularly worked with cyberneticians, such as Norbert Wiener, and corresponded with Jerome Lettvin (of frog's eye fame).

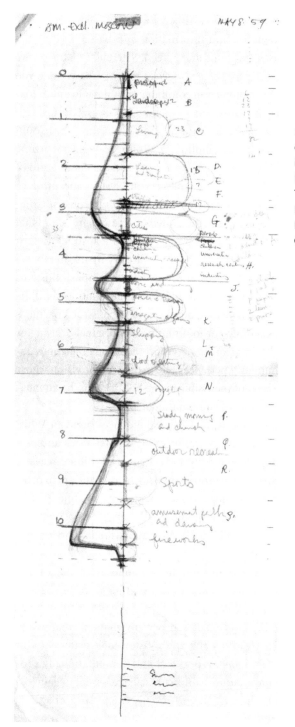

FIG. 4.5__Final editing flow chart for *Glimpses of the USA*, installation by Charles and Ray Eames, Moscow (1959), The Work of Charles and Ray Eames, (C-12r), Manuscript Division, Library of Congress. © 2013 Eames Office, LLC.

The movie was carefully thought out, timed to the moment, seemingly grounded in theories of cognition and information and based on a model of iterative feedback between spectator and screen. The chief editor of the seven-channel presentation, *Glimpses of the United States*, was the avant-garde moviemaker John Whitney Sr., a pioneer in computer graphics and animation and famous for using cybernetic concepts in making his films.[38] In discussing his editing tactics he wrote of producing a "liquid architecture" that would create "structured motion that begets emotion."[39] Whitney sought a machine cinema no longer in the realm of image and index to produce a new world of entropic potentials that would directly tap the nervous system and produce a mobile space. "As early as 1957," he recalled, "I had begun to construct mechanical drawing machines. . . . I was not motivated to create representational images with these machines but, instead, wanted to create abstract pattern in motion . . . to evoke the most explicit emotions directly by its [the film's] simple patterned configurations of tones in time."[40] To accomplish this ambition of surpassing the image to directly induce emotion, Whitney used the remains of antiaircraft servomechanisms from the navy to construct machines that could produce graphics on film without filming any original drawing or live action.[41] Whitney sought an algorithmic vision that made machines authors in the production of human experience; a form of machine vision whose work did not operate at the level of the visible image but through the attenuation of the nervous system, by way of computational logic.

The production notes are copious, and correspondences between Whitney and Eames afterward indicate the centrality of ideas of feedback and psychology to Whitney's thinking about editing and cadencing. Whitney in a later letter to Eames cited Kenneth Clark, a prominent British art historian, to make his point about the changing cosmology of the "image" and to argue that he (Whitney) sought to make a less "anthropomorphic" image.[42] A less anthropocentric image for a new age, one Whitney inaugurated with his work on animation, starting in the late 1950s.

If the organizers of the USIA pavilions still thought in terms of images, and architectures of geographical space, Whitney discussed his work in terms of harmonies, fluids, sine waves, raw patterns, and "material abstractions," implying the production of a visual-acoustic sensory environment that would transform cinema and in fact territory (fig. 4.6). Vision took materiality for Whitney as a process to be designed, replicated, and computationally programmed. In design, as in the anatomies of neuroscience, visuality gained a new logic, to be encoded into the architecture of USIA Cold War propaganda and the nervous system of the spectator.

FIG. 4.6__Screen stills from *Catalog*, by John Whitney Sr. (1961). He used his adapted servomechanisms-camera machines to produce the patterns and animations without pre-drawing the stills and to create patterned mutations and movements. This was one of the first examples of purely machine animation, produced shortly after *Glimpses of the USA*. Image capture from YouTube, https://www.youtube.com/watch?v=TbV7loKp69s.

Of greatest interest to us in demarcating a historical shift in media strategy and the management of attention was the Eameses' and Whitney's attitude to cutting and the image. They viewed idiosyncratic changes of images as useless but viewed relational shifts, the changing of a series or set of images together, as facilitating information exchange. In *Glimpses* only sets of images could shift. Charles Eames argued that changing one image at a time was useless. Only changing images in groups of three or four so the eye could find a pattern was useful. Transfer would always start at the bottom of the screen, and a number of screens would change together to allow the eye to pick up patterns. He labeled this a new form of relational editing.[43]

If the Steichen show was laid out in a linear flow, driving users to choose a pathway through a historical life cycle centered on the drama of motherhood and child development, for Whitney and the Eameses the spectacle had to be produced through redundancy and repetition. Whereas Steichen only occasionally used landscapes, tending to focus on midlength shots or close-ups that clearly showed emotion, and allowed individuals to be singled out as unique, in the Eameses' films extreme long range and extreme close-ups were the standard, usually following one after the other. Whitney's editing style followed the logic of musical scoring, favoring repetition, cadencing, and harmony. *Glimpses* operated by repeating cycles of slowness, accelerating images, and then again slowness as a way to move viewers through a "day" that was mostly about showing repetitive patterns of infrastructure and activity. This speed was paralleled musically and in scaling images, the crescendos being the closest or furthest moments—many highways at once or Marilyn's face.

Machine Vision

It should perhaps be no surprise that it was the figure of the movie star's close-up that served as the locus of transfer from human to inhuman vision. The close-up, as the feminist film theorist Mary Ann Doane has written, is the moment when that which is human (the face), the very sign of subjectivity, condenses into raw gesture and form. Film critics obsess about the close-up as the very mark of the cinematic medium, the singular operation that separates classical cinema from other performative and spectacular media, such as theater or panoramas. The close-up, however, demonstrates a built-in contradiction. The close-up simultaneously demonstrates how the cinema exceeds language and human perception, the inhumanity of the camera, while also asserting the human body (the star) at its center. As Doane demonstrates, the discourse on the close-up in film theory is fraught with an effort to maintain this human at the center. At the same time, the theory of cinema must repress the features of time and scale that make the close-up operate not through identification with the star but through a spectatorial immersion into the image. The close-up is a perspective that, despite its content, cannot be seen or fully accessed by human vision, it is too large, its scale and proximity impossible for an un-aided human eye to actually see. It drives an effort, a desire even, to be able to image the medium, even as it makes it impossible for a human being to do so. The star is not available to be apprehended by the spectator.

Perhaps surprisingly, what results is an intensification of media consumption. This joint feature, the intimacy of the face and the size of the screen, drives a frenzied proliferation of screens—both very small and very large—in an attempt to both enjoy the cinematic spectacle (IMAX) and maintain control over the image through personalization (handhelds). At the zenith of Doane's argument may be the failure, entirely, of the cinema as a medium in the face of another media landscape; a failure the multiscreen installations, signifying a new form of spectatorship and medium, appear to embody in proliferating screens, scales, and media.[44]

This installation did appear to be an exercise in eliminating older forms of spectatorship. Utilizing the latest imaging technologies—aerial views, micro-scopic close-ups—the purpose was not to align the human eye with the ma-chine eye. There was no mis-en-scène in the piece. It was not that the Eameses saw everything through a camera, and that the view of the camera and that of the human being were being aligned, but rather that the camera took an au-tonomous role. Mechanical vision emerged because of the focus on relation-ships and scale providing the direct conduit to emotion, as the editors of the piece imagined, rather than a conduit to seeing discrete images. This was not training in seeing like a machine but in being part of one.

If Steichen argued for the "oneness of man," the Eameses argued that images, such as highway interchanges, would be universally "familiar" (which is questionable).[45] This assumption of universality appeared to imply that the installation provided a format of species unification, perhaps an alternative definition of population to Steichen's, one grounded not at the level of the organism, or the human body and subject as the basis of the collective, but at a cellular or even molecular level, based on sharing neural patterns and forms of attention rather than identifying with similar subjects.

This global nervous network was propagated through a conception of vision as a channel or threshold. Charles Eames spoke of the choice of seven screens in terms of producing "credibility," which he defined as offering people a sense that they had seen something they were familiar with and was real in a documentary sense, without necessarily allowing them to focus or fulfill a total identification with any one image. "We wanted," Eames said in a later interview, "to have a credible number of images, but not so many that they [the spectators] couldn't be scanned in the time allotted. At the same time, the number of images had to be large enough so that people wouldn't be exactly sure how many they have seen. We arrived at the number seven." (We might also wonder if seven was arbitrary.)[46] Eames suggests that vision is a threshold operating between the "credibility" of large data sets and the scanning capacity of the human sensorium. This perceptual architecture insisted on an eye capable of finding patterns in vast data flows. This eye, however, could never be fully "sure"; it had to be never stable, always available—as in the new epistemology of cognitive science and cybernetics—to anticipate and assimilate more data.[47] This lack of "surety" mirrors Miller's expunging of paranoia and delusion. All three figures gesture at an epistemic transformation in the definition of observation and authenticity. Credibility, here, was not about knowledge but about capacities.

Most times, Charles argued, people confused multimedia with multi-images. In that case, this logic implied, any film is multimedia. But, he continued, the work of the Eames Office was different: "it had not only multiple images, including the relationship between still and motion pictures, but also sensory things. . . . We used a lot of sound, sometimes carried to a very high volume so you could feel the vibrations. . . . We did it because we wanted to heighten awareness."[48] The language alerts us to a new site of technical articulation. Awareness itself takes on a materiality, to be modeled and encoded as a form of media. Unlike their predecessor, Herbert Bayer (who had provided inspiration for the Steichen layout), the Eameses created architectural diagrams of their installation that showed no observer; they rendered the installation as itself an eye, perhaps a cognating one. The spectator had disappeared (fig. 4.7).

Perversely, the cinematic tactics of the Eames installation, along with earlier USIA films already shown at the world's fair in Belgium in 1958, more closely aligned themselves to Soviet Constructivist conceptions of an autonomous and machine vision than to the classical Hollywood organization of spectacle, where the spectator is safely comforted into an alignment with the camera in the hope of reasserting the place of the human body as the measure of the screen image.[49] But if Constructivism was linked to both behavioral psychology and utopian politics, this new multimedia practice linked to the cognitive and computational sciences had different understandings of truth and human subjectivity; gone were the notions of absolute truths or teleological progress, replaced by circuits of information.

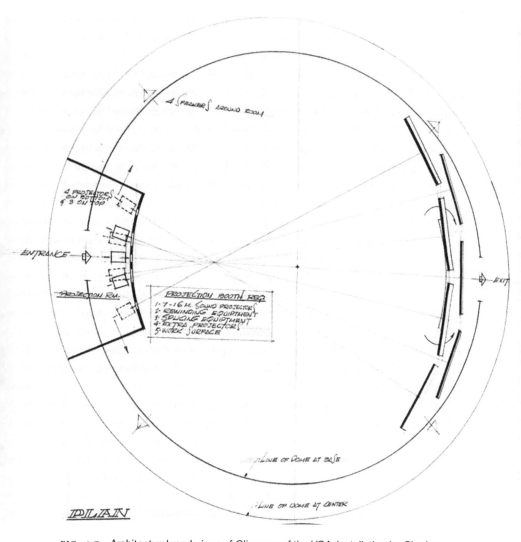

PLAN

FIG. 4.7 __ Architectural renderings of *Glimpses of the USA*, installation by Charles and Ray Eames, Moscow (1959). Viewpoints are taken from the perspective of the apparatus-projectors, not the observers; the piece is rendered as an "eye." The Works of Charles and Ray Eames, ADE unit 2833 no. 3, Photography and Prints Collection, Library of Congress. © 2013 Eames Office, LLC.

Theirs was a world emerging through the careful timing of data delivery, "We have always been committed to information," Charles recalled.[50] The Eames motto: information, not representation. For the Eameses, there was only communication. Sense made a channel that merged with cognition and was reformulated in terms of capacity and surety, always available to be re-coded in the terms of cognitive science.

Whether pattern recognition and subjectivization should be considered equivalent is unclear either at the time or in our present. This architecture of interactivity introduced in the 1950s was not the architecture of distraction put forth by Walter Benjamin and postulated by architectural histori-ans as applying to 1959. The concepts of the individuated subject produced through normative images of the self, or of propaganda assaulting psychol-ogy, do not describe this new media practice. Domesticity, often associated with mid-century design and considered by design and architecture theo-rists like Beatriz Colomina, long the hallmark of theorizing Eames, needs to be rethought. It was Steichen's show that facilitated affiliation and identifi-cation with clear subjectivities. The forms of spectatorship produced by the multichannel installation utilized the observer's presumed familiarity with the image for another purpose—to attune the spectator to finding patterns and consume information. This spectator, therefore, was no longer linked to norms and populations through identifying with ideal forms like the nuclear family or a stable subjectivity. The relationship between the anatomo-politics of the body and the population, posited by Foucault as the foundation of mod-ern biopower had been severed.[51]

Instead, this was an architecture of the network, producing a new form of spectator who was simultaneously hyperindividuated and linked into a broader circuit, whose very nervous system was already conceived as a part of an interface. Perhaps this was not even an anthropocentric form of spectator-ship. Certainly it was not one necessarily linked to the assumption of a stable human subject linked to a national population. A French press write-up covering the exhibit was titled "Le Cinema Prend L'Oeil De L'Insecte" and argued that a new form of vision, a challenge to cinema, had emerged in the cyclorama of the Moscow pavilion. This journalist equated the "vision" of the cinema with that of an "insect." The inventor, Charles Eames, had, accord-ing to the article, produced a new optic, utilizing the rules of psychology and physics to produce, for the first time in the history of cinema, a new view of the world: bifurcated into multiple points of view but synchronized tempo-rally. These multiple viewpoints that were still synchronically coordinated the reviewer equated with the eye of an insect, which also has multiple lenses and does not see the world through the cyclopean (to use language from chapter 1) synthesis of stereoscopic vision. Eames was thus credited with producing a radical, perhaps nonhuman, point of view.[52]

It is also worth noting that USIA-administered exit polls at numerous exhibitions at the time demonstrated that the "soft" message of U.S. cultural initiatives often originated because of the rather diffuse and unclear message of the installations. Counter to the Soviet installations, which offered clear points on the virtues of Communism and the technical prowess of the state, viewers tended to articulate pleasure at American exhibitions but lack of clarity as to the message.[53] Spectators were offered affective sensation without clear representation and without point of view.

Archive

But if the past was forgotten, it was still stored. For there was another (curious) feature of these installations that is never commented upon. Efficiency or speed was not the only temporality of this perceptual architecture. These installations had an archival sensibility. Absolute storage, the ability to save everything, was the unconscious desire structuring this form of visuality. However, the nature and organization of this storage system, and the identities it creates concerning race, nature, sex, human, and non-human—those questions that defined the nineteenth-century archival impulse in the work of Foucault, and so much colonial, postcolonial, and postmodern discourse, were repressed in the interest of producing these relational forms of seeing. There was no stable ontology or concern with recording here. Perhaps this attitude inflected itself in the cannibalization of the imagery of civil rights and human diversity into this architecture. It was a smooth space for global integration where the autonomy of vision, the interactivity of attention, and the absolute recordability of the world were givens.[54]

The relationship between the two displays, however, both as differing historical forms of visuality and as temporal experiences, at juxtaposition within the same space and operating in collaboration and competition is worth noting. This emergent perceptual territory of the American exhibition in Moscow relied, therefore, on maintaining an ongoing tension between multiple forms of spectatorship and perception. Observers' senses vacillated *between* these two different modes of interaction. Shuttling viewers between identification and a logistics of assimilation/substitution, the affective field wavered. While the Eames display targeted circulation and arguably consumption, Steichen's show resolutely focused on the human condition as separable from the material or consumer habits of its subjects. For while the Eames show may have espoused choice, Steichen also believed in choice, but still affiliated to individual subjects and to relationships between individuals who recognized each other. For Steichen choice was about politics, not consumption. While much has been said to critique Steichen's exhibition as Fred Turner has recently noted, the *Family of Man* exhibited a deeply democratic impulse critical of the forms of nationalism and identity that had supported totalitarianism, fascism, and Nazism during the war. The democratic subject was one who could choose how and where to look. Within the historical situation Steichen's impulses were in many ways, progressive. He, in fact, often warred with the USIA over the specific content and fought to have the United States represented in some of its diversity—of class and race—even if his dominant life narrative was heteronormative.[55]

The only element unifying the field between the two displays was the specter of the Cold War itself—the Bomb. In *The Family of Man* the atomic bomb figured as a color image only in the hardcover catalogue; it had been removed from the actual installation.[56] But its logic was central to the notion of a unified but differentiated humanity. The Bomb had no direct representation (but then, nothing does) in the Eames installation, but the idea of a perfectly communicated affective environment spoke to averting such a technical disaster. The entire space operated, therefore, to defer one possible future (nuclear annihilation) through the technical manipulation of the image archive and the modulation of attention. This then was a curious form of futurity, whose imaginary was both radically nihilistic and abundantly optimistic. The viewer experienced the universality and potentiality of a not-yet-realized human vacillating with the inevitability of technical disaster and extinction.[57]

In cybernetics the tension between storage and circulation continued to animate the production of endless interfaces. Between the Eameses' autonomous machine vision no longer linked to geographical space or humanity and the vision of a biologically threatened human species presented by Steichen lay the infrastructure for our current data-filled and sensory environments. This relationship between older forms of spectatorship and subjectivity linked to recognition and identification in the image and the proliferation of interfaces where data inundation and pattern-seeking are the dominant modes of observation are the two poles that substantiate our contemporary aesthetic and political situation. These two different architectures of perception drive a contradictory recentering on identity and subjectivity at the same time that the human measure of the screen is effaced in the name of another discourse of direct neural and cognitive interaction and manipulation. This contradiction emerges in our present through the ubiquity of computing interfaces and global social networks, while reactionary and identity politics are resurgent in many forms.

Violence?

Traveling from the interior of nervous systems to global media spectacles, I am left to ask about the implications of this situation for the remains of older political orders. As one USIA officer commented, this was the greatest piece of "psychological warfare" ever waged.[58] At the same time, Soviet journalists and spectators kept asking: where were the "technical" exhibits, industry, and science?[59] The exhibition seemed to lack any examples of national might or military capacities. The consumer aesthetic did not seem to befit a superpower in the midst of a global conflict framed by nuclear weapons and conducted at the level of postcolonial civil wars. One might extrapolate from these overheard comments that Soviet viewers wanted to know where are the weapons, where is the violence? What state shows up to prove its power by showing winking girls and playing children organized through circulative networks?

Charles Eames also appears to have had a number of reservations. Our own work, he recalled twenty years later, "has come back to haunt us." He went on to say that "franticness of cutting tends to degenerate the information quality. We have always been committed to information: it's not a psychedelic scene in any way."[60] We were not, Eames implied, inducing nonrational mental states (although in a world where eyes can cognate, the definition of "rational" should not be clear). Implicitly, he understood that his installations were drug-like. While seeking to distance himself from the counterculture and the Happenings of the (historical) day, this statement is also an implicit confession of affiliation on Eames's part. These massive installations bombarding their spectators with data produced a space where the boundaries of the body, perception, and environment might be reconfigured; but also a space where the ability to create legibility, meaning, or signals was threatened. Eames's statement can be read as an oblique confession that his own search for industrial efficiency, scientific authority, and legibility had been undercut by the very environment and techniques he generated.

The American National Exhibition in Moscow, therefore, poses critical questions about today's new forms of governance through media, and the infrastructures of psychology, perception, and cognition that underpin them. How do we reconcile the seeming disappearance of violence with these new models of psychic manipulation? What do these statements about the loss of a centered viewing subject, and the invisibility of violence, on the part of newspaper journalists and designers say about the perseverance or denigration of identity, geography, and recognition in the context of an emerging interactive media practice? In this attentive environment that integrated observers' neural nets and their governments, their marketing mechanisms and their optic nerves, where difference—racial, national, biological—itself was deferred as a question (I didn't say eliminated) and rendered politically impotent through consumption into an interactive architecture of hypervisualization, what still haunted, in the words of Eames, this machine?

The USIA pavilion was certainly a performance in making the infrastructures of American life stunningly visualized yet impossible to comprehend. If anything, this pavilion cannibalized older structures of vision and gaze in the interest of consuming the possibility of evidence or witnessing altogether. I might ask if this is the genealogical underpinning to what the anthropologist Rosalind Morris has argued is the "narcissistic economy" of contemporary warfare and torture?[61]

In our present, despite a ubiquity of violent images and performances of ethical horror at torture, there appears to be no scopophilia, or pleasure in looking. It is possible that looking is not even possible. Images appear to no longer prompt identification or desire—whether in love, hate, or disgust. Morris argues that what has disappeared is any concept of a social structure that organizes vision and judges witnesses or participants. This is a "shamelessness," to use Lacan's terms, that emerges because the always recorded world is always available to be personally replayed in the very near future. The censure of surveillance by anything but the self is obliterated in the self-referential circuit of images.

Under such conditions, circulating images do not produce evidence, proof, or emotional attachment (even if negative) but only an imperative to circulate more images. Thus, soldiers who torture prisoners continue to circulate images of their work, despite potential judgment by military tribunal, without, in her words, "satisfaction," on the Internet, and we as a public "see" them, but only as an incentive, perhaps, to use Facebook or YouTube, and not, as one might hope, as an invitation to action or commitment to stop these actions.[62] In the ability to always already feed the image of ourselves to ourselves in the near future, we make it impossible to feel shame or remorse. Morris stunningly argues that we cannot encounter difference in the field of vision. The imperative to encounter is renegotiated toward an imperative for interactivity and informational circulation.

It is possible to read the consumption of racial iconography into these architectures of 1959 as servicing such a narcissistic economy of torture. In these architectures one can envision that the relationship between the subject, the body, population, and territory had been severed and remixed, consuming identities and differences into a new logic of a global species-attentive field, where histories of inequality continue to operate but without recourse to representation or voice, thus posing a terminal threat to older forms of civic life (not to mention the civil rights movement, whose very iconography it has consumed) in the name of avoiding thermonuclear conflict. The elimination of civil rights in the name of global love. In this world where eyes speak but no longer in any language of translation, only in action, the resurrection of threat from within the network is always a possibility. Morris's account aligns the reformulation of desire and encounter from chapter 3 with the media strategies and psychologies of this chapter.

Histories Should Multiply

But the death drive of the contemporary media system is also suspect. I return to Eames's concern with psychedelic states as a double-coded concern about Communism and the counterculture, one that weakly confesses to an affiliation while recognizing that the very architectures being designed service as an apparatus for perpetual war. It is a concern for the fate of the species, now potentially homogenously dictated in the neurophysiological language of drugs or technology—the result of a paradigm where dedication to "information" has produced a global attentive infrastructure.

In response it may be worth returning to the earlier moment in this book when I discussed cybernetics and the archive in order to recall Marx's statements on language and translation. For Marx circulation is contingent on translation, and thus money and ideas can be made analogous.[63] The bourgeoisie and structuralist response to this assumption is to fantasize a global language, in this case of sense, to overcome the resistances within capital to circulation.[64] This is a fantasy we could extrapolate into the obsession with information and circulation; a dream once articulated in the massive multichannel installation in Moscow. If we believe Marx's dictum, then we will lose all possible forms of encounter, and even future, to the seamless coupling of the nervous system into the interface, resulting in an orgy of "narcissistic violence."

This is not, however, a predetermined fate. Earlier I turned to a "foreignness" within representation that is the source of difference and futurity. Here I wish to turn to an alienness that emerges from within the image. This radical exteriority is the source of a complex, but possible, form of encounter with differences—both among subjects and in thought. This encounter emerges at the moments of internal contradiction between circulation and identification that parallels Derrida's concept of différance, and Deleuze's notion of encounter in the time-image.

This alienness lies within the cybernetic discovery of a vision capable of destabilizing the boundaries of the human subject. And it lies in the space between older archival orders of memory and visuality and informational regimes—the haunting that troubles and inspires Charles Eames. How we define and maintain the temporal and spatial separation between the archives of visuality and the interface is part of this struggle. Does communication and translation automatically assume homogeneity and convergence between all mediums and entities? And times? The study of the past demonstrates that the field of vision is never coherent, and always multiplicitous. Circulation demands resistances; empires are affective and vacillating entities.

It is perhaps worth contemplating more seriously, then, the differentiations between Eames and Steichen and, further, the multiplications of the image emerging from this media condition. At that moment, when discourses of ideology and consciousness shifted to those of technology, cognition, and communication, we must ask, what alternative possibilities were never realized? Could the reformulation of bodies, identities, and screens have been reattached to a global human imaginary in different ways that would not lead to a Vietnam War or American racial apartheids, for example?

In 1966, the famous figurehead of American avant-garde cinema Jonas Mekas published a special issue of *Film Culture* dedicated to "Expanded Arts." In the issue, George Maciunas included the Charles Eames's multimedia installations as part of an image titled the "Expanded Arts Diagram."[65] This work was presented along with (ironically) mostly psychedelic works. Artists, movements, and other agents of aesthetic transformation were listed. The diagrams create creative genealogies that stretch from Walt Disney to, in the words of one title, "Anti-Bourgeois Popular Art," from the rather staid figures of Charles Eames to radical feminist artists like Carolee Schneeman. These figures were all diagramed creatively with all sorts of new categories and histories. As for the "expanded cinema" with which Eames and Whitney are credited, histories of "electronics, optics," "Fairs," "Disneyland," and "Collage, Junk Art, Concretism" lead to it, and emerging from it is "pseudotechnology," which makes "expanded cinema" a sibling, perhaps even mother or father, of "Kinaesthetic Theater" and "Neo-Baroque Happenings." It's not known what Charles Eames thought about all this, considering his concerns about confusing good information design and psychedelic art.[66] It is all patently absurd, and incredibly logical. The diagram is a flow chart leading to an alternative art history; perhaps a diagram to another future? Soon, following the "expanded arts" diagrams of Fluxus (fig. 4.8), came the pathbreaking text *Expanded Cinema* by Gene Youngblood—a manifesto for a new type of image making in the 1970s. Youngblood prominently mentions Charles Eames and particularly gives homage to John Whitney Sr., as a pioneer in computer animation.

In his work beginning in the late 1950s, John Whitney Sr., had used the machinery of servomechanical antiaircraft defenses to achieve the direct input of the image into the nervous system. Whitney lovingly recalled that these experiments in autonomous animation were initially made because antiaircraft servomechanisms had become available to him as a result of their obsolescence. These machines had been thrown out by the military because they had been replaced by other technologies—mainly computers—or were no longer needed. In these scrap heaps of Los Angeles, Whitney found desire and inspiration. For a moment, it might be interesting to take the obsolescence, and even untimeliness, of the antiaircraft gun sight as a form of seeing seriously— perhaps as a way to contemplate the idea that we don't actually see through the gun sight in our present; that something *has* changed in the field of vision.

His first film compiled from outtakes of all his experiments with the M-5 and M-7 Anti-Aircraft Directors, *Catalog* (1961), is in fact a "catalogue." The film is composed of all the different tactics Whitney developed using his machines. The film is thus an archive of pure optical strategies, set to the music of dissonant strings, somewhat recalling the sound of an Indian sitar.

In the film (fig. 4.6), numbers dissolve into circular spirals that turn into waves, and then return to circular spirals. The transitions have no cuts, no seeming edits. There is no montage, just the movements of machines inscribing themselves on film. Although Whitney did edit, in connecting the disparate sequences together, he conceals those cuts. He presents the sequences as series, mutating in forms rather than scenes.

As the spirals of the films made by John Whitney Sr., unfold, they turn slowly into new forms, unfocused vacillations and movements, the result of the very machines responding to their own positive feedback and oscillating, thus transforming the patterns of animation. If Norbert Wiener had originally proposed that vision must be a process of abstraction in order to maintain homeostatic equilibrium, here abstraction itself becomes the source of emergence, as the absolutely random patterns generated by servomechanisms start to imbalance themselves through feedback, transforming the image, directly imprinting it on the retina.

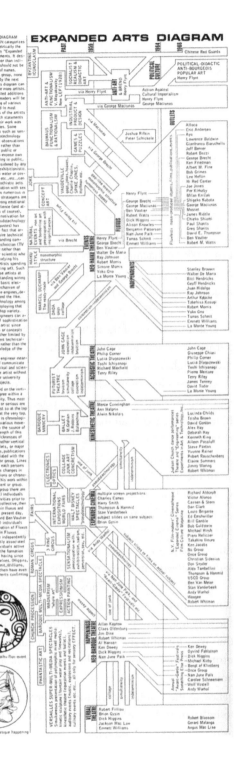

FIG. 4.8__Expanded Arts Diagram, by George Maciunas (1966). © All Rights Reserved, George Maciunas Foundation, Inc., 2013. The captions under the two illustrations on the lower left are: (top) "monomorphic neo-haiku flux event"; (bottom) "mixed media neo-baroque happening."

For Whitney this image was no longer, perhaps, even linked to the cinematic image. Instead it was music that he named as inspiration. However, unlike Hans Richter or Oskar Fischinger, with whom he worked, who had also aspired to produce a visual music with animation, Whitney had a different idea. Music for him was not a separable sense; it was rather a different method or approach to organizing perception:

> people talk about abstraction in graphics as being cold or inhuman. I just don't see that at all. What is a musical note? It's totally abstract. That's the essential point and that's why I use the *musical analogy*. The essential problem with my kind of graphics must resemble the creative problem of melody writing. . . . Music really is the art that moves in time. The many statements about architecture being frozen music notwithstanding, here we are truly looking at another art that moves in time. Someone once said about musical compositions: "Time and tone completely fill each other . . . what the hearer perceives in the tones and rests of a musical work is not simply time but shaped and organized time . . . music is a temporal art because, shaping the stuff of time, *it creates an image of time*." I like that idea very much, so I ask myself, what can be essentially the image of time for the eye to perceive?[67]

In these time-images, as labeled by their makers, architecture and space are undone. Whitney envisioned a new territory of sensory reformulation. He wrote of his later films that they contained patterns, which produced "words," but not in any language of speech; rather, they produced serial resemblances, structural and syntactical similarities that would also differentiate in meaning and experience. This language he analogized again to music: "I am moved to draw parallels with music. The very next term I wish to use is 'counterpoint.'"[68] Counterpoint for Whitney denoted the layering of graphics, superimposed in time—backward and forward—on animation cells. Whitney insisted, in fact, that his thinking about music started with the image.

In watching Whitney's films, they operate through a layering of perception. This layering is topological, it does not emerge from dialectical relationships between images and sound. It is flat, but immersive and active. It is depth through time. The forms move through screens by way of mutation, usually in rotation, or through changing shape, not through a literal movement across a screen.

When Whitney speaks of "counterpoints," what is actually experienced is a structure, like the corner of two walls, which joins two senses in a relationship. The sound does not operate to produce movement; it is not precisely timed to the animation. Rather, the soundtrack is a bit off in timing. The strings are a distinct experience producing immersion into the image (sound of course has long been considered adding dimensionality to cinema), but without actually offering coordinates spatially.

The temporal, rather than spatial, logic of this cinema also operates through an absence of devices of scale. There are no establishing forms that changes are compared to, no moments of stoppage in the image transform. Certainly this cinema is absolutely absent of figural devices, close-ups, or establishing shots or devices of recognition or affiliation—the reverse shot, for example. It is not the only cinema to attempt such strategies, but it was the first to do it through the operations entirely of the machine.

But these were dynamic machines. While the analogue machines used required the input of an initial pattern, the results would always be surprising. The outcome was structured by the original input but never fully predictable, differentiating from within the actions of machines. Gene Youngblood wrote of John Whitney's films that they possessed a "seriality":

> Second, is the quite noticeable seriality of the composition, the unified wholeness of the statement, although it is composed of discrete elements. In defining "serial" in this context I should like to quote the art critic John Coplans: "to paint in series is not necessarily to be serial. Neither the number of works nor the similarity of theme in a given group determines whether a [work] is serial. Rather, seriality is identified by a particular interrelationship, rigorously consistent, of structure and syntax: serial structures are produced by a single indivisible process that links the internal structure of a work to that of other works within a differentiated whole.[69]

These wholes emerge in the films through a pushing of vision and sound into singular relationships through a similar process. Each data point, or input, may be different, but the method is not. But here, rather than culminating in homogeneity, we see the opening up of the image; a reorganization of affective forces. In fact, while starting with analogue, Whitney's interest in the digital emerged because he continued to believe that the complexity and interactivity available through formal algorithms would become emergent, and unpredictable, when they were actually responding to their own feedback.

Cybernetics opens from within these circuits and channels. The teleology culminating in the destruction of the enemy, the original purpose of gun direction, is eclipsed by a logic of capacities and thresholds. These images participate in producing a sensory infrastructure with no teleological endpoint, whose seeming aftereffect is the proliferation of ever more images.

One can imagine, therefore, that in these oscillations subscribing to a pure algorithmic logic, these images break from all purpose. One might even imagine these images pushing the very logic of military vision to the point where the necessary alignments between sentiment and action (particularly aggression) necessitated by our contemporary wars on terror, for example, are not invoked.

At the same time, when contemplating the screen-filled environments of such sensorial consumer landscapes as Songdo, the place of such a purely affective image must be understood as related to our contemporary architectures of responsiveness and interactivity. Whitney created a new image that reconfigured the senses, and even bodily relations. The question is whether such maneuvers need to be channeled into a relentless need to "use" interfaces in the interest of circulating data, or whether they can produce other forms of relationality. In his day, Whitney hoped it would be a "human" image.[70]

What makes Whitney's work uncanny, in the way that circuits vibrate, and porpoises suddenly perform novel tricks, is its resemblance to another discourse. In *Cinema 2: The Time-Image*, Gilles Deleuze seeks to recuperate the "image" for "thought." While the language of encounter rarely appears in the text, it is implicit to the discussion. For Deleuze certain cinematic operations force encounter with the unthought, the virtual, that which is not apparent or available to our limited imaginations in the present. These operations create an encounter with an alienness within the image, something that exceeds what we think we know or any image we already possess for thought.

This encounter bases itself in the de- and rerealization of the perceptual field; the interruption of homogeneity and circulation in an otherwise seamless channel. Deleuze argues that in rare works of art, "there is no longer any movement of internalization or externalization, integration or differentiation, but a confrontation of an outside and inside independent of distance, this thought outside itself and thus un-thought within thought."[71] Here Deleuze repeats a longer running theme in his work. Already in *Difference and Repetition*, Deleuze stated that the condition for thought must emanate from within a system, from within thought, and relies on the destruction of any coherent image. The image of thought, Deleuze argued, must be opposed to "recognition" and can only be "sensed."[72] In insisting that this "inside" and "outside" are not a matter of subjects linked to individual bodies, but another type of encounter altogether, Deleuze repeats the injunction that thought cannot arrive through identity or object relations. Encounter is, therefore, not a matter of "distance" or space. It is the production of thought through the self-produced actions of systems; perhaps the oscillations of machines.

Deleuze then turns to a term that preoccupies his work on cinema—temporality. He specifies that time-images are capable of "a perceptual re-linkage. Speech reaches its own limit which separates it from the visual; but the visual reaches its own limit which separates it from sound. So each one reaching its own limit which separates it from the other thus discovers the common limit which connects them to each other in the incommensurable relation of an irrational cut, the right side and its obverse, the outside and the inside . . . the visual image become stratigraphic is for its part all the more readable in that the speech-act becomes an autonomous creator."[73] While time is not mentioned here, implicit is the idea that these thresholds and folds are ongoing processes, and that this reintegration of the perceptual field is a durational operation. This differentiation and then encounter between the senses in the *time-image* does not occur because senses are discretely separated and atomized, but rather through the pushing of thresholds to absolute capacity, a sort of "counterpoint," to use Whitney's phrase, where sense deterritorialized is reorganized.

Deleuze appears to seek a moment of differentiation and reintegration in the midst of new proximities between sentiments and technologies. Difference, here, an encounter with the "unthought," comes only in delay, and only through the radical incongruity but simultaneous reassembling between different forms of image and sense. These encounters with the "unthought" that produce "thought" emerge from the internal multiplications of the system, the generative feedback loops of the now cybernetically infected cinema.[74] Perhaps this is the "seriality" Youngblood deploys to contemplate the new computer animation? This incongruity sees itself perhaps mirrored in the splits between an identificatory image and the circulative channels embedded in our contemporary infrastructures of sense. But for Deleuze it is only the rare practice that can unleash this possibility to create an "image" of thought. What Deleuze points out to us is that the organization of affect into encounter is the central dilemma for both politics and philosophy; perhaps it always was.

In 1959, the new biopolitics of reconfiguring population through data inundation and the reformulation of perception had already begun to emerge. Our historical vantage point allows us to understand the heterogeneity of that moment, and the possibility that these new techniques could have been (but largely were not) attached to different historical and spatial configurations. At that moment, a brief interlude of détente in the Cold War, the affective field wavered between global identification, circulative consumption, and individual identity, between species being and hyperpersonalization. Older histories of vision and documentation supported the emerging computational and algorithmic visions. The image itself continued to multiply—computational, representational, neural. This image had not yet formalized into a memory capable of attacking the present, as in the image of torture. In the example of Abu Gharib images of war are replayed and reenacted not as a form of working through but as a form of operant conditioning and integration into the media network. There is nothing alien and no encounter at the interface in the instance of torture, only the seamless redirection of attention into the circuits of war and capital.

But history demonstrates that this state was not, and is not, inevitable. It is the work of critique in the present to explore and remember these instabilities and contests over how perception and cognition would be organized, integrated and modulated through the built environment, and used for political and economic purposes. In contemplating this history, we realize that the vast wavering space of that Moscow pavilion was a moment of potential, and missed, encounters.

Perhaps it is worth recalling George Miller's initial comments about delusion and paranoia. Miller also wondered about the surprising changes that happen from inside of circuits. He contemplated a thought game. Perhaps, instead of acting like Senator McCarthy and assuming that that which emerges from within the mind is an external and persecuting force, if we began to investigate the way our own networks haunt, trouble, and delude us, then perhaps we would end with something different from what we started with. Patterns can emerge that are not necessarily static or eternal but are arbitrary and chancy. These patterns may teach us something about how our own systems work and can be enhanced without necessarily falling to a reductive history. This is what the frog's eyes and the magical and mystical number seven, a biblical and mythological number, can foretell—a future that we may approach and anticipate without fully knowing it. McCulloch's "ignorance" that produces knowledge perhaps? Miller closes his essay:

> and finally, what about the magical number seven? What about the seven wonders of the world, the seven seas, the seven deadly sins, the seven daughters of Atlas in the Pleiades, the seven ages of man, the seven levels of hell, the seven primary colors, the seven notes of the musical scale, and the seven days of the week? What about the seven-point rating scale, the seven categories for absolute judgment, the seven objects in the span

of attention, and the seven digits in the span of immediate memory? For the present I propose to withhold judgment. Perhaps there is something deep and profound behind all these sevens, something just calling out for us to discover it. But I suspect that it is only a pernicious, Pythagorean coincidence.[75]

There are no enemies. But these older histories also remind us that there are no fully built, or entirely familiar and known, machines and minds. But our paranoia has taught us that we can recode our memories, transform our cognition, embrace chance without nihilism. Cybernetic eyes and minds have many different forms of time and truth simultaneously operating—mystical and predictive, archival and historical; the trick is to use them to push the system into the future, and to "withhold judgment" on the present. This perhaps was the ethical lesson we could have learned, and continue to forget, in 1959—but one of the many violent political tragedies emerging from the Cold War confrontation.

If today we think we can know our minds and each other because our brains work like genetically programmable computers and our environments have been automated to modulate attention, we may wish to remember that there was a time when people considered machines, eyes, and minds to be far less knowable and far more capable. A moment when the very logic of paranoia or embodiment was subverted and rechanneled into another discourse of capacity, and where the internal differentiations of the image were available for multiple uses. Humanitarianism, pure affect, nonhuman vision—all these multiple forms of imagining the world constantly erupt out of the translations between our myriad databanks and interfaces.

The ethics, and politics, of that transformation are still being negotiated. This is the nature of politics now, negotiated at the level of attention and nervous networks, structured into our architectures of perception and affect; feedback providing the opening to chance and the danger of repetition without difference. Forget me not—both a promise to rethink difference, life, and our relations to each other and a warning that we will not.

Conclusion

I opened with an image of a city: the fabled greenfield city of Songdo, a lush, verdant, and simultaneously sterile space, part of a new network of territories that crisscross the globe. These images (fig. C.1) precisely mirror the marketing materials published by Cisco and the South Korean government. The advertising uncannily resembles or anticipates reality. In these advertisements, these cities are imagined in any number of locations around the world. They are to be rolled out as the infrastructures for economic expansion and sustainable life. The South Korean government and Cisco already have projects across Asia in Vietnam, Malaysia, China, and India.

What is so curious about all the images in the marketing for smart cities is their resistance to being seen—their raw, indifferentiable amorphousness. What is noticeable is the pure aesthetics of computation. Sleek glass. Pure transparency. The ubiquity of nonstructures. This is the territory of nonarchitecture. The location of the city, the site, is unimportant. It is hard to know precisely what is being marketed, except some concept of greenness and the fluidity of life itself as rendered by a computer. The only thing visible is the latest software rendering programs' conventions. That is perhaps what these images make us "see." It is clear that one does not need the distinctive hand of an auteur in these spaces. What is even more curious in the standard visions of these spaces is that engineers confess that they have little interest or concern with the spatial form. The conduits and telepresence services, the satellite networks, and integrated data analytics instruments that are purveyed by Cisco can be applied to any city.

These structures, lacking an image or a definitive architecture, actualize something that the USIA installations of the late 1950s already suggest—the collapse of the visual field, in the name of an attentive and affective global information-consumer space. Returning to these environments "enclosed by images," in the words of the architectural historian Beatriz Colomina, I am forced to conclude that they are anything but.[1] These spaces are occupied only latently by what might be identified as an image *of* anything; rather, what they actually do is attenuate the nervous system. These structures en-

FIG. C.1__**The New Utopia**.
"Global Business Utopia," Songdo,
Incheon Free Economic Zone, South
Korea. Photo credits, 1-r: author,
July 4, 2012; Jesse LeCavalier
(2012); author, September 1, 2013.

courage the proliferation of ubiquitous computing infrastructure to the point
where vision loses any function in producing identification or recognition between or within subjects. This transformation in population and subjectivity
is about the shift from an observing subject to a "user." A user is not consolidated in identity but rather operates through the logic of units of attention,
and bandwidth, consisting of roving populations of action in the network.
"User" is the discourse for a collapse of the individual subject as a spectator,
listener, or reader. No longer can there be an individuated subject distracted
by an overwhelming environment. Rather, there is only a sensorium imagined
as infinitely extendable. As with so many other of the techniques of biopolitics, what was once supplemental—the limits of representation, legibility, or
perception—has now become the imagined substrate for technical intensification.[2]

This resistance to imageability is symptomatic of a historical condition no
longer defined by the image of the commodity. Where once capital was accumulated to such a degree that "it becomes an image," in our present, perhaps,
as the architectural critic Anthony Vidler notes, "spectacle is an image accumulated to such a degree it becomes capital."[3] And in being so, it ceases to
provide a concealing spectacle at all but folds into the logic of the interface, as
demonstrated through this book.

The literary critic Fredric Jameson labels this nonimage infrastructure
"nonutopian" and calls on Rem Koolhaas's term "junkspace." In these junkspace global territories, there is no longer any architecture; all elements in the
environment are perfectly convertible and interchangeable. It is the world of
the shopping centers in their many forms, a world where images of buildings,

or even architectural renderings and plans, say nothing about the space or the systems they serve. It is a nonarchitectural space lacking any authorship, with the exception of a few flagship buildings that only serve as vestigial odes to a now long gone concept of space and value.[4]

What concerns both Jameson and Vidler is that this nonimageability perpetuates a violence against the future. Returning to the nonimage of the ubiquitous computing landscape, its impermeability to visibility forecloses the very possibilities within these structures to produce radical technological environments. This was the same concern first articulated by Derrida in *Archive Fever*, involving the automation of recording to the point of death.

These "future cities," unsurprisingly, lack for Jameson a utopian image. While there is much to be maligned in the modern ideal of utopia, for Jameson, "utopia" is also the only name for the imagination of a time and future type of life that does not exist in the present. What both Vidler and Jameson seem to argue in the early twenty-first century is a strange, sudden desire for an image of thought to return. A sort of revalorization of being able to "envision" and representationally apprehend something beyond or outside of us; to be offered some visualization that allows action, not merely reaction.

But this lack of imageability, to also use Kevin Lynch's term, is not only the province of massive corporations and technologists. It is also the result of contemporary critiques of computation. If the latest real-estate development and frontier of "soft" computing is failing in its mission to produce alternate futures (which perhaps may only be expected, considering the purpose of real-estate speculation), then the critics are doing no better.

It is, in fact, an understatement to say that few projects provoke as much debate among urban planners, architecture circles, and urban sociologists as Songdo and its sister "smart" cities such as Masdar, in Abu Dhabi, another superexpensive, free-trade zone, sustainable, space-station-looking complex.[5] Prominent sociologists such as Richard Sennett argue that these cities are "stupefying," as a result of their corporate parentage, speed of development, and homogeneity.[6] These sentiments are repeated daily by the designers and architects I work with at New School Parsons School of Design in our research to develop alternative models for visualizing and thinking about urban digital infrastructures. My colleagues almost universally express disdain for these large, "corporate," quantitatively driven, nonqualitative sensorial spaces. Arguments regularly circulate about how retrograde the assessment and measurement instrumentation of corporations like IBM are, how the data structures deployed to monitor and organize urban traffic, environment, and population are overly taxonomical, and how unsophisticated is the approach of these large technology corporations like SAP, IBM, Cisco, and Siemens to data gathering, human experience, and community. Even such data-rich moguls as Google are guilty of similar sins (even as all anyone does is build apps to use their system). Counter to such monoliths of autonomous data gathering, ethnographers and designers espouse the merit of ethnographic richness, unstructured data, and some amorphous fantasy of flexibility in databases. No one seems to doubt the need for data.

The reverse side of the coin is a faith in the power of networks to facilitate democratic social action. Thus major figures such as the prominent political economist Saskia Sassen speak of "open source urbanism," critiquing the Masdars and Songdos of the earth, while at the same time demanding an intensification of technology in the urban environment. She writes that technology must be "urbanized" (an unclear statement, since technology and cities, or cities as technology, have a long history) and must be put at the hands of users, open to be hacked and otherwise propagated through the ever-popular do-it-yourself movement.[7] According to another apostle of the urban digital future, the Media Lab professor Carlo Ratti, who heads MIT's Senseable Cities Lab, these "smart" city developments are morally repulsive because they are the products of "top-down" planning.[8] He compares such planning to Twitter platforms and other social networks that are "bottom-up." As the website of his lab proclaims, "the real-time city is here,"[9] and emerging sensory technologies are making new forms of life possible. For Ratti, the "Arab Spring" and other similar political moments of the recent past have been enabled by social networking technology, reductively understood as telecommunication and Internet infrastructure. His lab's greatest successes have been such projects as putting radio tags on trash to create visualizations of where our refuse moves. Refuse itself is now a source of knowledge. The very supplement incorporated into the economy of information. The lab thus encourages the ongoing exploration of the place of environmental sensors and lighting to make visible phenomena like water pollution, and assumes the virtues of embedding sensory technology into everyday devices to allow the regular monitoring of human activity, and of course, to encourage social interaction.

This is not all negative. There is little doubt that planners need to know where the trash is going. The work is openly stated to be in the legacy of Gyorgy Kepes and Kevin Lynch, although it lacks the sense of time, density, or even sadness that marked those earlier projects. And often the visualizations produced are very beautiful. The brightly lit sensors floating in environments, the vacillating maps of transport systems modulating by the day, week, month, the fluctuations of pollutants in the air—these maps all produce sentiments, and visualizations, even if they do not always produce ways to comprehend different technical futures or diagnose structural concerns. But most of these projects are dedicated to such concerns as linking local food producers to markets, visualizing commuter behavior, and other efforts to enhance "efficiency" understood as bandwidth—amount of information traveling in a channel in specified times.[10]

The irony of these discourses is that the global territory envisioned by Cisco and the localized technological demos envisioned by Ratti's Senseable City project and the many critics of large-scale corporate projects are largely the same. It would appear that we live in a culture of disavowal. Feedback as a democratic virtue makes a critical stance difficult. The critique of the smart city and the suggested resolution of the problem both largely foreclose the actual question of technology, assuming the necessity, even the demand, for attentive, even nervous stimulation as an answer to social questions. As Ricky Burdett, the head of Cities Program at the London School of Economics and the lead planner for the city of London for the 2013 Olympics put it, quoting the famous British architect Cedric Price, a member of the famous 1960s avant-garde architectural group Archigram: "technology is the answer. But what is the question?"[11]

What Burdett then proceeds to neglect is the other part of Archigram's project, which was to generate questions. To experiment with forms. To build impossible fantasies of a world where technical rationality was taken to its extremely logical endpoint. Instead Burdett proceeds to assume the necessity of technology without asking: how do we want to live? We are back to the discourse of "smart planets" and sustainable technical management. The future city, for Burdett and any number of other prominent urban planners in our present, is correlated with democracy, open source, and other similarly black-boxed terms that, ironically, all base themselves on the same technology, techniques, and corporations that are rolling out the much-loathed greenfield smart cities.[12]

What is solidly missing from all these fantastical debates about smart cities and smart urbanism, and all the smartness that will soon envelop us in the cocoon of ubiquitous computing, is any sense of historical contingency or possibility. It is precisely the imperatives for a future founded in circulating data that make it so difficult to produce alternative conceptions of life. But this fetish in the eternal revolution supplied by bandwidth and interactivity is also illusory. Every age, Deleuze suggests in all his work, has its own "image" of thought, which is another way of saying that every age has its forms of perception and attention, its modes of visuality and sensation.[13] The temporariness and fleetingness of our modes of perception should be encouraging.

What then would it mean to really take the incredible reformulations of territory and population that such developments as Songdo denote to their zenith? What does an experiment in possible futures look like? What kind of histories would we wish to write that might both situate and reimagine our present?

In 1960, the prominent architectural critic Reyner Banham already anticipated this question of how we might create utopian images under the conditions of computational technology. As the architectural theorist Anthony Vidler notes, Banham called for a new program at the time to ask what computation and architecture might have to say to one another, if anything. Banham called for "anti-building": a "zone of total probability, in which the possibility of participating in practically everything could be caused to exist"; to produce this zone of probabilities, the mandates of operations research and reductive cybernetics would have to be disbanded in the interest of a new "image" of the city. To provide such images, Banham turned to Archigram.[14]

In a series of journals, filled with fantastical, never built constructions, the group paraded prototypes and blueprints of sometimes hilarious, often touching, images of future possibilities. The group became famous through the production of a series of journals, titled *Archigram*, that they published throughout the 1960s in order to provide a space for students and junior architects in the field to showcase their ideas. Merging the term "telegram" with architecture, their purpose was to bring urgency to the demand for rethinking space in a technical world. Borrowing from such figures as Buckminster Fuller and the emerging drug and countercultures and, most important, engaging cybernetics, computing, and environmentalism, their work became the inspiration for many architects, from Philip Johnson to Rem Koolhaas.[15]

Sometimes patently absurd, the journals all resonated with a certain sadness for the present, a beauty of the unbuilt, and a creativity that could only come from the almost total embrace of a particular technical condition. Archigram had an image of the future without a clear philosophy guiding their program. It was a program produced through these experiments in "what if" . . . what if we are all consumers? What if everything is disposable? What if architecture is over? What if programmability is the dominant logic of space? What would it take to experiment in architecture?

What would it be like, Archigram asked, to be guided not by preconceived structures but through programs, to truly take the logic of computing to its extreme, to start with certain algorithmic rules and run them until they start to oscillate? The first issue of *Archigram* in 1960 opens with a poem by David Greene:

> The love is gone.
> The poetry in bricks is lost
> We want to drag into building some of the poetry of countdown,
>   orbital helmets, discord of mechanical body transportation
>   methods
> and leg walking
> Love gone . . .
>
> A new generation of architecture must arise
> With forms and spaces which seem to reject
> The precepts of "Modern" yet in fact
> Retain these precepts. WE HAVE CHOSEN TO
> BYPASS THE DECAYING BAUHAUS IMAGE
> WHICH IS AN INSULT TO FUNCTIONALISM.[16]

We have chosen, in other words, to take modernism to the limit. To truly embrace our technical condition, to enact critique not through supposition but through a practice. Their practice is not unlike Derrida's deconstruction, one of finding the supplements and excesses that generate new meanings and forms. Or perhaps it's more like Deleuze's idea of the fold, pushing the internal inconsistencies until systems fold into new forms.

Walking Cities, Plug-In Cities, Computer City, Underwater Cities, Instant Cities, Sin Centres, Fun Palaces (figs. c.2 and 3). Each imagined design took the condition of consumption, technology, and pushed it to the extreme. In

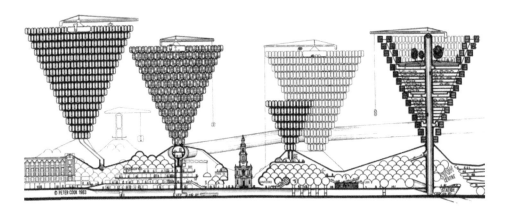

FIG. C.2__**Plug-In City**: "Within the big structure, almost anything can happen. This is in effect, the brief from which Plug-In city develops. . . . The big structure in Plug-In City is at the other end of the scale in that it incorporates lifts and services within the structure tubes. It controls the discipline of the whole city, but on a very large scale." Peter Cook, *Plug-In City, Housing for Charing Cross Road*. Archigram Archive © Archigram 1963.

Plug-In City, disposability was taken to its logical endpoint . . . an entire city full of throwaway, interchangeable parts. Peter Cook, one of the leaders of Archigram, described *The Plug-In City* as the result of a series of discussions that decided to embrace "expendable buildings: and it was then inevitable that we should investigate what happens if the whole urban environment can be programmed and structured for change."[17] As an extension of the Plug-in City project, a fellow member of the group, Denis Compton, also envisioned a "Computer City." These "cities" were planned for density, disposability, movement—different strata of the city were organized by the speeds at which things moved—interchangeability, and networking. Everything was assumed disposable, and it was assumed that only "urban, or architectural, or mechanical or human mechanisms thrive on being stirred together." Like Lego pieces, they can be built on top of, added to, plugged in to existing networks and cities and of course, moved around and reattached to other cities and networks. Another proposed design for a computer city "suggests a continual sensing of requirements throughout the city and, using the electronic summoning potential, makes the whole thing responsive on the day-to-day scale as well as on the year-to-year scale of the city structure."[18] The plans call for high-intensity spatial usage—everything is plug-and-play; office units, automated shopping centers, and rail links can be moved around and intermixed with one another according to feedback systems of control.

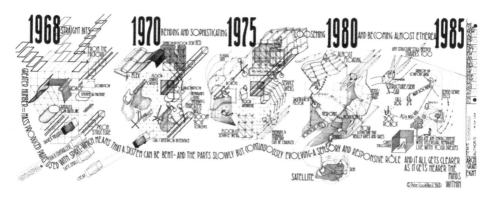

FIG. C.3__*Control and Choice, Metamorphosis,* by Peter Cook. "GREATER NUMBER=MASS PRODUCED PARTS USED WITH SPIRIT–WHICH MEANS THAT A SYSTEM CAN BE BENT–AND THE PARTS SLOWLY BUT CONTINUOUSLY EVOLVING–A SENSORY AND RESPONSIVE ROLE AND IT ALL GETS CLEARER AS IT GETS NEARER THE MINDS WITHIN"; *Archigram,* issue 8, 1968, 3. Archigram Archive © Archigram 1968.

According to Banham, these cities were the opposite of operations research and cybernetics, not because they disavowed the methods but because they provided images instead of banal operation manuals. "Archigram can't tell you for certain whether Plug-in City can be made to work, but it can tell you what it might look like."[19] This effort to produce an aesthetic that might drive technology, an aesthetic perhaps out of time with the speed of computing—sometimes slower, and often faster, than that of computing—was for Banham a form of encounter with a different form of thinking. It did not merge architecture with computing but asked what each might give (if anything) to the other, and established the differentiation between the two.

This radical possibility, of course, is not a politics of opposition. It should be noted that many Archigram members thanked such figures as Charles Eames as mentors, for making them think seriously about Disneyland, for opening them to play and computation. These projects, unlike other counter-cultural movements, had little counter within them. As the opening poem denotes, what was fantasized was taking functionalism to its logical extreme, to embrace the rationality of cybernetic systems until its sparkling underbelly was revealed. And, most important, to genuinely experiment with the new forms technical life could take.[20]

Archigram was not, of course, an ideal prototype to be emulated in our present without question. When buildings were built, they were sometimes banal; most were never constructed. And today so much of this ironic play is clearly assimilated into the contemporary landscape of consumption, marketing, and corporate architecture. Finally, when it comes to encounter with different classes or people from one's own, there is none in these diagrams and journals. But that is not why I have offered these examples. Rather, it is to remember that Archigram did *not* succeed. This project has yet to be realized, and therefore leaves something to our imaginations. For all its performativity, for all of Archigram's insistence on the event of architecture, these images are powerful because of their untimeliness—their obsolescence, even in their own day. These were not the architects pioneering the latest computer-aided design programs at MIT or IBM.

Rather, it is about how these cutouts and montages of images from popular culture and archives of different architectural imaginings are strung together in handmade journals. They are seemingly quaint, today, to us, because it is so easy to create the same effects with Adobe suites. But the radicality of these projects exists in reminding us, in the present, of a moment when computing was something to be encountered, a radical externality that would become the very heart of architectural and design practice and that prompted new "images" to proliferate. As the architectural historian Simon Sadler writes, "what remains compelling about Archigram's work for progressive architects was the possibility of an architecture without architecture, organizing experience without incarcerating it, . . . Archigram's work . . . has resonated too with the dream of escaping the conventions of space, as it is organized around the clutches of the market, the family, the state, and other hegemonies."[21] These leftover images are, therefore, not documents of a reality but traces of a moment when the encounter with the radical unknowability of cybernetic systems prompted a program of questioning and reinvention. There is a shock here of encounter with this archive of possibilities of what a computational future would look like. What might appear ludicrous, funny, amusing, perhaps merely aesthetic was also a genuine effort to envision a world that was not yet known.

Archigram's project asks critical questions, therefore, about our response and analysis of the present. One wonders: what might it mean to find in the possibility of a Songdo an incentive to experiment in the future? Balancing emergence and learning, organization and structure, some cyberneticians enjoyed reflexively performing their logic until the miscalculation, error, and unpredictability in the system became visible. What spectacles are still capable of such action today? The circuits and diagrams of the late 1950s and 1960s enacted deconstructive visualizations of systemic failures. In the debates between policy analysts, in the reformulations of subjectivity, race, and territory were articulated contesting imaginaries of perhaps more plural or perhaps ever more homogenized systems. Cybernetics was many things to many people. I have attempted to show the perversity of our contemporary situation.

The two decades after World War II saw a radical reconfiguration of vision, observation, and cognition that continues to inform our contemporary ideas of interactivity and the interface. The reformulation of the observer as simultaneously networked, decentered, and multiplied came adjoined with new notions of environment and interactivity that are the infrastructures for our contemporary models of economy, technical, and aesthetic practice. What I also hope to have demonstrated is that even as the body of the individual subject ceased to be the central site of intervention for knowledge production or the extraction of value, new forms of life emerged, and new types of population were produced. These forms are multiple. On one hand there is the surprising reformulation and reterritorialization of sense into older formations of emotion and identity, but now bereft of any modern sense of an empowered body or agent. On the other hand there is the centrality of population as a defining site of intervention and interest. Like the markets that Cisco envisions for telepresence services, population is no longer linked to subjectivity or discrete bodies but is composed of agglomerations of nervous stimulation; compartmentalized units of an individual's attentive, even nervous, energy and credit.

From these divergent paths, there are also new possibilities for different forms of affiliation, even care, to develop. The images of gardens, Archigram cities, cinematic machines all emerge here, as in Bateson's porpoise's performance, as untimely possibilities. These images are not in any way representative directly of a condition or subjectivity. Nor are they the images of realized utopias or successful projects. Rather, these performances offer a challenge to continue a never realized project of engaging differently with our sociotechnical networks. Of being forced to thought through encounter with these systems beyond and outside of direct representation. Bridging Deleuze's hope for an encounter unencumbered by disciplinary strictures of thought; free of the weight of historical determinism. An encounter with things that have not yet been thought.

Perhaps it is fitting to end with a little ditty written by Archigram member David Greene, in a little essay and project called the *Gardener's notebook*. The notebook is full of technical things made to look like natural things. He wanted to build outlets for a mobile life, so people could plug in to the electrical and information grids anywhere they went. So he envisioned and drew fake logs and rocks that would grow algae and moss, but serve as outlets:

For the present we have to wait, until the steel and concrete mausoleums of our cities, villages, towns, etc. decay and the suburbs bloom and flourish. They in turn will die and the world will perhaps again be a garden. And that perhaps is the dream, and we should all be busy persuading ourselves not to build but to prepare for the invisible networks in the air there. . . . Read a bit of this incredible poem—it's all watched over by Machines of Loving Grace

I like to think
(right now please!)
of a cybernetic forest
filled with pines and electronica
where deer stroll peacefully
past computers
as if they were flowers
with spinning blossoms . . .[22]

Computer blossoms resonant of the spinning mechanisms of Whitney's cinema, the vertigo-inspiring sensations of the Eames installations, the shiny beauty of data flickering across our interfaces, serving as beacons and warnings of a future we have yet to envision—these are the gardens we have yet to cultivate.

Epilogue

History is never linear. So its study seemingly welcomes repetition. Cybernetics, in particular, drives this historical imperative; inviting feedback and repetition until one's own thinking starts to oscillate. Scales are ruptured. Different "images" emerge.

Epilogues offer such possibilities. They are heterotopic spaces, not really part of the book, not yet a new program or project. Perhaps spaces that can produce other visions of our present. So I want to return one last time to a theme recurring in this text—the binding and unbinding of history and meaning to territory.

Of all the stories I came across when attempting to do a history of cybernetics and its many influences on the cultural landscape, the one that remains most poignant for being both exemplary of a post–World War II transformation in aesthetic sensibility and irreducibly specific to historical accidents and conditions is that of two gardens both designed by one of the preeminent sculptors and environmental artists of the time—Isamu Noguchi. Perhaps these are the fabled cybernetic gardens presupposed by Archigram. Or maybe they offer something less ironic, less dialectic, less humorous but more empathetic. A little less boys with their toys. Far less familiar to those of us studying computing or cybernetics, for sure. If there is one thing that is certain it is that these gardens evoke a different set of emotions than the cute and churlish diagrams of agitprop architects.

The first is a sculpture garden built in 1962–1963 in the midst of the new corporate headquarters of IBM in Armonk, New York, a far western suburb of New York City. The second is the sculpture garden that was built, at roughly the same time, to house the new collection of the recently opened Israel Museum in Jerusalem. These rock gardens, like all gardens, are fertile in their function, and they tell a tale of transformations in aesthetics, economy, and politics. These gardens unlike the magical palaces and instant cities of Archigram, were actually built, but perhaps never realized, and so they offer material sites of excavation into alternative futures of the past. They offer, perhaps, in Jameson's account, visions of the future. They are also deeply personal, and deeply flawed. But they exist, and therefore provoke a challenge to how we write history itself, and how we negotiate the present.

Their creator, Isamu Noguchi, would not immediately appear a likely candidate to build a national or for that matter corporate monument. He had an ambiguous relationship to nation, materialism, and identity. Born of an American mother and a Japanese father who was a prominent poet, he had long struggled to fit in within any national space. When Isamu was an adolescent his mother returned to Japan with him, but his father spurned him, and he had found himself an alien outsider in pre–World War II Japan. On his return to the United States and completion of high school, he briefly attended Columbia University in New York for premedical studies, but he left to begin arts training. In 1927 he received a Guggenheim Fellowship to travel to Paris, studying with the famous sculptor Constantin Brancusi. He then returned to New York. He briefly returned to Japan in the early 1930s, where he was again not welcomed by his father, but his uncle showed him kindness, and he began to introduce himself to the practices of Japanese gardening and pottery, which he had previously spurned. He also traveled to Beijing and studied Chinese line drawing.

Returning to the United States in the 1930s, like many artists he became deeply invested in using sculpture for social benefit, and he worked under the influence of politically active artists such as the Mexican muralist José Clemente Orozco, who was producing a major mural in Mexico City, *History Mexico*, which merged the dramatic scenes of racial and labor oppression with a technological hope for science. The 1930s also marked Noguchi's entry into dance, starting a lifelong collaboration with Martha Graham in set design. Perhaps most prominently, Noguchi created a 1930s sculpture, *Death (Lynched Figure)*, which was one of his few expressly figurative and violent denunciations of racial violence for a show on the theme. The piece was denounced by the New York art establishment, and lost him Works Progress Administration contracts during the Great Depression.[1]

The war, however, shaped a turning point in Noguchi's career. Having considered himself American, it was the internment of the Japanese that suddenly prompted a consciousness of his alien subjectivity. Denouncing the forced internment of West Coast Americans of Japanese descent, Noguchi started the Nisei Writers and Artists Mobilization for Democracy, and volunteered, as a political act, to enter the Colorado River Relocation Camp in Poston, Arizona, in 1942. Having entered, he was not allowed to leave for numerous months, and the camp disillusioned him mightily about both nation and human spirit.

In his frustration and despair he wrote a never published essay, "I Become a Nisei (Japanese American)."[2]

Noguchi opens this essay: "to be hybrid anticipates the future. This is America, the nation of all nationalities, . . . For us to fall into the Fascist line of race bigotry is to defeat our unique personality and strength."[3] Before the war, he related, he had never come to see himself as Japanese. He was plagued by a "haunting sense of unreality" as events unfolded, forced to take an identity he did not want.[4] Only after the war had started in 1940 had he first heard the name "Nisei," when traveling to Hawaii. And only after Pearl Harbor had he come to actively associate himself with this term.

His affiliation with this name, however, was not as a matter of separation from other Americans, or identification with a stable identity or group. Rather, Noguchi viewed the adoption of this term as a democratic act that recognized the predicament of those who have no clear identity or territory. This name, wrongfully applied by Americans to those of Japanese genealogy, was appropriated by Noguchi to stand for precisely that which does not have a set origin.

He related: "how strange were my reactions on entering camp. Suddenly I became aware of a color line I had never known before. . . . Along with my freedom I seemed to have lost any possibility of equal friendship. I became embarrassed in their [white administrators] presence." This embarrassment of the self, and seeming sense of inferiority, did not, however, make Noguchi automatically identify with his fellow Nisei; "their background seemed too different, or does imprisonment make also a prison of one's mind?"[5] For Noguchi also recalled that one of the things marking the Nisei was their "Americanism," their discomfort with being around so many Japanese faces, their lingual isolation from their parents. Noguchi, too, spoke little or no Japanese and had a fragmented and bi-polar relationship to his and his mother's relationship to Japan, and to Japanese history.

Alienated from identity but forced into a closed community, Noguchi turned to his art and design in an effort to reconfigure the overdetermined architecture and psychology of the camp. This effort ended in despair and failure. There would be no better conditions for life in the camps. Futilely, but repeatedly, Noguchi designed playgrounds, gardens, and architectural improvements to camp buildings in a seemingly desperate set of acts to maintain an imaginary about the human capacity to exceed historical conditions. These designs were, of course, never executed by camp authorities, but Noguchi continued to produce them.

Recollecting this effort, he wrote that he designed these imaginary spaces in the hope of preserving "the arts of peace." Perhaps these designs for better condition and this condemnation of racism might be part of the global war effort? He hoped that "we will build a seed reservoir for the future . . . teach Asia the meaning of democracy." Such hopes were in vain. The artist recalled that "moments of elation [were] only to be defeated by the poverty of our actual condition."[6] All the efforts of the communities in the camps to produce self-governing institutions, to demonstrate the desire for Japanese Americans to participate in democracy and build community, Noguchi writes, were ultimately thwarted by the oppression, poverty, and difficult conditions of the camps. While seeking the benefits that might be learned from these small hopes of making do with less, ultimately the lessons of Poston did not hearten Noguchi.[7] They were also prophetic; Noguchi would never build a playground, although his archive is littered with imaginative designs of what he labeled "adventure" playgrounds. His fields of joy and potential for children were thwarted time and again in the name of concerns about security and safety (of children). Although, usually other reasons were to blame for the project failures. In the case of his plan for Riverside Park in Manhattan, a collaboration with Louis Kahn, to be built in 1967–1968, concerns over race and an emerging fiscal crisis destroyed the project.[8]

Leaving the camp after six months, he returned to New York, where he turned to abstraction. Arguably he turned away from politics, but this is disputable; perhaps he had a new concept of abstraction, as a mode of expression, outside of identity or nation, an aesthetics that might serve global society and that might articulate the unnamable and unspeakable personal and human catastrophe wrought by the war. His immediate works were landscapes imagined to be produced by aerial bombings and final sculptures that might leave remnants visible from space, like the Great Wall of China, so that some other species and civilization might know we were here.

He created an image of a face in the desert (fig. E.1), an imaginary picture to be seen from Mars after humanity had destroyed itself with atomic weapons. His sculptures also turned from the social murals of the 1930s inward, to the personal and psychic. While abstract, the forms appear to hang in space, and are violently cut apart. The sculptures take the form of figures whose innards are being punctured by bones and shards of foreign materials; seemingly barely balanced figures that hang out of place, testifying, perhaps, to his own sense of dislocation.[9]

Increasingly individuated while recognizing the global, if not cosmic, scale of conflict; Noguchi transformed this crisis in biography into a concern with environments and into the manipulation of scale in a manner not unanalogous to that of his colleagues (he knew Eames and Kepes) in other design

FIG. E.1__*Model for Sculpture to be Seen from Mars* (originally titled "Monument to Man"), by Isamu Noguchi (1947). Courtesy of Noguchi Museum and Sculpture Garden. © 2013 The Isamu Noguchi Foundation and Garden Museum, New York/Artists Rights Society (ARS), New York.

FIG. E.2__Bridges in Hiroshima, Japan, by Isamu Noguchi. Ikiru—*To Live* (1953). Courtesy of Noguchi Museum and Sculpture Garden. © 2013 The Isamu Noguchi Foundation and Garden Museum, New York/Artists Rights Society (ARS), New York; Shinu—*To Die* (1953). Image: author, August 2010.

fields (see chapter 2). It was out of this context that Noguchi turned to producing actual gardens and landscapes, embracing a plethora of corporate and institutional benefactors that came to prominence after the war and were ready to pay for such designs.

At this moment, as older dialectical political contests of identity, race, and nation were folded into emerging information organizations and postcolonial and postwar territories, we find a curious merger of contesting fantasies about what the future of space might be and what its relationship to history would become. Noguchi, critical of identity politics, always concerned with social welfare, and originally socialist in proclivity, would find his major benefactors and greatest site of creativity and ingenuity in the landscapes of the new institutions of the postwar world—corporations, the UN, national entities and governments emerging from colonialism and the War.

His first major effort at an environmental scale was to be in Hiroshima, Japan. He built two bridges (fig. E.2) (that are still there), one to death and one to life, and initially designed a memorial, one that faintly resembles the existing structure: an arch with a large underground space beneath it, but much heavier and more austere. "It was," Noguchi wrote in his memoires, "to be a mass of black granite, glowing at the base from a light from beyond and below." Noguchi described it as a concentration of spiritual energies, a space where the future (death, return to the earth, the first atomic dead) would meet the future, for it would also be a "womb" for those not yet born. After months of work it was rejected. The excuse from the authorities was that it was "too abstract." Noguchi sadly remembered that perhaps "it was just density not to get the point—a lack of delicacy . . . where nothing can be said, sculpture may indeed not be needed, not wanted. However abstract, in sculpture is a meaning. . . . Though humanity protests, officially, it seems, no lessons may be learned, no warnings given."[10] Perhaps, he wrote in a letter to a friend, it was just too "shocking," heavy and ugly (figs. E.3 and E.4).[11] For Noguchi abstraction had never been self-referential or disembodied but was always a form of expression that might take place when other forms failed. The rejection of his sculpture was a rejection, perhaps, of the possibility of producing imaginaries even in the face of historical disasters. Even if nothing can be said, history must be contended with.

Perhaps it was too soon to have an American, even a partially Japanese one, build such monuments. Noguchi thought perhaps his memorial was too brutal and sad, too weighty and monumental, to act as a postwar restorative. The bridges, however, still stand; strange monuments, almost unremarkable unless one approaches them very closely to see that their surfaces have been produced with steel emulating the burning of the bomb; faint reminders of both technical catastrophe and reconstruction.

Throughout his career Noguchi always imparted a faint sadness of the future in his work. But he believed in this possibility of a future, perhaps the virtual, even if his optimism about its emergence was not infinite. In his work it is often a spare and stark future.

Not long after his failure in Japan, however, he received a major commission to build the UNESCO gardens (1956) in Paris, his first success and the start of a career as one of the more significant voices in postwar landscape and sculpture art. Soon thereafter he received the commissions for the two gardens for IBM and the Israel Museum.

Beautifying the Machine

The first garden tells a story of the aesthetics of communication and computing, one that refracts the story of beautiful data and aesthetic informational infrastructures told throughout this book. In 1964 Noguchi was approached by James Watson Jr., the chairman of IBM, to design a garden for the new corporate headquarters (fig. E.5). Watson recalls personally calling the sculptor

FIG. E.3_*Memorial to the Dead, Hiroshima*, by Isamu Noguchi (1952). Composite photograph of plaster model. Courtesy of Noguchi Museum and Sculpture Garden. © 2013 The Isamu Noguchi Foundation and Garden Museum, New York/Artists Rights Society (ARS), New York.

FIG. E.4_*Memorial to the Dead, Hiroshima*, model, plaster, by Isamu Noguchi (1952). Courtesy of Noguchi Museum and Sculpture Garden. © 2013 The Isamu Noguchi Foundation and Garden Museum, New York/Artists Rights Society (ARS), New York.

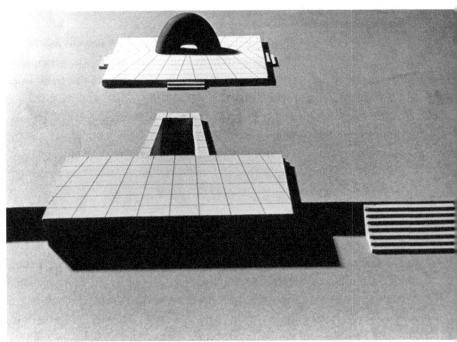

in Jerusalem to ensure that he would come and create this space for the company, unready to accept that the famous artist might deny one of the world's foremost organizations in power, capital, and influence.[12]

Watson's personal attention to the building could more appropriately be classified as an organizational obsession. As discussed earlier, one of his abiding legacies and most explicit programs was the integration of a cohesive design vision as a corporate strategy. The building and the garden were parts of a radical reconfiguration of IBM itself as a corporation.[13]

If form follows function, to cite the famous Bauhaus adage, then the low, rectangular office buildings, with their gleaming courtyards, light-filled walkways, and self-enclosed habitus speak volumes about the intended function of the organization housed within. The headquarters, built in 1963 by the firm Skidmore, Owings and Merrill, had the low-lying block structure we now identify with so many suburban office parks. The headquarters was a quadrant, whose entirety could not be visualized from the outside approach; the size, scale, and rectangular structure made it impossible.

As a structure, the building is focused on its own interior, which is where the best offices face. In this interior was an expansive (and expensive) garden, crisscrossed by glass-encased walkways, so that human movement in the building was highly visible and illuminated by the surrounding offices.

This internal courtyard broadcast to those within the building an aesthetics of circulation and internal environmental control that doubled the machine rooms that housed the computers, which had to be kept inside the building, insulated from real light, in the interest of keeping these vast machines cool and running. Those computer rooms, however, like the transparent courtyards and walkways, were ringed with glass, transparent to those within the building, even if the apparatus linking this building to a global network of operations was opaque to the outside world. This was a building produced through a contrary interplay of transparent surfaces and self-referential optics focused on circulation and communication—between departments, people, and machines.

In the midst of this redesign of the headquarters to house this emerging powerhouse of computer technologies, Noguchi planted his gardens— gardens that refract a world of self-referential and gleaming data and reflexive and responsive organizations.

Composed of two courtyards, the North garden, in Noguchi's words was to face "science and mankind's future," and the South garden was to "represent mankind's past."[14] The past was an ode to the geological landscape of the site, consisting of the blasted and fractured bedrock from the building site's

FIG. E.5__*Garden of the Future*, IBM Armonk, by Isamu Noguchi (1964). Photo by
Jon Naar. Courtesy of Noguchi Museum and Sculpture Garden. © 2013 The Isamu
Noguchi Foundation and Garden Museum, New York/Artists Rights Society (ARS),
New York.

parking lot and composed of a small pool in a condensed landscape of grass and raked gravel, a small pathway, and a number of large rocks jutting out of the sculpted lawn in an idiosyncratic manner, reminding the viewer of the remains of some archeological excavation or some violent geological event, their roughness and lack of hew offering a disjunction with the manicuring and sculpting of the gravel and grass. Noguchi labeled these stones "ugly," but useful. Apparently these pieces of humanly disturbed earth, thrust out of place from the parking lot and put together with small trees and grass, would present "nature."[15]

For the future Noguchi had another vision. A large black dome, off-center, "emerges from the earth to explore the universe." Dotted with circuitous lines, it has inscriptions taken from John von Neumann and Herman Goldstine's book *Planning and Coding*, showing circuits of memory and reverberation; let us imagine that they are the faint traces of the first efforts to imagine a stored program computer and an information society into being. Noguchi sought diagrams that might "present the scientific frontiers of the world" and "keep up with changing concepts of nature."[16] This new nature also included diagrams of nuclear formations and stellar constellations. Near this dome is a pyramid dissected by geometrical lines, and opposite is a dial-like marble semicircle inscribed with more scientific formulas. Scaling from the smallest particles to the circuits of networks to space, Noguchi's garden expresses an optic of repeating diagrams and algorithmic space.

This garden was also austere, embodying an aesthetic of abstraction. Noguchi did not even want trees planted (trees were added, but only under pressure from the architect, Gordon Bunshaft of Skidmore, Owings and Merrill, and at the behest of his corporate clients, who felt a somewhat warmer environment might be necessary). Instead Noguchi sought to maintain the future in the space of diagrams, circuitry, and formulas, which were now both abstract and stunningly materially embedded into the sculptural material of marble, basalt, and bronze. His attitude to medium seemingly materialized in the diagrams that decorate his basalt relief. As in communication theory, the medium should speak for itself. If Noguchi resisted the trees, it was not because he disliked organicism but rather that he embraced the liveliness of diagrams—their capacity to evoke sentiment and affect without further explanation or adornment by other artifacts. These algorithmic and informatic patterns were beautiful in his estimation . . . and generative. In this garden the future emerged directly from the stark interplay of shapes with the building—from the organization of forms and the structure of the space.

Like the communication theories and computational diagrams Noguchi mimed, this was also a space of predictive time emerging from the recombination of many data fields. He merged the iconography of computational and scientific diagrams with forms he adopted from his visits to Japan and his study of Zen gardens, and his travels to India and the astronomical observatories of Mogul kings; particularly the eighteenth-century observatory created by Maharajah Jai Singh in Jaipur.

This interiority that produces the future out of examining a self-contained past might speak to an entire new relationship to temporality and history encoded through space viewed as an interface—a passage point between times, no matter how short. The garden might serve as a mirror world refracting back the material aesthetic of abstraction that substantiates our belief in information as plentiful, beautiful, and virtuous. This is a mirror world made more poignant by the fact that the gardens at IBM were not to truly be used; or at least the documentation of the garden is always bereft of humans. While there is a walkway, there are no benches. Noguchi treated these landscapes as "sculptures" perhaps demanding some new form of interaction with viewers: not one of casually strolling but perhaps one of encounter with alien space, or unearthing patterns in the forms. Noguchi wrote of the garden: "to me, it's very important to keep up with changing concepts of nature. I don't think there's enough of it."[17] Noguchi further reflected that sculpture needed "new forms," of which the landscape was now one: "nature and non-nature. There will come other gardens to correspond to our changing concepts of reality [induced by technology]: Disturbing and unbeautiful gardens to awaken us to a new awareness of our solitude. Can it be that nature is no longer real for us or, in any case, out of scale?"[18] For Noguchi, these spaces revealed the malleability of human being and nature through a technology of both environment and aesthetics. Scale was a bridge, a mode of transformation of time and space, through a new optic embodied in material. The sculptor suggests that the scale of life has changed, and this new situation in the organization of time and space demands new forms. The IBM garden refracts this aesthetic of materialized abstractions and a scalable optics that has been traced throughout this book; making nature itself no longer of the appropriate proportion.

In structure and aesthetics, the gardens produce a strange folding in time and space. Encased within buildings that evade visibility, whose totality as a structure is not available to human perception, sculpture, Noguchi implies, comes to be about ongoing interactions, not gestalt moments, with the viewer. Structure is about time and environment instead of isolated or pure forms. The visitor in this garden is projected into the future through experiencing the sculptural traces of many pasts—of art, nation, and knowledge. This aesthetic infrastructure, supporting a new territory that was global, flexible, and recombinable, found expression within the confines of an organization, mirroring the time space of its gardens. These garden images are, therefore, the vestigial, almost disappeared, traces of the emergence of an aesthetic infrastructure of communication that reorganized not only the senses of human beings but those of organizations.

The Aesthetics of Territory

The second garden also speaks of territory and scale, but it tells a slightly different tale. In 1960 Noguchi was approached by the New York–based Broadway showman Billy Rose to make a sculpture garden for the new National Museum being built in Jerusalem. Rose had jumped at the opportunity to sponsor an entirely new space for contemporary art at a large scale. His proclivities, however, were hardly toward sponsoring Jewish art or Israeli craftspersons. Rather, Rose turned to the stars of abstract expressionism and modern art.

The Israel Museum garden would become Noguchi's largest commission, and one of his most important works. Judging from his archive, it was also one of his most beloved works. For the rest of his life until his death in 1988, he would visit it, nurture it, and treat it as the apex of his effort to make the environment itself into an aesthetic and abstract space, contradictorily capable of sentiment. He wrote of this garden: "perhaps I had a notion that I was building something in the Bible retroactively. Jerusalem is an emotion shared by all of us. It gains new meanings and it is my hope that the garden and the museum, of which it is a part, will come to be a very integral part of this new image—an acropolis of our times."[19]

Initially, Noguchi declined, on the ground that he was not Jewish. But the donor, Billy Rose, did not accept such spurious excuses. Noguchi recalled, "this he would not accept, contending that one who had voluntarily incarcerated himself in a War Relocation Camp could not refuse such a challenge. An argument more to be appreciated were I Jewish. But why not Jewish? The man in the Foreign Office in Tokyo had said, 'For purposes of art, you are Japanese.'"[20] Rose reminded Noguchi that at many times and in many places, he had served different identities. Rose stated that the Jew was like an artist, a "continuously expatriated person who really does not belong anywhere."[21] Noguchi acceded, saying that perhaps for the purpose of art he might even be considered Jewish, and that as a self-proclaimed exile from many nationalities, it would be appropriate to build this space for those who also had had no place in the world.

This garden stands to this day a monument to the contested visions for Israel. The artist recalled that the hill, Neve Shaanan—"tranquil habitation"—was an entry into "the timeless world of antiquity," but it was to house the newest of art at the time. As such, he envisioned it, like the IBM gardens, as "the homage of one age to another (what else could it be?), or our new world and its civilization to the primal space of its birth."[22] He envisioned it as a recombinable space where the past could be remade in the name of the future (fig. E.6).

The garden was also an opportunity to sculpt earth at a scale never before achieved by the artist. He was enthralled with the "drama," in his words, of the site, bridging the hills of Judea, between the future site of the Israeli Knesset, the building that was to house the Dead Sea Scrolls, and the Valley of the Cross. To accomplish his mission of unifying the past while seeking a new geography for the world, he produced five curving retaining walls, utilizing the stones from the site, and the curve of the hills. From the university, he hoped that the site would have the "evocative power of great wings." To accomplish this effect of his words of sanctity, he sought to create a "dialogue" between earth and sky, using only the structure of the hills to form the garden. The result of this dialogue is an asymmetrical garden composed of a series of plateaus that offer numerous scales by which to encounter the sculptures, which include works by Sol LeWitt, Picasso, and many other modern and contemporary artists, such as Richard Serra and James Turrell.

Scale was a critical technology in the garden. Noguchi built simultaneous individuated and personal spaces with partial walls to permit the viewer to encounter a sculpture in isolation, while leaving the space simultaneously open and offering vantage points where the topography could be viewed. In the interest of producing this effect of both intimacy and connectedness, he created a system of terracing and walls to produce enclosures, while maintaining links between all the elements of the garden, a system of open "rooms" that provided a smaller scale in which to present the work while maintaining the flow and openness of the landscape (fig. E.7). The garden had elements similar to those of his other works: highly finished pyramids and a fountain, producing multiple levels and strata. The garden was to have no arbitrary paths to break the curve of the landscape but instead would have different strata that would illuminate different dimensions of the topography, some hierarchical—the pyramids and fountains—some horizontal, along the paths of the hills; all in juxtaposition to the blocks of the museum (which could be understood as Bauhaus or Oriental, depending on the commentator).[23]

Since the construction of the garden was separate from the building of the

FIG. E.6__*Model for Billy Rose Sculpture Garden*, plaster, wood, cardboard, by Isamu Noguchi (1960–65); final model, scale: approximately 1/20" = 1'0". Courtesy of Noguchi Museum and Sculpture Garden. © 2013 The Isamu Noguchi Foundation and Garden Museum, New York/Artists Rights Society (ARS), New York.

FIG. E.7__*Museum Garden*, by Isamu Noguchi (1960–65). Courtesy of Noguchi Museum and Sculpture Garden. © 2013 The Isamu Noguchi Foundation and Garden Museum, New York/Artists Rights Society (ARS), New York.

museum, Noguchi also approached it architecturally as a scalable space for two functions: as a self-enclosed and isolated display of artifacts and as a networked ecology of related objects. While Noguchi argued that the garden was in some sense architectural in creating a total environment, he was clear that there was also a separation, in his mind, between architecture and sculpture. Architecture, he argued, was utilitarian, and these gardens and sculptures were not. His sculptures were always related to the buildings and spaces they were placed in, but their function was to serve no architectural or commercial function outside of making beautiful the purpose for which the building was built. Noguchi's desires were never to serve functions but rather to produce experiences. Abstract as his work was, his language of sculpture is one of awareness and sensitivities. As if to assert the nonfunctional aspect of the garden, one of the main critiques of it is that he failed to put enough shade or benches in the garden for human occupation.

This non- or afunctionalism is also an aesthetics of aspiration and desire for that which does not yet exist. While the garden purportedly speaks to the religious and deeply symbolic landscape within which it is set, Noguchi also framed the space in terms of a global future. "What forms must sculpture aspire to, to conform to our new nuclear, non-Euclidean, non-base-lined, relativistic world? It is hoped that the ancient hill of Neva Shaanan will become a vital part of our new awareness of the world without, I hope, losing its significance to the past, with the present making the past even more meaningful." He sought for this garden to be an "acropolis" for our times, a space that would extend "all the way to China." The amorphous territory envisioned here traversed time, using the past without returning to it, while producing a new stratified topography that could scale, expand, and contract linking localized geographies to a global aesthetic territory. Noguchi would return to this garden all his life, mentoring it, growing it, overseeing its development.[24]

When placed in relation to the other spaces on this hill, the Knesset, completed in 1966, stands in stark contrast. Returning to the architectural memory of the Second Temple, the Knesset asserts an architecture of stable identity, religion, and blood that appears almost at odds with the global and loose definition of "Jewish" assumed by Noguchi. These hillsides thus see themselves at a strange site of emergent differentiating imaginaries that negotiate in a shared lingua franca of recombinable territory, global identity, transformed environment and space, and governance, to very different effects, playing out tensions between occupation and circulation.

While the area was and has never ceased to be fraught with political tensions, Noguchi was more consumed with producing a global imaginary for sculpture, a form for the future, a "new image" by way of the past, than with addressing the immediate political uses and abuses of the site.[25] Dallying in many religious and historical metaphors, analogies and concepts, Noguchi seemed married to none, even if his patrons were. Here a curious conflation between the global and local occurred with the force to interrupt both.

These two gardens, hybrids influenced by European modernist sculptural movements, American postwar design, and histories of Japanese Zen gardening, produce a small network of aesthetically linked spaces whose connections offer us beautiful and terrifying insights into a striking mid-twentieth-century transformation in attitudes to aesthetics, vision, technology, and environment. Tied together by a globalizing aesthetic infrastructure of international-style modernism and information design, while simultaneously fragmented by specific territorial and spatial contexts and histories, the two spaces carry the residual traces of a moment of transformation in the relationship between vision, power, and territory that continues to inform our present but has, perhaps, now been rendered invisible through its ubiquity in the worlds we inhabit.

The architectural critic Eyal Weizman has written about the international style in Jerusalem and the effort in the 1960s to both "Orientalize" and internationalize that city. His analysis of Israel's use of architecture and design as a governmental tactic to control the space is important. Noguchi was not familiar or knowledgeable about the environment he worked in. This is true of most modern architecture and design; it is global but rarely local. In this, his efforts to deal with the landscape without dealing with the Palestinian inhabitants might be seen as the folding of political conflict into aesthetic technology, to the bereavement of history and futurity.[26]

Weizman's analysis is important, but it is a perspective only from within Israel. One wonders: if these other global influences as demonstrated by Noguchi, had been more potent, would a different organization of space or population have been possible? Or of identity? Noguchi was one of the major aesthetic influences on a global style closely integrated with the rising information economies. Noguchi's gardens, when one walks through them, feel modern, and are based on Japanese garden planning, and other aesthetic influences that are not "Oriental," in the sense of deployed by the museum's advertising as denoting Arab or Middle Eastern in origin, in their genealogy. The plan of the garden is one of delicate networks between open rooms, where one confronts a sculpture in isolation, closed only into the work, although surrounded by sky, and vistas, which are, however, incomplete. The garden never offers spaces of absolute surveillance over the territory. It is difficult to locate yourself in this space; you must trust the guidance of the planner, who walks you through the garden with carefully laid out paths that move through the terraces. In many ways, it is space where one encounters the difference of the artist's vision, unamenable to immediate appropriation or change. It is a space of depersonalization, not personalized recombination.

I call attention to these gardens because they are about the binding and unbinding of control over space and over meaning. In these gardens are performed a series of acts of repression and commencement that uncannily mime what the anthropologist James Siegel has said of the work of nations at moments of revolution. In the moment of revolution, Siegel recounts, there are contradictory forces at work, and their organization can either become that of an opening to the world and a new network of affiliations or a closure into self-reference and violence, the "fetish" of revolution, where the past is resurrected as the future and national identity consolidated into localized territory.[27]

In these gardens, we are returned to the problem I first opened with: in another space, in our present, in Songdo, how do these pasts transform the future? At the most limited level, we are confronted with questions of whether recuperating such histories—of moments when what it was to be of a particular nation, territory, or population could be remade through contests over aesthetics, computation, and communication—offers us a lens into how fetishes of territory and identity were produced retroactive to this deterritorialization. I would argue that these gardens pose unique attributes because they bring, in the most physical form possible, messages from elsewhere, and possibilities for affiliation not yet present in the space. These gardens intervene in the tight binding of form and content. They could, if allowed, offer encounters with the untimely arbitrariness and the often dense and heterogeneous networks that produce a space. This does not mean that being forced to encounter Noguchi's sculpture garden automatically leads to political change, but it opens the possibility of asking what, in the future, might produce territories of encounter with different visions of space, identity, and humanity. These gardens pose questions of scale and framing. Would rewriting this history of localized conflicts in a more global framework, as affiliated with the broader transformations in aesthetics, technology, and economy, transform the present? The answer is unknown. But the question recalls Cedric Price's challenge that we do not yet know the questions to which we answer, and therefore have yet to fully encounter other possibilities for life.

This may all seem far away from the worlds of data, information, and cybernetics, but it is not. Noguchi's closest confidant and most regular interlocutor throughout his life was Buckminster Fuller, the techno-idealist and cybernetician, the high sage propagating the integration of science, technology, and design. The designer, Fuller argued, as I mentioned at the start of this book, must become a "design scientist," ready to appropriate all the tools of technology to improve the state of the planet. Noguchi did not share the automatic assumption of technological progress, being a melancholic voice, but he always shared in the interrogation of technology, and the search for forms to address the age. He recalled Fuller as a "prophet" but a misunderstood one. Noguchi wrote, with some sadness, that "his [Buckminster Fuller's] thinking which started out with much the American Dream of material progress would inevitably come to question the means by which this has come about. Bucky himself is without acquisitiveness excepting possibly with boats. Believing in man's essential rationality, Bucky remains the supreme optimist—there is a way out. This must be his great appeal to the young. He is a true believer, a prophet for our times. The ultimate machine is no machine—a little black box he calls it—no machine but the knowledge and control of the forces of nature that bind us all in mutual dependence."[28] No machine but the apparatus of care and relations. These gardens, so fragile and shining, perhaps new acropolises, perhaps disasters of abstraction and informatic disembodiment, loom before us; they call to us, and they remind us that there are no machines, only little black boxes from which we may derive potential forms of encounter or destruction. They provide us with different images of alternative forms of life. These fragments of partially built, partially fantasized, landscaping are utopian promises or warnings that have never been fully realized. Creativity? Torture? Love? Here history must multiply, divide, and reenvision the pasts to produce that which is never known . . . the future.

Notes

PROLOGUE. **Speculating on Sense**

1 Arthur, "Thinking City," 56.

2 The research conducted in Songdo, South Korea, was done in collaboration with two architect-designers, Nerea Cavillo and Jesse LeCavalier, and with the excellent assistance and translation help of the electronic arts curator Dooeun Choi. I am indebted to all of them for the inspiration and education they provided.

3 "Korean Free Economic Zones," FEZ Planning Office, Republic of Korea, accessed October 1, 2012, www.fez.go.kr/en/incheon-fez.jsp.

4 Taken from promotional materials and interviews with the IFEZ officials Kyung-Sik Chung (director of cultural affairs) and Jongwon Kim (director of marketing and development for IFEZ Authority), on July 4, 2012. "Incheon Free Economic Zone," Wikipedia, accessed October 1, 2012, http://en.wikipedia.org/wiki/Incheon_Free_Economic_Zone#Yeongjong_Island, and the IFEZ website, www.fez.go.kr/en/incheon-fez.jsp.

5 Schelemetic, "Rise of the First Smart Cities," last modified September 20, 2011, accessed October 12, 2012, http://news.thomasnet.com/IMT/2011/09/20/the-rise-of-the-first-smart-cities/; Arthur, "Thinking City," posted January 11, 2012, accessed October 12, 2012, http://www.songdo.com/songdo-international-business-district/news/in-the-news.aspx?d=360/title=The_Thinking_City.

6 Keller Easterling labels these cities "spatial products." Easterling, *Enduring Innocence*.

7 Engineers, planners, and marketers all agreed that "smartness" and "ubiquity" were not really definable terms, but they were useful as a goal and ideal. From interviews with Tony Kim, senior vertical manager at Public Sector Internet Business Solutions Group, Cisco Systems, July 6, 2012, and Gui Nam Choi, client solutions executive, Cisco, July 6, 2012.

8 IBM has an entire new management consulting service branded around "smart planet" services; see the website of this service, accessed August 7, 2012, www.ibm.com/smarterplanet/us/en/?ca=v_smarterplanet. Cisco is planning to also retrofit itself into a more consulting service oriented company rather than only or mainly selling hardware like routers for digital infrastructure. In building these cities, Cisco's role is largely as the management consulting and strategy firm for high-tech services. The conduits, routing systems, sensors, telecomm towers, and other hard portions of the infrastructure are built by telecom companies Cisco is partnered with, in this case Korean Telcom (KT), and the build-

ings are built by GALE International, affiliated with the engineering and architecture firm Kohn Pedersen Fox Associates, located in New York City. Kohn
Pedersen Fox did the master urban plan. The master plan for the buildings,
however, was rarely worked on in consultation with the technology infrastructure groups. One of the outstanding features of this "digital" city is how its built
form has little relationship to its technical infrastructure. Form and function are
seemingly disassociated. From interviews with Tony Kim, July 6, 2012, and Gui
Nam Choi, July 6, 2012.

9 Lindsay, "The New New Urbanism," 90.

10 Definitions of terms like "information" and "data" are kept rather black-boxed
   in this endeavor or reduced to behavioral concepts like bit rate.

11 Interviews with IFEZ officials, July 4, 2012, and interviews with Tony Kim,
   Senior Vertical Manager at Public Sector Internet Business Solutions Group
   Cisco Systems, July 6, 2012. Both officials detailed and demonstrated marketing
   videos showing the rollout of telemedicine applications.

12 This situation offers sustenance to Deleuze's claim that "individuals" have become "dividuals," and masses have become samples, data, markets, or "banks."
   Deleuze, "Postscript on the Societies of Control."

13 As theorist Patricia Clough puts it, there is presently an "affective turn" that
   "marks these historical changes . . . indicative of the changing global processes
   of accumulating capital and employing labor power through the deployment of
   technoscience to reach beyond the limitations of the human in experimentation
   with the structure and organization of the human body, or what is called 'life
   itself.'" Clough and Halley, *Affective Turn*, 3.

14 There are numerous debates about biopolitics and digital media. Many theorists follow a more Agambenian approach that views contemporary capital and
   politics as leading to an inexorable transformation of life into capital. I take a
   less disciplinary and perhaps more optimistic approach and follow the lead of
   feminist philosophers like Rosie Bradotti, Donna Haraway, and Elizabeth Grosz
   in insisting on the productive elements of these assemblages. Haraway, "Cyborg
   Manifesto," Braidotti, *Nomadic Subjects*, Grosz, *Nick of Time*.

15 Ben Rooney, "Big Data Demands New Skills." Wall Street Journal, February 10,
   2012, http://blogs.wsj.com/tech-europe/2012/02/10/big-data-demands-new
   -skills/?mod=google_news_blog.

16 Seagran and Hammerbacher, *Beautiful Data*, Segel and Heer, *Beautiful Data*,
   Edward Tufte, *Beautiful Evidence*.

17 Benjamin, *Illuminations*, 238.

18 Easterling, *Enduring Innocence*, 1–13.

INTRODUCTION. **Dreams for Our Perceptual Present**

1 Bacon, *New Atlantis and the City of the Sun*.

2 Bentham, "Panopticon."

3 Le Corbusier, *City of To-morrow and Its Planning*, 173.

4   Fishman, *Urban Utopias in the Twentieth Century*, 186–87.

5   Anable, "Architecture Machine Group's *Aspen Movie Map*," 500.

6   Negroponte, *Architecture Machine*, 1.

7   Counter to other building management design programs at the time, such as those developed by Skidmore, Owings, and Merrill, an architectural firm responsible for many of the largest and most impressive post-war corporate buildings around the world, Negroponte and his colleagues dreamed of an ecology, formed out of constant feedback loops of machine human interactions, one that evolved and changed, grew "intelligent." Negroponte, *Architecture Machine*, 7.

8   Negroponte, *Architecture Machine*, 7.

9   Negroponte, *Architecture Machine*, 3.

10 While there has long been a discussion of architecture as media, or reflecting media, it should be critically noted that the Architecture Machine Group was later absorbed into today's Media Lab, eliminating the language of architecture entirely. Negroponte has been remembered as regularly posting on blackboards that TV, computers, and publishing had to merge "now." The dominant philosophy was the integration of media and a cybernetic philosophy that design must not be merely computer aided but actually a matter of interactivity coconstituted through the integration between humans and machine systems. The lab grew from a collaboration between the Architecture Machine Group, the Center for Advanced Visual Study (founded by Hungarian artist and designer Gyorgy Kepes), and two other groups working on visual language and video and digital cinema. Interview with Michael Naimark, one of the members of the Aspen Movie Map team, August 12, 2013. See also other works by Negroponte, including *Soft Architecture Machines*.

11 Negroponte, *Soft Architecture Machines*; Negroponte, *Architecture Machine*.

12 Norbert Wiener, *Ex-Prodigy*, 63.

13 Wiener, *Ex-Prodigy*, 63.

14 Wiener, *Ex-Prodigy*, 130.

15 I open with this memoir because we are arguably still negotiating the legacy of this transformation. The choice of Norbert Wiener is, of course, not arbitrary. The foremost preachers of the new gospel of networked and smart cities, such as the former and current directors of the Media Lab, William Mitchell and Nicholas Negroponte, respectively, all came from MIT (the Media Lab was originally involved in the Songdo project but withdrew due to supposedly conceptual, but probably monetary, concerns) and all laid claim to cybernetic influence. Mitchell labeled these "cyborg" producing cities. See Mitchell, *Me++*, Negroponte, *Soft Architecture Machines* and *Architecture Machine*.

16 Such statements rethinking the role of representation, memory, and perception were repeated in many fields at the time, ranging from anthropology to biology, to sociology, to computing, to architecture. For more information on the influence and use of information theory and communication science on a variety of fields see: Keller, *Refiguring Life*, 89–99. For an opposing argument on the role of information theory in the history of molecular biology see Kay, *Who Wrote the*

Book of Life? Keller views cybernetic and information theories as providing the possibility to view life in its complexity, while Kay posits an account that argues that these notions of codes are ultimately reductive.

On postwar attempts to build a unified theory of science see: Bowker, "How to Be Universal." And on the cultural impact of cybernetics see: Edwards, *The Closed World Computers*. For the impact of communications theory, cybernetics, and digital media architecture, design, and the arts see: Pamela M. Lee, *Chronophobia*.

17 Kepes, *New Landscape in Art and Science*, 24.

18 Kepes, *New Landscape in Art and Science*, 20.

19 Kepes, *New Landscape in Art and Science*, 22–24.

20 Fuller, *Synergetics*.

21 Fuller, *Synergetics*.

22 Deutsch, *Nerves of Government*, ix.

23 Deutsch, "Review: A New Landscape Revisited by Gyorgy Kepes Manuscript," April 1960, Karl Deutsch Papers, HUGFP, 141.50, manuscripts and research materials ca. 1940–1990, box 1, Harvard University Archives.

24 Deutsch, "Review: A New Landscape Revisited by Gyorgy Kepes Manuscript," April 1960, Karl Deutsch Papers, HUGFP, 141.50, manuscripts and research materials ca. 1940–1990, box 1, Harvard University Archives.

25 Hayles, *How We Became Posthuman*; Manovich, *Language of New Media*; Galloway, *Protocol*; Galison and Daston, *Objectivity*; Crary, *Techniques of the Observer*; Schievelbusch, *Railway Journey*.

26 Galloway and Thacker, *Exploit*, 67.

27 Arendt, *Human Condition*, 4–6. It is worth noting, of course, that there were other attitudes to the productivity of mass mediation at the time, including Walter Benjamin.

28 Adorno, "Culture Industry Reconsidered," 32, 37.

29 Melley, "Brainwashed," Marks, *Search for the "Manchurian Candidate."* The famous document NSC-68, establishing containment as an American policy, framed Soviet Communism as a "slave state" lacking diversity or freedom, forwarding the idea of a mass homogenous group. The document, written for President Truman, included in its authors Secretary of State Dean Acheson, and was an extension of George Kennan's strategy of containment, but with a narrowly military focus. The document makes statements such as: "the implacable purpose of the slave state to eliminate the challenge of freedom has placed the two great powers at opposite poles. It is this fact which gives the present polarization of power the quality of crisis. . . . [On the other hand] the free society values the individual as an end in himself. . . . From this idea of freedom with responsibility derives the marvelous diversity, the deep tolerance, the lawfulness of the free society." NSC Study Group et al., *NSC 68*. The document clearly defines Communism as opposed to liberal subjectivity, and makes this connection by envisioning Communist political movements and political actors as faceless hordes or swarming masses who lack reason.

30 Crary, *Techniques of the Observer*, 3.

31 Sterne, *Audible Past*, 1–2.

32 Crary, *Techniques of the Observer*, 6.

33 Crary, *Techniques of the Observer*, 6.

34 *Oxford English Dictionary* (Oxford: Clarendon Press, 2013).

35 *Oxford English Dictionary*.

36 SAS software, accessed May 6, 2013, www.sas.com/data-visualization/overview
.html?gclid=CKHRtpP6hbcCFYef4AodbEcAow.

37 This summary is taken from the "Smart Planet" website and the "Graphics
and Visualization" website of IBM, accessed January 2, 2013, www.ibm.com
/smarterplanet/us/en/overview/ideas/index.html?re=spfesearcher.watson.ibm
.com/researcher/view_pic.php?id=143.

38 A number of theorists and scholars have informed this understanding of "visu-
alization," even if not directly speaking about data visualization practices. See:
Mitchell, *Reconfigured Eye*, Manovich, *Language of New Media*, Beller, *Cinematic
Mode of Production*.

39 Silverman, *Threshold of the Visible World*, 131.

40 Oetterman, *Panorama*, Braun, *Picturing Time*, Dickerman, *Inventing Abstraction*.

41 There is an extensive literature on the complex histories and ontologies of
"vision" in Western thought. Some critical works that have critically engaged
histories of vision, ideas of oculocentrism, and of the relationship between
vision, knowledge, power, and difference, and concepts of abstraction and representation to which I am indebted: Krauss, *Originality of the Avant-Garde and
Other Modernist Myths*; Rose, *Sexuality in the Field of Vision*; Jay, *Downcast Eyes*;
Bryson, *Vision and Painting*.

42 These definitions are taken from the website of Farlex, Inc., accessed March 6,
2013, www.thefreedictionary.com/vision.

43 This discussion is indebted to Rajchman, "Foucault's Art of Seeing."

44 Rajchman, "Foucault's Art of Seeing," 91.

45 Rodowick, *Gilles Deleuze's Time Machine*, 5.

46 For a sampling of work on standards and measures as related to the production
of new forms of perception, observation, and governance see: Sekula, "Body and
the Archive"; Gould, *Mismeasure of Man*; Gould, *Mismeasure of Man*; Braun,
Picturing Time; Dagognet, *Etienne-Jules Marey*; Cartwright, *Screening the Body*;
Bowker, *Sorting Things Out*; Star and Bowker, "How to Infrastructure"; Lampland and Star, *Standards and Their Stories*; May, "Sensing"; Mumford, *Technics
and Civilization*; Carson, *Measure of Merit*; Dumit, *Drugs for Life: How Pharmaceutical Companies Define Our Health*.

47 Canales, *Tenth of a Second*, 215–19.

48 Wiener, *Human Use of Human Beings*, 15.

49 Rabinow, *Ethics*, 67.

50 Foucault, *Birth of Biopolitics*, 20.

51 Knight, *Risk, Uncertainty, Profit*.

52 For a comprehensive treatment of the theme of "uncertainty" and the logic of
smart cities please see: Halpern et al., "Test-Bed Urbanism."

53 For an elaborate discussion on territory in Foucault's late lectures see: Foucault,

Birth of Biopolitics; Foucault, *Security, Territory, Population*; Elden, "Governmentality, Calculation, Territory."

54 Clough and Willse, *Beyond Biopolitics: Essays on the Governance of Life and Death*, 1.

55 Galison and Daston, *Objectivity*, 49, 50.

56 I call on Deleuze and Guattari's concept of "assemblage" here. In *A Thousand Plateaus* Deleuze and Guattari speak of the nature of assemblages as sites of concentration, or density, where bodies, discourses, agents, technologies all accumulate, each with their own histories. Assemblage denotes a site of historical accumulation. Assemblages are topographical and temporal entities, that accumulate in Deleuze and Guattari's language "content" and "expression," which is another way to say that an assemblage is a marriage of historical content with forms, "on one hand it is a *machinic assemblage* of bodies, of actions and passions, an intermingling of bodies reacting to one another; on the other hand it is a *collective assemblage of enunciation*, of acts and statements, of incorporeal transformations attributed to bodies." Deleuze and Guattari, *Thousand Plateaus: Capitalism and Schizophrenia*, 88.

57 See for example Jackie Orr's work on panic and the rise of "psycho-power"; Orr, *Panic Diaries*. On shifts in planning and governance to ideas of ecologies and networks see: Graham and Marvin, *Splintering Urbanism*; on the emergence of risk and environment see: Beck, *Risk Society*.

58 Chun, *Control and Freedom*; Deleuze, *Anti-Oedipus*; Liu, *Freudian Robot*; Massumi, *User's Guide to Capitalism and Schizophrenia*.

59 Hayles, *How We Became Posthuman*; Golumbia, *Cultural Logic of Computation*; Amadae, *Rationalizing Capitalist Democracy*. For an opposing account see: Mirowski, *Machine Dreams*; Daston, "Rule of Rules."

60 From interviews with Tony Kim, senior vertical manager, Public Sector Internet Business Solutions Group, Cisco Systems, July 6, 2012, and Gui Nam Choi, client solutions executive, Cisco, July 6, 2012.

61 The planners and engineers of Songdo, and other smart cities in South Korea, regularly mention Tokyo and Japanese engineering as models. In Japan, the subway systems, for example, are run on decentralized data clouds. The network is not centralized; rather, each station speaks to the next, like a hive or the Internet's packet-switching architecture. Decisions about directing trains are made locally, and the network fluctuates according to localized demands that are regulated through a decentralized system of stations. As the anthropologist Michael Fisch notes in his study of the Tokyo infrastructure, these systems no longer rely on centralized cognition or command centers but diffuse the functions of regulating feedback into the network. In his words, the new form of autonomous traffic control "is its operation beyond capacity, whereby the entire focus of railroad operators is not on maintaining precision per se but rather on maintaining the precision of the margin of indeterminacy. The margin of indeterminacy is the space and time of the human and machine interface. Put differently, it is the dimension in which bodies and machines, with their incommensurable qualities (technicities), intersect with the time and space of institutionalized regularities

to produce a metastable techno-social environment of everyday urban life. In other words, it is a dimension of perpetual precarity that is held together by collective investment in the security of repetition and routine." The Seoul subway system that extends to Songdo operates on a similar technical architecture, possessing a decentralized and autonomous traffic control system for which human beings are only necessary in cases of extreme disruption, but in which daily disorders—of traffic, accident, and so on—are managed through the autonomous network. The idea, therefore, of increasing control finds its corollary in the networking and dispersal of power and sense, and what is remarkable about these systems is their approach to the indeterminacy and unknowability of the systems' actions to human observers. Fisch, "Tokyo's Commuter Suicides and the Society of Emergence," 329, 331–38.

62 Wendy Hui Kyong Chun has a complex discussion of memory and transparency that bears on this conversation and that I will return to in chapter 1. See: Chun, *Programmed Visions*.

63 Stoler, *Along the Archival Grain*, 1.

64 As Keller Easterling has noted, these architectures are charmingly "enduring(ly) innocent," masquerading in innocence and experimentation while smoothing space for the renegotiation of politics by way of aesthetics. Easterling, *Enduring Innocence*, 2–3.

65 When Le Corbusier produced his models for this utopian machine city, retrofitted for mobility and providing machine-buildings that would perfectly regulate human life, he was imagining redesigning the French colonial city of Algiers. While the colonial administration denied his plan, he never realized that his work both aided and abetted, even if through a different vision of the independent artist, a regime that was violently racist. For a discussion of Le Corbusier's utopian vision in relationship to colonialism see: Fishman, *Urban Utopias in the Twentieth Century*, 244–52, Corbusier, *City of To-morrow and Its Planning*. Felicity Scott in *Architecture and Techno-utopia* makes a similar argument about the potential in modernism, and the failure of some of the more radical efforts—like ant-farm—to actually produce alternative visions of space and territory. Scott, *Architecture and Techno-utopia*.

66 Steiner, *Beyond Archigram*.

67 Caillois, "Mimicry and Legendary Psychasthenia."

68 Some major discussions about the ideologies of communication and the body, history, and situation: Hayles, *How We Became Posthuman*; Golumbia, *Cultural Logic of Computation*; Keller, *Refiguring Life*.

69 See, for example: Fred Turner, *From Counterculture to Cyberculture*.

70 I take my lead here from Elizabeth Grosz's work on the Nietzschean "untimely" as a force for emergence. Grosz, *Nick of Time*.

71 Lury et al., "Introduction." The sociologists Celia Lury, Tiziana Terranova, and Luciana Parisi have recently written about "topology" as a central concern for discussions of the social, as it is a form of thinking linked to the relations between surfaces without depth, and where "movement" in their terms is not an aftereffect but constitutive of social processes.

72 Jazairy, "Toward a Plastic Conception of Scale," 1. Other urban theorists, like Ash Amin, are also attempting to think about urban space in terms of probability distributions and conditions of possibility. Amin and Cohendet, *Architectures of Knowledge*.

73 Wiener, *Human Use of Human Beings*, 24, 33.

74 Wiener, *Cybernetics*, 44.

75 Parikka, *Insect Media*, xx, xxi.

76 Bruno Latour is interested in accumulation and densities as producing effects and agents. While the term "immutable" is contestable, the idea that forms are made by accumulating agents, documents, and inscriptions is a suitable analogy to the way I will be dealing with density and scale in this text. Latour, "Visualization and Cognition," 6. Foucault, *Discipline and Punish*; Rajchman, "Foucault's Art of Seeing."

77 Definition of "speculate" from the *Oxford English Dictionary*.

CHAPTER 1. **Archiving**

1 The virtual is used throughout this chapter to denote that which does not yet exist but is being brought into being. It serves as both an operation and a field for conditions of possibility. I am not using the term, however, in the sense of a simulation or a simulacra. The virtual cannot exist as a materialized form in the present.

2 Wiener, *Human Use of Human Beings*, 7, 11.

3 Wiener, *Human Use of Human Beings*, 15.

4 Arguably, after the war, this "complex" is never dismantled. The reorganization of research, economy, and military formalized during the war is fortified as part of the Cold War by the GI Bill with its emphasis on higher education, and by the Cold War infusion of scientific research funds through the Department of Defense and later the National Science Foundation as a central strategy for safeguarding national security. In fact, science funding never returns to prewar levels in the period between the mid-1940s and the late 1960s, and far exceeds even World War II funding levels during the Cold War. For example, after the launching of *Sputnik* in 1957, defense funding of scientific research was higher than peak wartime levels, and fifty to sixty times prewar levels. Leslie, *Cold War and American Science*, 1–10. See also: Akera, *Calculating a Natural World*, 181–222; Hughes, *Rescuing Prometheus*.

5 Lettvin, cited from Simon Garfunkel, *Building 20: A Survey part of Jerome B. Wiesner: A Random Walk through the 20th Century Website at MIT*, accessed October 16, 2012, http://ic.media.mit.edu/projects/JBW/ARTICLES/SIMSONG .HTM.

6 Brand, *How Buildings Learn*.

7 Weaver also oversaw research in operations research, demonstrating the close integration between the fields. Originally this work was considered part of operations research.

8 Wiener, *Cybernetics*, introduction. Wiener also repeats this idea in many of his

writings. For example in a short essay for a lecture "The Nature of Analogy" written in September 1950, Wiener argues that the purpose of cybernetics is to "compare phenomena" in "widely different fields which nevertheless show common properties or behaviors." "The Nature of Analogy" [September 1950], Norbert Wiener Papers, MC 22, box 29B, folder 655, Institute Archives and Special Collections, MIT Libraries, Cambridge, MA.

9   Galison, "Ontology of the Enemy," 231.

10   The ideas about "closed world" and the imaginary of cybernetic systems come from Paul Edwards's work. See Edwards, *The Closed World Computers*.

11   What defines this form of engineering is feedback and the integration of previous histories of engineering into a coherent assemblage with the sciences of the mind, process philosophy, and logic. David Mindell and others have noted that feedback, a core and central aspect to the ideas of cybernetics, computing, and information theory, has numerous earlier origins. Feedback already played a role and was formalized as an engineering concept earlier in the twentieth century as an answer to problems in spreading telephony and communications. Control and homeostasis was also formulated as a problem for both organisms and machines in industrial systems. See: Mindell, *Between Human and Machine*; Beniger, *Control Revolution*; Rabinbach, *Human Motor*.

12   Quoted in Heims, *John von Neumann and Norbert Wiener*, 127.

13   *Cybernetics and Society*, 136.

14   Wiener, *Human Use of Human Beings*, 33.

15   Rosenblueth et al., "Behavior, Purpose and Teleology," 18.

16   Rosenblueth et al., "Behavior, Purpose and Teleology," 24.

17   Wiener clearly articulates his new concept of communication, which adjoins Gibbsian theories to communication, and in doing so reworks the idea of information. In his postwar work, Wiener summarizes these ideas: "information is the real clue to a study of causality. In the Newtonian system, where everything is determined and the whole present suffices for the knowledge of the whole future, it is not a significant question to ask, 'what causes what?' The really interesting questions of causality are always of a more or less nature." As an extension of this observation, Wiener argues that something cannot be information if there is no choice between options. Wiener, "Problems of Organization," 227.

18   The above synopsis is based on the foundational article put forth by Norbert Wiener in collaboration with Arturo Rosenblueth and Julian Bigelow from Harvard Medical School in 1943. See: Rosenblueth, "Behavior, Purpose and Teleology."

19   For further information and explanation about definitions of behavior, communication, and purpose see also: Rosenblueth, "Behavior, Purpose and Teleology." Rosenblueth and Wiener, "Purposeful and Non-purposeful Behavior."

20   Wiener worked closely with Weaver, who produced the mathematical theory of communication in which these assumptions are explicitly stated. As Weaver notes, "Dr. Shannon's work roots back, as von Neumann has pointed out, to Boltzmann's observation, in some of his work on statistical physics (1894), that entropy is related to 'missing information,' inasmuch as it is related to

the number of alternatives which remain possible to a physical system after all the macroscopically observable information concerning it has been recorded. L. Szilard (Zsch. f. Phys. Vol. 53, 1925) extended this idea to a general discussion of information in physics, and von Neumann (*Math. Foundation of Quantum Mechanics*, Berlin, 1932, chap. V) treated information in quantum mechanics and particle physics. Dr. Shannon's work connects more directly with certain ideas developed some twenty years ago by H. Nyquist and R. V. L. Hartley, both of the Bell Laboratories; and Dr. Shannon has himself emphasized that communication theory owes a great debt to Professor Norbert Wiener for much of its basic philosophy. Professor Wiener, on the other hand, points out that Shannon's early work on switching and mathematical logic antedated his own interest in this field; and generously adds that Shannon certainly deserves credit for independent development of such fundamental aspects of the theory as the introduction of entropic ideas." Shannon and Weaver, *Mathematical Theory of Communication*, 1.

21 Shannon and Weaver, *Mathematical Theory of Communication*, 8–9.

22 One of the readers of this book insightfully noted that the conflation of making decisions, or binary choices, with the encoding of data is a remarkable action on the parts of cyberneticians; this move effectively evades the difference between making a decision and data storage. Two activities that, as the rest of this chapter will attest to, are quite separate.

23 Quoted in Bowker, "How to Be Universal," 111.

24 See Heims, *Cybernetics Group*.

25 For some examples of cybernetics impact on contemporary media, life sciences, and digital technologies see Packer and Jordan, *Multimedia*; Manovich, *Language of New Media*; Johnston, *Allure of Machinic Life*; Shanken, "Cybernetics and Art: Cultural Convergence in the 1960s"; Chun, *Programmed Visions*. For a thorough excavation of the Macy Conferences and debates concerning cybernetics, posthumanism, and embodiment see: Hayles, *How We Became Posthuman*.

26 Heims, *John von Neumann and Norbert Wiener*. See also Mara Mills, "Deaf Jam."

27 Hacking, *Taming of Chance*.

28 For an excellent summation of McCulloch's relationship to Wiener and his philosophical influences, see Michael Arbib, "Warren Mcculloch's Search for the Logic of the Nervous System."

29 Wiener, *Cybernetics*, 42–44.

30 Hacking, *Taming of Chance*.

31 Wiener, *I Am a Mathematician*, 327–28.

32 A number of scholars have seized on this discursive shift to information, communication, and process as signaling larger transformations in governmentality, economy, and subjectivity that identify "postmodernity"; in fact Wiener himself regularly argued for an age and economy of "servo-mechanisms" to replace industrial systems: Wiener, *Cybernetics*, 42–44; Lyotard, *Postmodern Condition*; Haraway, "Cyborg Manifesto"; Castells, *Rise of the Network Society*.

33 Barthes, *Rustle of Language*, 143.

34 For more about the relationships between sociobiology and information theory, particularly in relation to complex systems emerging from nonconscious and agents in both theory and history, see: Deleuze, "Postscript on the Societies of Control"; Haraway, "Cyborg Manifesto"; Lyotard, *Postmodern Condition*; Gould, "Dance Language Controversy"; Wenner and Wells, *Anatomy of a Controversy*; Kay, *Molecular Vision of Life* and "Cybernetics, Information, Life"; Wilson, *Sociobiology*, 176–79.

35 Barthes, *Rustle of Language*, 143.

36 Barthes, *Rustle of Language*, 146.

37 The film theorists Kaja Silverman and Mary Ann Doane have both argued that one of the central critiques of media in modernity has been the alienation from ideals of reality, usually encapsulated in discourses of representing and capturing the present or "experience." Silverman, *Threshold of the Visible World*; Doane, *Emergence of Cinematic Time*.

38 Wiener, *I Am a Mathematician*, 327–28.

39 Bergson, *Matter and Memory*, 22.

40 Freud, "Note Upon the 'Mystic Writing-Pad,'" 228.

41 Doane, "Freud, Marey, and the Cinema," 315.

42 Doane, *Emergence of Cinematic Time*, n. 26.

43 See Kittler, *Gramophone, Film, Typewriter*, and *Discourse Networks 1800/1900*.

44 Bergson, *Creative Evolution*, 332.

45 Early cinema actually did not have as standardized a speed for projection as even Bergson presupposed. Most films were projected at sixteen frames per second or faster, sometimes faster then the standard of twenty-four frames per second that became the standard with sound. Card, "Silent Film Speed," 55–56.

46 Bergson, *Matter and Memory*, 219.

47 Deleuze, *Bergsonism*, 63, 60.

48 Deleuze, *Cinema 1*, 2–3. Deleuze explicitly argues that Bergsonian thought anticipates the later terms of the cinema, in which the apparatus, as an information system, allows a mobility of the eye, and a freeing of viewpoint. This perspectival mobility recuperates Bergson's understanding of the relations between matter and memory as put forth in his previous work.

49 Arbib, "Warren McCulloch's Search for the Logic of the Nervous System," 204, and "How We Know Universals," 46–66.

50 Bergson, *Creative Evolution*, 210–12.

51 Deleuze, *Cinema 1*, 23.

52 Deleuze, *Cinema 1*, 58–59.

53 Hansen, *New Philosophy for New Media*, 5.

54 Deleuze, *Cinema 2: The Time Image*.

55 Wiener, *Cybernetics*, 122.

56 Wiener, "Problems of Organization," 396–97.

57 Heims, *John von Neumann and Norbert Wiener*, 304–29.

58 Wiener, *Cybernetics*, 135, 24.

59 Mills, "On Disability and Cybernetics."

60 Julesz regularly cites famous studies by cyberneticians and computer scientists

such as Jerome Lettvin, Warren McCulloch, Humberto Maturana, and J. C. R. Licklider, to name a few. Julesz, *Foundations of Cyclopean Perception*, 381.

61 Julesz, "Binocular Depth Perception of Computer-Generated Patterns."

62 Rutgers University: Center for Cognitive Science, New Brunswick, New Jersey, Second Julesz Lecture on Brain Research, accessed May 9, 2013, http://ruccs .rutgers.edu/ruccs/index.php/news/58-second-julesz-lecture-on-brain-research. Similar attitudes to vision as linked to pattern seeking and environment emerged at the time in important work of psychologist James Gibson: Gibson and Gibson, "Perceptual Learning" and *Senses Considered as Perceptual Systems*.

63 Julesz, *Foundations of Cyclopean Perception*, 153.

64 Julesz marks the start of the computer modeling of vision in mid-1956 by Bert F. Green, Alice K.Wolf, and Benjamin W. White. The work of Donald McKay is also prominently discussed in attempting to use stochiastic parameters for the study of vision. Julesz, *Foundations of Cyclopean Perception*, 153–54.

65 Julesz, "Binocular Depth Perception of Computer-Generated Patterns."

66 Bowker, "How to Be Universal," 114–15.

67 Siegel, "Choices."

68 Jonathan Crary makes this suggestion in the introduction of his work on attention. See Crary, *Suspensions of Perception*.

69 Trifonova, "Matter-Image or Image-Consciousness," 81.

70 Doane, *Emergence of Cinematic Time*.

71 Wiener, *Cybernetics*, 121.

72 Von Neumann architecture called on ideas of memory and neural nets from McCulloch and Pitts, the same ideas that Wiener is working from, to build the stored program architecture. See von Neumann's initial paper on electronic computing instruments: Burks et al., *Preliminary Discussion of the Logical Design of an Electronic Computing Instrument*, 34–79. Also see von Neumann. *First Draft of a Report on the* EDVAC.

73 These ideas are substantiated in the further work of Warren McCulloch, Walter Pitts, Gregory Bateson, and the Macy Conferences: Bateson, *Steps to an Ecology of Mind*; McCulloch, *Embodiments of Mind*; the transactions of the Macy Conferences on Circular, Causal, and Feedback Mechanisms in Biological and Social Systems, particularly the transactions for the Sixth Conference (March 24–25, 1949, New York, NY), and the Eighth Conference (March 15–16, 1951, New York, NY) in Pias, *Cybernetics*.

74 Wiener, *Cybernetics*, 147.

75 Rosenblueth et al., "Behavior, Purpose and Teleology," 21.

76 See Wiener's other autobiographical and philosophical writings, *Ex-Prodigy* and *Human Use of Human Beings*.

77 Wiener, *Cybernetics*, 149–50.

78 Packer and Jordan, *Multimedia*, 48.

79 Wiener, *Human Use of Human Beings*, 7–37.

80 Wiener, *Cybernetics*, 149.

81 Kittler, *Gramophone, Film, Typewriter*, 3.

82 Kittler, *Discourse Networks 1800/1900*.

83 Doane, "Freud, Marey, and the Cinema," 316.

84 Doane, "Freud, Marey, and the Cinema," 316, and *Emergence of Cinematic Time*.

85 See the chapter "Feedback and Oscillation" in Wiener, *Cybernetics*. Shannon and Weaver's work is also largely dedicated to problems of system stability and noise control; Shannon and Weaver, *Mathematical Theory of Communication*.

86 Wiener, *Cybernetics*, 123.

87 Throughout this chapter "real time" should be understood as denoting a desire for immediacy in interactive exchanges without assuming an external referent to those communicative exchanges.

88 Wiener, *Cybernetics*, 124.

89 McCulloch quoted in von Foerster, *Transactions of the Sixth Conference*, 163.

90 Daston and Peter, "Image of Objectivity."

91 Vannevar Bush, "As We May Think," *Atlantic Monthly*, July 1, 1945. http://www .theatlantic.com/magazine/archive/1945/07/as-we-may-think/303881/.

92 Lacan, *Seminars of Jacques Lacan*, 296–300. Some serious treatment of the relationship between Lacan and cybernetics, and between poststructuralism and cybernetics, has been done by a number of authors. However, none of them discuss the epistemological relationships and contexts, or what it is about cybernetics over other computational theorems that made it so amenable to these conceptual programs. Of these authors, only Johnston recognizes the emergent properties within cybernetics; Liu treats cybernetics as a logic of empire but says little of how Lacan reveals this logic through analysis. Geoghegan demonstrates the influence of cybernetics visible in Roman Jakobson's work. Johnston, *Allure of Machinic Life*; Liu, *Freudian Robot*; Geoghegan, "From Information Theory to French Theory."

93 Derrida, *Archive Fever*, 17.

94 Derrida, *Archive Fever*, 15.

95 Derrida, *Archive Fever*, 19–20.

96 Marx, *Grundrisse*, 163.

97 Morris, "Review of James T. Siegel, *Fetish, Recognition, Revolution*," 165.

98 Morris, "Review of James T. Siegel, *Fetish, Recognition, Revolution*," 165.

99 This term and my method are clearly influenced by Benjamin's essay, "Theses on the Philosophy of History," in Benjamin, *Illuminations: Essays and Reflections*, 253–65.

CHAPTER 2. **Visualizing**

1 Scott et al., "Civilian Action Goes into Action," 10. This chapter is indebted to Rheinhold Martin, who first alerted me to this citation, and the figure of Gyorgy Kepes. I am extending Martin's observations about "pattern-seeking" and the history of control in architecture to discuss knowledge, governmentality, vision, and "biopolitics." Readers are advised to read his excellent book, Martin, *Organizational Complex*.

2 Kepes, *New Landscape in Art and Science*, 17.

3 Eames et al., brochure for "A Sample Lesson." The Work of Charles and Ray

Eames (WCRE), Box 230, Folder 1, Publicity File 1953, Manuscript Division, Library of Congress.

4 Eames et al., brochure for "A Sample Lesson."

5 Lynch, *Image of the City*, 1–3.

6 Kepes, *Language of Vision*.

7 Lemov, "Towards a Data Base of Dreams"; Masco, "'Sensitive but Unclassified'"; Martin, *Organizational Complex*; Lee, *Chronophobia*; Edwards, *Closed World Computers*; Colomina, *Domesticity at War*.

8 Gyorgy Kepes to Norbert Wiener, September 27, 1951, Norbert Wiener Papers, MC22, box 9, folder 141, MIT Institute Archives and Special Collections. Norbert Wiener to Gyorgy Kepes, August 1, 1951, Norbert Wiener Papers, MC22, box 9, folder 140, MIT Institute Archives and Special Collections.

9 "Gyorgy Kepes, Founder of CAVS, Dies at 95." *MIT News*. January 16, 2002. http://newsoffice.mit.edu/2002/kepes.

10 "Gyorgy Kepes Interview." MIT 150. Accessed April 14, 2014. http://mit150.mit .edu/multimedia/gyorgy-kepes.

11 Kentgens-Craig, *Bauhaus and America*; Forgács, *Bauhaus Idea*.

12 Kepes, *MIT Years*.

13 Lynch, *Image of the City*.

14 Kepes, *New Landscape in Art and Science*, 17.

15 Moholy-Nagy, *New Vision*, 8.

16 Kepes, *Visual Arts Today*, 187; *Education of Vision*; *Structure in Art and in Science*. Material on Kepes's interactions with many pioneering and cybernetically in-formed art and design practices are in his papers. He has letters to Buckminster Fuller, and was participating with the Experiment in Art and Technology group, for example. Gyorgy Kepes Papers, Archives of American Art, Smithsonian In-stitution, reel 5305; for founding documents on the Center for Advanced Visual Study at MIT, reel 5303–0168–170.

17 Kepes, *New Landscape in Art and Science*, 80–82, 108–12. Kepes also discussed these ideas in his notes from plans for a course at MIT, 1945 (?), Gyorgy Kepes Papers, reel 5312, Archives of American Art, Smithsonian Institution.

18 Kepes, *New Landscape in Art and Science*, 371.

19 Martin, *Organizational Complex*, 42–79.

20 Kepes, *Language of Vision*, 23.

21 Lupton and Miller, *ABC's of [Triangle, Square, Circle]*; Golec, "Natural History of a Disembodied Eye." The architectural theorist and historian George Vrach-liotis argues that Kepes occupies a privileged space in the shift in architecture and design between the gestalt orientation of the Bauhaus and a new technical-scientific structural perspective. Vrachliotis, "Popper's Mosquito Swarm."

22 Kepes's attitude was a subtle reconfiguration of older histories of psychology and perception. He opened *Language of Vision* by quoting one of the founding members of Gestalt psychology, Wolfgang Köhler, "as so called optical illusions show, we do not see individual fractions of a thing; instead, the mode of appear-ance of each part depends on not only upon the simulation arising at that point but upon the conditions prevailing at other points as well." Kepes, *Language of*

Vision, 17, quoting Wolfgang Kohler, *Physikal Gestalten, 1920*. For Kepes, Gestalt psychologies' innovation was to replace a psychological concern with sensation to one with perception. Gestalt psychologists focused on these "modes of appearance" and the forms allowing sensations to be organized cognitively into perceptions. For more information on the histories of cognitive psychology and the use of gestalt see: Galanter, "Writing Plans," 41. Kepes continued to work with Kohler, who contributed to a number of the Vision+Value books. He engaged with geneticist John Haldane, physicist John Desmond Bernal, biochemist and historian Joseph Needham, and geneticist C. H. Waddington during his two years in London from 1935 to 1937. Roach, "Positive, Popular Art," 86–89, 109.

23 Haraway, *Crystals, Fabrics, and Fields*, 62; Roach, "Positive, Popular Art," 86–89, 109.

24 Roach, "Positive, Popular Art," 86–89, 109.

25 One that began with the Bauhaus but found itself reformulated and reinvigorated in the milieu of postwar America. For the members of the Bauhaus design school, the production of a universal language for vision is a dominant aspiration, and this vision for a universal design applicable to anyone, anywhere, based itself in a belief that all humans shared basic psychological processes as defined by Gestalt psychology. For negative critique of the specific relationship between Gestalt psychology, Bauhaus, and design see Lupton and Miller, eds., ABC's of [Triangle, Square, Circle]. Kepes also worked in a milieu of critics who regularly used the concept of gestalt, such as Rudolf Arnheim; see Arnheim, *Art and Visual Perception*, 3, 5, 1.

26 Kepes, *Language of Vision*, 15. My account of Kepes departs from the accounts of individuals, like Ellen Lupton and Michael Golec, who treat Kepes and the sciences of gestalt he worked from as disembodying and reductive. See Golec, "Natural History of a Disembodied Eye"; Lupton and Miller, "Language of Vision," 64.

27 Kepes, *New Landscape in Art and Science*, 104.

28 Kepes, *New Landscape in Art and Science*, 104.

29 This interest was further spurred by the appearance of a series of texts in the late 1950s and very early 1960s, influencing artists and media producers, by Reyner Banham of the British Independent Group and Marshall McLuhan, announcing the rise of a second or electronic machine age, accompanied by works such as Thomas Kuhn and George Kubler on the changing nature of time, and introducing epistemic changes as historical models. For artists and designers, these discourses fed an ongoing belief that the postwar era was fundamentally different and demanded new institutional and organizational responses: Banham, *Theory and Design in the First Machine Age*; McLuhan, *Understanding Media* and *Gutenberg Galaxy*; Kubler, *Shape of Time*; Kuhn, *Structure of Scientific Revolutions*.

30 Goodyear, "Gyorgy Kepes, Billy Klüver, and American Art of the 1960s," 613–14. There is a great deal of work on art and technology at this period and preceding the war. Particularly, the relationship between modernism and the technological imaginary, theories of relativity in physics, and the interaction between biology and artistic and filmic production is much discussed. However, there has been very little examination of differences in artistic or design approaches. For work

on the first part of the twentieth century see: Braun, *Picturing Time*; Doane, *Emergence of Cinematic Time*; Galison and Jones, *Picturing Science, Producing Art*; Crary, *Suspensions of Perception Attention, Spectacle, and Modern Culture*. For postwar surveys: Shanken, "Cybernetics and Art"; Lee, *Chronophobia*; Jones, *Machine in the Studio: Constructing the Postwar American Artist*.

31 Gyorgy Kepes to Mrs. Jerome Rubin, March 28, 1967, Gyorgy Kepes Papers, reel 5305-0169/0170, Archive of American Art, Smithsonian Institution.

32 Correspondences between E.A.T. and Kepes, however, were warm, and there are also many affiliations between the groups and the forms of vision and practice Kepes was encouraging: Brian O'Doherty to Gyorgy Kepes, February 26, 1967, Gyorgy Kepes Papers, Archive of American Art, Smithsonian Institution, reel 5305-0127-0133.

33 Goodyear, "Gyorgy Kepes, Billy Klüver, and American Art of the 1960s."

34 Kepes, MIT *Years*, 14.

35 Kepes, "Lost Pageantry of Nature."

36 Kepes, "Light and Design," 25.

37 Otto Piene, *In Memoriam: Gyorgy Kepes, 1906–2002. Leonardo Journal*. Accessed October 12, 2012. http://www.leoalmanac.org/in-memoriam-gyorgy-kepes-1906 -2002-by-otto-piene/.

38 Kepes, MIT *Years*, 14.

39 Bacsó, *Gyorgy Kepes and the Light Art*.

40 Kepes, MIT *Years*, 10–15.

41 Kepes, "Toward Civic Art," 71.

42 For information on Rand's artist residency programs in the early 1960s, see Lee, *Chronophobia*.

43 For background on MIT's history with CAVS see the founding history of the arts at MIT at the website of CAVS, accessed March 30, 2014, http://cavs.mit.edu /about/id.3.html.

44 Goodyear, "Gyorgy Kepes, Billy Klüver, and American Art of the 1960s"; Leslie, *Cold War and American Science*.

45 Eames et al., brochure for "A Sample Lesson."

46 Eames and Gingerich, "Conversation with Charles Eames," 337.

47 Charles Eames, "Language of Vision: The Nuts and Bolts," 14–16.

48 Norbert Wiener worked on the film emerging from this class. Eames corresponded with the mathematician John von Neumann and most significantly with Claude Shannon, the inventor of the mathematical theory of communication, in the course of preparing this class, and many of the subsequent Eames Office projects. "Letters to Oscar Morgenstern," box 70, folder 3; letter, "Notes for IBM Museum," WCRE, box 148, folder 1, Manuscript Division, Library of Congress; and the credits, where Norbert Wiener and Claude Shannon are listed for *Communication Primer* (1953) available at The Prelinger Internet Archive, accessed March 30, 2014, https://archive.org/details/communications_primer.

49 Shannon and Weaver, *Mathematical Theory of Communication*, 8–9.

50 Eames, "Language of Vision," 14.

51 Eames, "Language of Vision," 17, 20, 14. This was a regularly repeated theme for

Eames. See also: Eames, Norton Lectures at Harvard, WCRE, folder 10, box 217, and folder 7, box 202, Manuscript Division, Library of Congress.

52 Shown slides of Eames lectures and their collection prepared potentially for Sample Lesson, and also for other slide shows, Charles often reused slides. Slides contained a range of images, usually in "sets" demonstrating a theme, the images included for example set of animal claws, hands, and other ways to grasp, different forms of flight-birds, planes, etc, and assortments or collections of artifacts, such as button series, toy series of children's toys such as dolls, boxes, and tops, and examples of design work with pure color blocks and also prototypes of corridors and mazes. Only a few of ths slides remain according to David Herstgaard the archivist, August 16, 2009, Eames Office, 850 Pico Boulevard, Santa Monica, CA (as of June 2014, the office is closed and moving to a new location).

53 Tagg, *Burden of Representation*; Stoler, *Along the Archival Grain*; Canales, *Tenth of a Second*.

54 Colomina, *Domesticity at War*; Kirkham, "Humanizing Modernism"; Carpenter, "Eames Lounge: The Difference between a Design Icon and Mere Furniture."

55 Lynch, *City Sense and City Design*, 11–12.

56 Lynch, *City Sense and City Design*, 12–18.

57 Fishman, *Urban Utopias in the Twentieth Century*.

58 Lynch, *Image of the City*, 13.

59 Kevin Lynch, folder "Environmental Perception: Research and Public Policy, May 1976," Kevin Lynch Papers, 1934–1988 (MC 208), MIT Institute Archives and Special Collections.

60 Kevin Lynch headed major research projects for UNESCO in Argentina, Poland, Australia and Mexico, on the future of the city, as well as directed policy and planning commissions for numerous American agencies, and for the city of Boston throughout the 1960s and 1970s. For some samples of a very extensive portfolio: folder "Environmental Perception: Research and Public Policy, May 1976"; folder "The Possible City (1967) AIP Conference: 'The Next Fifty Years,' October 1967"; folder "Man and the Biosphere (MAB) Project 13—The United States National Committee for Man and the Biosphere, September 17, 1975," Kevin Lynch Papers, MC208, Box 1, MIT Institute Archives and Special Collections.

61 Licklider was a psychologist, and trained in psychoacoustics; he is best known for his work in interactive computing and his contribution in important ways to conceptualizing the Internet. He served as the appointed head of the Information Processing Techniques Office of ARPA and his ideas of time-sharing and networking machines led to ARPAnet, often considered a precursor to the Internet. ibiblio: The Public's Library and Digital Archive, accessed March 30, 2014, www.ibiblio.org/pioneers/licklider.html; Waldrop, *Dream Machine*.

62 Licklider, "Man-Computer Symbiosis."

63 Lynch submitted the original proposal for the project to Licklider in September 1954, asking about his opinions concerning the method, and the analysis of the information. Licklider responded on two fronts, which were incorporated into the project design in interesting ways. The first question Licklider had concerned Lynch's preference for "imageability" in urban space, and wondered if

there might not be a more ethical or morally neutral way to consider the material. The second topic of conversation involved computational approaches to form. Licklider proposed, for example, that Lynch "consider the silhouette of a sky line. Except for open work in electrical signs, etc. this is a single function of a single spatial variable. A possibly interesting hypothesis is that the Fourier transform of the sky line silhouette, which one might call the spectrum of the skyline, is relatable to the artistic acceptability, or perhaps to the average city dweller's reaction." Licklider to Lynch, September 13, 1954, 2; Licklider to Lynch, September 13, 1954; Lynch to Licklider, October 12, 1954; Licklider to Lynch and Kepes, March 3, 1955; Lynch to Licklider, April 8, 1955. Kevin Lynch Papers, MC 208, MIT Institute Archives and Special Collections, available at Dome, digitized content from the MIT Libraries' collections, accessed January 10, 2012, http://dome.mit.edu/bitstream/handle/1721.3/35615/KL_002005.pdf?sequence=1.

64 Lynch to Licklider, April 8, 1955, 1. Kevin Lynch Papers, 1934–1988, MC 208, MIT Institute Archives and Special Collections, MIT Libraries, available at Dome, digitized content from the MIT Libraries' collections, accessed March 30, 2014, http://dome.mit.edu/bitstream/handle/1721.3/35615/KL_002005 .pdf?sequence=1.

65 Lynch, *Image of the City*, 1.

66 For information about Negroponte's initiatives to transform architecture and planning see Negroponte, *Soft Architecture* and *Architecture Machine*.

67 Jacobs, *Death and Life of Great American Cities*; Gans, *Urban Villagers*.

68 Lemov, *World as Laboratory*.

69 Lynch, *City Sense and City Design*, 4–5.

70 Schorske, *Fin-de-Siècle Vienna*; Fishman, *Urban Utopias in the Twentieth Century*; Lopez, *Building American Public Health*; Shanken, *194x*, 41–44; Colomina, *Privacy and Publicity*.

71 See Lynch's course lectures for information on his disciplinary interests: folder "Visual Plan (Seminar 4.572)" (1964 and 1965); folder "Urban Landscape, Articles, Clippings, Photographs" (1966–67[?]). Kevin Lynch Papers, MC 208, box 3, MIT Institute Archives and Special Collections.

72 Here the distinction Arjun Appadurai makes between risk and uncertainty might be useful. For planners there is a great deal of uncertainty in the outcomes of their experiments, but what has increased is a faith in "techniques of calculatability," which operate without the clear-cut endpoints and numbers of risk, but rather through ongoing methods of assessment and documentation. Appadurai, "Ghost in the Financial Machine," 528.

73 Jen Light notes in her study of Rand and urban planning that nuclear deterrence and security was largely subsumed by other concerns about race throughout the 1960s, and in fact largely dropped from the discussion of urban planning. Light, *From Warfare to Welfare*.

74 "1964 New York World's Fair," in Wikipedia, accessed February 14, 2011, http://en.wikipedia.org/wiki/1964_World%27s_Fair. Rosenblum, *Remembering the Future*; Samuel, *End of the Innocence*. It should be noted however that the fair

failed to get the attendance desired, and barely broke even, the result of the conflicts Robert Moses had with numerous groups—racial, religious, and also international. The Bureau of International Expositions boycotted due to the fees Moses was requesting for use of the grounds. Caro, *The Power Broker*, 1093.

75 "Urban League Lays Bias to '64–65 Fair," *New York Times*. February 9, 1962; "World's Fair Urged to Employ Negroes," *New York Times*, June 15, 1961.

76 "Integration 'Drive-in' Set for Fair Grounds," *New York Times*, July 11, 1963; "U.N. Is Picketed on Jobs at Fair." *New York Times*, August 13, 1963.

77 Samuel, *End of the Innocence*, 14–15, 27, 28; "City Rights Panel Summon Unions to Racial Inquiry," *New York Times*, August 14, 1963.

78 For information on the original exhibit see the website of the IBM Corporate Archives, accessed April 15, 2006, www-03.ibm.com/ibm/history/exhibits/vintage/vintage_4506VV2085.html.

79 Colomina, "Enclosed by Images," 20.

80 Script of the IBM film *View from the People Wall*, shown at the Ovoid Theater, New York World's Fair, 1964, transcribed April 17, 2009, from movie reconstruction, courtesy of Eames Office, Santa Monica, CA.

81 IBM, "IBM Pavilion at World's Fair Features 'Information Machine,'" press release, April 21, 1964. Courtesy IBM Corporate Archives, Somers, NY.

82 IBM was early to move corporate headquarters, but this trend would accelerate throughout the decade and into the early 1970s. New York City's employment was stable at about 3.5 million in 1950 and 1960, increasing to 3.8 million in the late 1960s. After 1969 there was a relentless decline, with overall employment reaching lowest levels in 1977, with a fiscal crisis, at 3 million (including government jobs). By 1987, when employment figures rose again, the entire structure of the economy had changed, with manufacturing dropping from 1 million jobs in 1950 to 387,000 by 1987, and finance, insurance, real estate, and services replacing these sectors. This transformation was accompanied by the flight of Fortune 500 corporate headquarters. In 1965, New York City had 128 of these; in 1976 there were 84; and in 1986, 53. Sassen, *Global City*, 200–202.

83 Watson, "Good Design Is Good Business." "The New IBM Research Center in Yorktown," project prospectus, 1959–60, 1, 6. Courtesy IBM Corporate Archives, Somers, NY.

84 Johnston, "From Old IBM to New IBM," 48–53.

85 The soft-sell approach of IBM addressed "the need for science to be better understood by the American public." The company's sponsorship of films, exhibitions, and books was intended to foster a climate in which computers would be perceived as acceptable and benign. In its ongoing film program in the 1970s, IBM directed the Eameses to convey three points: "IBM as an international company; technology is a basic economic resource—like land or water—for a country; and computer technology is being applied to improve the quality of life for people around the world." Albrecht, "Design Is a Method of Action," 38, quote n. 45; Jane P. Cahill to Charles Eames, June 16, 1972, folder 4, box 48, WCRE, quote n. 46; Charles G. Francis to Charles Eames, June 16, 1973, WCRE, folder 4,

box 48, Manuscript Division, Library of Congress. For further reading on IBM and the design process see also: Harwood, "White Room"; Watson, "Good Design Is Good Business"; Friedman, *Nelson, Eames, Girard, Propst*.

86 Saarinen described his "unusual" design in an interview with *Architectural Forum*, "Labs and offices today depend on air conditioning and fluorescent lighting for their air and light not on windows . . . Windows are like fireplaces, nowadays: they are nice to have, but rarely used for their original purpose." Saarinen appeared to conceive of the lab as a closed but self-organizing system, his placement of the external windows also made the building "expandable" and encouraged collaboration and efficiency between the labs, from "Research in the Round," 81–82. In "The New IBM Research Center in Yorktown," project prospectus (1959–60) IBM stipulates that the plan is "flexible," "aesthetic" and "efficient," 6–8. Courtesy of IBM Corporate Archives.

87 "IBM Pavilion at World's Fair Features 'Information Machine,'" 2.

88 I received the full reenactment, viewed the storyboard, and received the script when visiting the Eames Archive in Santa Monica, California, August 16, 2009.

89 Eames and Gingerich, "Conversation with Charles Eames," 334.

90 Eames, "Language of Vision," 24.

91 Charles Eames to Ian McCallum, September 3, 1954, WCRE, folder 3, box 218, 1, Manuscript Division, Library of Congress.

92 Charles Eames to Ian McCallum, September 3, 1954, WCRE, folder 3, box 218, 2, Manuscript Division, Library of Congress.

93 Charles Eames to Ian McCallum, September 3, 1954, WCRE, folder 3, box 218, 1, Manuscript Division, Library of Congress.

94 Charles Eames, Norton Lectures Transcript, 1970–71, WCRE, box 217, folder 10, Manuscript Division, Library of Congress.

95 Charles Eames to Ian McCallum, September 3, 1954, WCRE, folder 3, box 218, 1, Manuscript Division, Library of Congress.

96 Szanton, *Working with a City Government*; Light, *From Warfare to Welfare*. Cyberneticians also worked on urban dispersal and networking: Kargon and Molella, "City as Communications Net."

97 Hamilton, "Films at the Fair," 36–41.

98 Kepes, *Visual Arts Today*, 4

99 Kepes, *Language of Vision*, 217.

100 Kepes, *Language of Vision*, 217.

101 Deleuze, *Cinema 2*, 279.

102 Taken from video interview of Kepes done in 2011, available at MIT+150, accessed March 30, 2014, http://video.mit.edu/watch/gyorgy-kepes-6728/.

103 Kepes was very close with artists like Robert Smithson. He was, however, often not a champion of the artists of his time. In an excellent survey of CAVS and Kepes and the relationship to art movements of the time see Melissa Ragain, where she discusses Kepes's formation of organizational models, institutionally based versus the more corporate and project oriented version of E.A.T. and the other models of resistance and counterculture that artists like Smithson took. She particularly highlights how Kepes's readiness to work with institutions and

governments urging subtle critique from within rather than radical opposition did put him at odds with other artists in the counterculture of the 1960s and early 1970s. Ragain, "From Organization to Network."

104 Roach, "Postivie, Popular Art," 15–16.

105 Lupton, *The ABC's of Design*, 29–32. Behrens, "Art, Design, and Gestalt Theory."

106 Lynch, *What Time Is This Place?*, 1.

107 Lynch, *What Time Is This Place?*, 189.

108 Lynch, *What Time Is This Place?*, 242.

109 Deleuze, *Cinema 2*, 278–80.

110 Martin, *Organizational Complex*, 8.

111 I would also like to mention the important work of Eden Medina on Chilean cybernetics and Salvador Allende here. She also demonstrates alternative potentials and spaces afforded by cybernetics in relation to politics and nation. Medina, *Cybernetic Revolutionaries*.

CHAPTER 3. **Rationalizing**

1 "The General and Logical Theory of Automata," 391–431.

2 McCulloch would argue that schizophrenia was the dominant diagnosis for psychosis. He also complained in letters and talks that its ubiquity made it almost useless, and too general, since it was, apparently, overly utilized to cover the whole spectrum of psychotic disorders in the 1950s. This is enclosed in an unspecified talk or unpublished paper, "The Modern Concept of Schizophrenia, Warren S. McCulloch Papers, B:M139:ser. 3, Warren," ca. 1950s, American Philosophical Society, Philadelphia, PA.

3 The history of the pathology of schizophrenia is both long and constantly changing. The disease was first formally identified as dementia praecox in the 1890s; the term "schizophrenia" and the aforementioned symptoms were formalized by Eugene Bleuler in 1908 to describe a "split," or cognitive dissonance, between personality, thinking, memory and perception. The definition of the disease continued to evolve; at the time McCulloch was writing there was no *Diagnostic and Statistical Manual*, and his definitions adhered to those of earlier in the twentieth century. McCulloch himself was critical of the term, and often thought it was used too often to catalogue too many psychiatric pathologies, particularly psychotic ones. Like other practitioners at the time, he classified schizophrenia into multiple subtypes, of which only one—paranoid—was prone to violence and to the regular imagination of threat and enemies to the self. Bleuler, *Dementia Praecox*; McCulloch, Warren S. McCulloch Papers, "Physiology of Thinking and Perception," June 22, 1954, B:M139, Series III: Creative Engineering, 1, American Philosophical Society.

4 Warren McCulloch and Walter Pitts, "A Logical Calculus of Ideas Immanent in Nervous Activity," 21.

5 McCulloch and Pitts, "Logical Calculus of Ideas Immanent in Nervous Activity," 38.

6 For other discussions about time and memory in cybernetics and early com-

puting see also Wendy Hui Kyong Chun, "The Enduring Ephemeral, or the Future Is a Memory"; Orit Halpern, "Dreams for Our Perceptual Present: Temporality, Storage, and Interactivity in Cybernetics."

7   There is a significant body of work concerning cybernetics and the behavioral and social sciences, for example see Andrew Pickering, *The Cybernetic Brain*; Steve Heims, *The Cybernetics Group*; Rebecca Lemov, "Hypothetical Machines: The Science Fiction Dreams of Cold War Social Science"; N. Katherine Hayles, *How We Became Posthuman: Virtual Bodies in Cybernetics, Literature, and Informatics*; Jean Pierre Dupuy, *The Mechanization of the Mind: On the Origins of Cognitive Science*.

8   For further discussion see Turing, "On Computable Numbers, with an Application to the Entscheidungsproblem"; Russell, *Principles of Mathematics*; Reck, *From Frege to Wittgenstein*; Goldstein, *Incompleteness*; Gödel, *On Formally Undecidable Propositions*; Turing, *Essential Turing*; A. M. Turing's ACE *Report of 1946*.

9   McCulloch, *Embodiments of Mind*, 359.

10 Foucault, *History of Sexuality*; Galison and Daston, *Objectivity*, 115–90, 304–5; Sekula, "Body and the Archive."

11 A number of authors have discussed the transformations in reason, probability, and epistemology; see Daston, *Classical Probability in the Enlightenment* and "Rule of Rules"; Hacking, *Emergence of Probability* and *Taming of Chance*; Matthews, *Quantification and the Quest for Medical Certainty*; Hald, *History of Parametric Statistical Inference*; Gigerenzer, *Empire of Chance*.

12 Mirowski, *Machine Dreams*.

13 Foley, "Strange History of the Economic Agent"; Schull and Zaloom, "Shortsighted Brain."

14 Dumit, "Circuits in the Mind."

15 McCulloch has a number of essays referencing his personal history and implying certain problems with, but also affinity for, religious and ideological stances. Warren McCulloch, "Of I and It" (original title "Of Eye and It"), 195?, series of manuscripts, Warren S. McCulloch Papers, B:M139: ser. 3, American Philosophical Society.

16 McCulloch regularly discusses wartime shocks as case studies in proving his neural theories. There are a number of radio interviews, personal recollections, and letters in the archive discussing such cases. McCulloch, "Of I and It" (original title "Of Eye and It"), 195?, series of manuscripts, and "Physical Treatments of Mental Diseases." Transcript of prepared comments for a BBC radio show to be recorded at Edinburgh, July 23, 1953 with Warren McCulloch, Warren S. McCulloch Papers, B:M139: ser. 3, American Philosophical Society.

17 For histories of the development of psychopharmaceuticals, of which new classes emerged in the 1950s along with new methods of drug testing, and placebo-controlled trials, which enhanced the ideas of targeting and specificity in drug action, see: Greene, *Prescribing by Numbers*, Healy, *Creation of Psychopharmacology*, Shorter, *Before Prozac*, Tone, *Age of Anxiety*.

18 McCulloch, *Embodiments of Mind*, 38.

19 McCulloch, "Physiology of Thinking and Perception," June 22, 1954, B:M139,

Series III: Creative Engineering, p. 1, Warren S. McCulloch Papers, American Philosophical Society.

20 McCulloch, *Embodiments of Mind*, 308–9.

21 Warren McCulloch, "The Computer and the Brain," 1968, unknown—lecture, Cambridge, MA, Warren S. McCulloch Papers, B:M139: ser. 3, American Philosophical Society. McCulloch recalls his early experiences in a number of unpublished essays and lectures in his archive; see McCulloch, Warren S. McCulloch Papers, "Of I and It" (original title "Of Eye and It"), 195?, series of manuscripts, B:M139: ser. 3, American Philosophical Society.

22 W. McCulloch to Dr. Francis Getty of the Illinois Psychiatric Council, March 17, 1950, folder "Illinois Psychiatric Research Council," Warren McCulloch Papers, B:M139, no.1, American Philosophical Society; this letter summates the past of McCulloch's research. W. McCulloch, Aeromedical and Physical Environment Laboratory, progress report no. 5, May 2, 1951, Warren S. McCulloch Papers, B:M139, no.2, American Philosophical Society; details long-running work, and purchase of further equipment for the study of nervous and chemical weapons, supposedly for therapeutic purposes under extreme temperatures. "Physical Treatments of Mental Diseases"; Arbib, "Warren McCulloch's Search for the Logic of the Nervous System."

23 McCulloch recounts, in an undated eulogy, meeting Wiener at dinner with Rosenblueth during the early 1940s, where he encountered the idea of "mechanized teleology." Arbib, "Warren McCulloch's Search for the Logic of the Nervous System"; Kay, "From Logical Neurons to Poetic Embodiments of Mind," 596–97; McCulloch, "Eulogy of Wiener (Unpublished?)," n.d. [1964?], 1, Warren S. McCulloch Papers, B:M139, no.1, American Philosophical Society.

24 For Warren McCulloch, who met Wiener through Arturo Rosenblueth in 1941–1942, cybernetic concepts were instrumental to his reformulation of how to treat the nervous system. For a decade between the mid-1940s and early 1950s the two men shared ideas, and students, in a curious barter of young men that included figures like logician Walter Pitts and neurophysiologist and psychiatrist Jerome Lettvin, both of whom were mentored and offered laboratory space, money, and assistance through McCulloch and Wiener's affiliations at MIT and earlier at the University of Illinois, Urbana-Champaign, and through their vast network of acquaintances inside and outside the university. Trading men, materials, and ideas, Wiener and McCulloch produced a space for investigation linking mind, perception, and communication.

The two men also participated in a conference on cerebral inhibition in 1942 at the behest of Frank Fremont-Smith, later chief administrator of the Macy Foundation, bringing biologists and engineers together in this event, which led to the infamous Macy Conferences, on circular causal and feedback mechanisms in biological and social systems, held in New York throughout the late 1940s and early 1950s, in which both men went on to participate (McCulloch chaired them). McCulloch later came to MIT, with his young protégé Walter Pitts, at Wiener's invitation, in 1952. McCulloch remained at MIT for the rest of his life, working along with Lettvin and others at the Research Laboratory for Electronics, an offshoot of the Radiation Lab, applying work in communications, computation, and signal processing to problems ranging from sound recording to neurophysiology. Arbib, "Warren McCulloch's Search for the Logic of the Nervous System"; McCulloch, "Recollections of the Many Sources of Cybernetics" and "Eulogy of Wiener (Unpublished?)," n.d. [1964?], 1, Warren S. McCulloch Papers, B:M139, no.1, American Philosophical Society.

25 The model has been reviewed elsewhere; here I am briefly outlining the work with focus on epistemology: Kay, "From Logical Neurons to Poetic Embodiments of Mind"; Abraham, "(Physio)Logical Circuits."

26 The logic used in the article was taken from Rudolf Carnap, with whom Pitts had worked.

27 Warren S. McCulloch and Walter Pitts, "The Logical Calculus of Ideas Immanent in Nervous Activity," 21–24.

28 McCulloch and Pitts, A Logical Calculus, 35.

29 Aspray, John von Neumann and the Origins of Modern Computing; von Neumann, "General and Logical Theory of Automata."

30 Markov, Theory of Algorithms, 1.

31 McCulloch, Embodiments of Mind, 10.

32 McCulloch and Pitts, "Logical Calculus of Ideas Immanent in Nervous Activity," 35.

33 The pair had derived their assumptions about how neurons work on what, by that time, was the dominantly accepted neural doctrine in neurophysiology. Using the research of Spanish pathologist, Ramón y Cajal, who first suggested in the 1890s that the neuron was the anatomical and functional unit of the nervous system and was largely responsible for the adoption of the neuronal doctrine as the basis of modern neuroscience, and the work of Cajal's student, Lorento de Nó, on action potentials and synaptic delays between neurons and reverberating circuits, McCulloch and Pitts had the neurological armory to begin thinking neurons as logic gates. Santiago Ramón y Cajal, Texture of the Nervous System of Man and the Vertebrates; Warren McCulloch, Embodiments of Mind, 52.

34 McCulloch and Pitts, "A Logical Calculus of Ideas Immanent in Nervous Activity," 22.

35 See also: Halpern, "Dreams for Our Perceptual Present: Temporality, Storage, and Interactivity in Cybernetics."

36 See for example interviews on treatment of soldiers coming from World War II done in Britain where McCulloch steadfastly spoke against narrative therapy, and proactively promoted drug treatment to rewire circuits in the brain: British Broadcasting Corporation, "Physical Treatments of Mental Diseases." Warren S. McCulloch Papers, B:M139: ser. 3, American Philosophical Society.

37 See also Peter Galison, "The Ontology of the Enemy: Norbert Wiener and the Cybernetic Vision"; Paul N. Edwards, The Closed World Computers and the Politics of Discourse in Cold War America.

38 Von Neumann, "First Draft of a Report on the EDVAC." Von Neumann confessed to being influenced by the McCulloch/Pitts model in conceiving machine memory.

39 Herman H. Goldstine and John von Neumann, "Planning and Coding of Problems for an Electronic Computing Instrument: Report on the Mathematical and Logical Aspects of an Electronic Computing Instrument, Part II, Volume I," 157.

40 Turing, "Proposal for Development in the Mathematics Division of an Automatic Computing Engine (ACE) (1946)," 21.

41 These concepts are inspired by Dumit, "Circuits in the Mind," 7.

42 Turner, *From Counterculture to Cyberculture*. Bergthaller, *Addressing Modernity: Social Systems Theory and U.S. Cultures*, 277.

43 Deleuze and Guattari, *Thousand Plateaus*, 21–22.

44 Bateson had a history with working on psychology and culture. He had worked with Margaret Mead on major studies on topics of schizophrenia before the war. Particularly the pair worked on a psychoanalytically informed study on "dementia praecox," or schizophrenia, in Bali throughout the late 1930s. His shift to discourses of communication, control, and cybernetics was made during the war when he, like most anthropologists, readily (a fact he later critiqued about himself) went to work for the Office of Strategic Services, serving twenty months in Ceylon, India, Burma, and China in service of the Allied forces as an interpreter of local custom and culture. While initially excited, he eventually found the work dull and accomplished little—with one exception, when he briefly operated a radio station undermining Japanese propaganda in Burma and Thailand. Studying the content of the broadcasts, Bateson decided to engage in acts of positive feedback. He related: "we listened to the enemy's nonsense and we professed to be a Japanese official station. Every day we simply exaggerated what the enemy was telling the people." Lipset, *Gregory Bateson*, 174. He recalled how his work in ethnography had made him recognize that psychology was a matter of interactivity, not an isolated aspect contained within individual subjects. Linking together his observations on the schizophrenic subject of the Balinese, the nature of propaganda and media, and the way cultures engage in conflict, Bateson came to reformulate mind in terms of communication and, in his words, "ecology." His future interest in cybernetics emerged out of these seemingly unrelated observations about culture, propaganda, and feedback. Bateson et al., "Towards a Theory of Schizophrenia"; Lipset, *Gregory Bateson*.

45 Heims, *Cybernetics Group*.

46 Gregory Bateson to Warren McCulloch, November 29, 1954, Warren S. McCulloch Papers, B:M139, no.2, American Philosophical Society.

47 Bateson et al., "Towards a Theory of Schizophrenia." Bateson, *Steps to an Ecology of Mind*, 177, 80, 89, 365.

48 Bateson, *Steps to an Ecology of Mind*.

49 Wilder-Mott and Weakland, *Rigor and Imagination*, Mental Research Institute, MRI.

50 Jackson, *Therapy, Communication, and Change*, 130–32; Ruesch, *Communication and Disturbed Communication*.

51 Festinger, *A Theory of Cognitive Dissonance*

52 Research Center for Group Dynamics, University of Michigan, accessed April 23, 2014, http://www.rcgd.isr.umich.edu/history/.

53 Festinger, *A Theory of Cognitive Dissonance*, 11, 145, 283, 284.

54 Festinger initially started work examining ecology and housing, which sparked an interest in interactions between individuals and environment. At MIT, as an assistant professor coming from the University of Iowa, he was finally exposed to communication theories that led to his theory of cognitive dissonance. However, he had already worked at Iowa and come to MIT with Kurt Lewin, one of the "founders" of social psychology, who always viewed individuals in relationship to environments. Festinger also did extensive statistical work for the Committee on Selection and Training of Aircraft Pilots at University of Rochester during the war (Stanley Schachter, "Leon Festinger," 99–105).

Along with many of the rising cognitive scientists of the time, including, for example, George Miller at Harvard, Festinger wanted to transform behaviorism, which in his mind took too linear and simple an approach to the relationship between stimulus and response (whether this is true or not is independent of the fact that many emerging social psychologists utilized a discourse involving behaviorism's reductionism and its ideal of an isolated subject): Festinger, *Theory of Cognitive Dissonance*, 11, 145, 283, 84; "Obituary, Leon Festinger, 69, New School Professor," *New York Times*, February 12, 1989; Festinger, *Conflict, Decision, and Dissonance* and *Social Pressures in Informal Groups*. For some of his predecessor studies to the book announcing the theory of cognitive dissonance see Festinger, *When Prophecy Fails*, "Theory of Social Comparison Processes," and "Informal Social Communication"; Lewin, "Action Research and Minority Problems."

55 McCulloch and Pitts, "Logical Calculus of Ideas Immanent in Nervous Activity," 38.

56 McCulloch and Pitts, "Logical Calculus of Ideas Immanent in Nervous Activity," 38; "Physical Treatments of Mental Diseases."

57 It is also not coincident that at this time "gremlin" is a term used by engineering crews for mechanical failures on airplanes. McCulloch, *Embodiments of Mind*, 374.

58 McCulloch, *Embodiments of Mind*, 377.

59 McCulloch, "Of I and It" (original title "Of Eye and It"), 195?, series of manuscripts, 558–59, Warren S. McCulloch Papers, B:M139: ser. 3, American Philosophical Society.

60 Thurschwell, "Ferenczi's Dangerous Proximities," 156–57.

61 Bateson, *Steps to an Ecology of Mind*, 135–36, emphasis mine.

62 Bateson, *Steps to an Ecology of Mind*, 139.

63 Von Neumann and Morgenstern, *Theory of Games and Economic Behavior*, 33.

64 Von Neumann and Morgenstern, *Theory of Games and Economic Behavior*, 8–9.

65 Aspray, *John von Neumann and the Origins of Modern Computing*, von Neumann, "General and Logical Theory of Automata," Poundstone, *Prisoner's Dilemma*.

66 As Kenneth Arrow, the Nobel laureate Stanford economist and leading proponent of rational choice decision-making theory, wrote in his critical 1951 text *Social Choice and Individual Values*, "in a capitalist democracy there are essen-

tially two methods by which social choices can be made: voting, typically used to make 'political' decisions and the market mechanism, typically used to make 'economic' decisions. In the following discussion of consistency of various value judgments as to the mode of social choice, the distinction between voting and the market mechanism will be disregarded." *Social Choice and Individual Values*, 12. Writing in support of that democracy in the same work, Arrow signals the compression of economics and politics by way of a discourse of "choice" and "information." The implications of these algorithmic logics, therefore, also spelled a transformation of politics and the erasure of political economy in the social sciences for a focus on unifying the methods across disciplines. But while Arrow was a clear supporter of a reasonable and liberal subject and an obvious proponent of American capitalism and anticommunism, not all those who were rethinking politics and economics were as antagonistic to Communism or as ready to equate freedom with functional consumption. Political scientists began to trace networks of communication and model organizations and networked intelligences.

67 Miller discusses the conferences at RAND and the influence of neural nets and cybernetics in early cognitive psychology: Miller et al., *Plans and the Structure of Behavior*, 3, 49. See also Miller, "Reviewed Work(s)."

68 Simon, "Behavioral Model of Rational Choice." See also: Crowther-Heyck, *Herbert A. Simon*; Simon et al., *Economics, Bounded Rationality and the Cognitive Revolution*.

69 Simon, *Sciences of the Artificial*, 6–7, 14, 24; Parikka, *Insect Media*, 135.

70 Clough, *Autoaffection*; Mirowski, *Machine Dreams*.

71 MacKenzie, *Engine, Not a Camera*; Crary, *Techniques of the Observer*.

72 Mirowski, *Machine Dreams*, 423.

73 MacKenzie, *Engine, Not a Camera*, 38, 123.

74 Cited in Foley, "Strange History of the Economic Agent," 90.

75 Deutsch wrote scathing reviews of her work, particularly *Origins of Totalitarianism*. See Papers of Karl Wolfgang Deutsch, "Notes on Arendt Manuscript (1949)," HUFGP 141.50, box 9, Harvard University Archives. A major figure in analyzing critical events such as the Berlin Blockade of 1948–1949, the occupation of Hungary by Soviet forces in 1957, and the fraught status of his own homeland, Czechoslovakia, for the U.S. government and military, Deutsch had thought hard about the sort of decision-making and forms of reason mandated to answer the imperatives of highly technical, global, and rapid politics. This text, coming immediately on the heels of the Cuban Missile Crisis, was written by a man who might very well have had a sense of just how nervous and potentially unreasonable governments really were.

76 Deutsch, Papers of Karl Wolfgang Deutsch, "Comments on Margaret Mead," September 1949, unpublished manuscript, 1949, HUFGP 141.50, box 2, Harvard University Archives.

77 Deutsch, Papers of Karl Wolfgang Deutsch, lecture, 195?, "Cybernetic Approach to History of Human Thought," HUFGP 141.60, box 4, Harvard University Archives. Emphasis mine.

78 Deutsch, Papers of Karl Wolfgang Deutsch, "Cybernetic Approach to Human Thought," speech, 195?, 4–5. Also see Deutsch, "Notes to an Approach to Communication Research in Political and Social Science," unpublished manuscript, October 18, 1952, HUGFP 141.50, Box 1, Harvard University Archives.

79 Deutsch, *Nerves of Government*, 87.

80 Deutsch, *Nerves of Government*, 75–109, 258.

81 Deutsch, Papers of Karl Wolfgang Deutsch, Lecture: "Race (1969–1970)," Papers Relating to Conferences and Symposia, 1944–1990, HUGFP 141.60, box 7, 2–6, Harvard University Archives.

82 Deutsch, Papers of Karl Wolfgang Deutsch, Lecture: Race (1969–1970), Papers Relating to Conferences and Symposia, 1944–1990, HUGFP 141.60, box 7, 2–6, Harvard University Archives.

83 Deutsch, Papers of Karl Wolfgang Deutsch, Lecture: Race (1969–1970), Papers Relating to Conferences and Symposia, 1944–1990, HUGFP 141.60, box 7, 2–6, Harvard University Archives.

84 See note 95.

85 Paul Lazarsfeld and Robert Merton, for example, two of the most prominent sociologists in postwar America, located at Columbia University, turned their attention to producing new methods for assessing individual psychological relationships to collective formations and media. Lazarsfeld began the Bureau for Social Research, a rather sinister name for a research center at Columbia that did work throughout the war, often at the behest of the U.S. government, on the impact of media on populations, by way of integrating qualitative and quantitative methods. Lazarsfeld then proceeded to attend the Macy Conferences on Cybernetics, where he was exposed to communication and cybernetic theories. Barton, "Paul Lazarsfeld as Institutional Inventor"; Lemov, "Hypothetical Machines"; Heims, *Cybernetics Group*; Lazarsfeld and Merton, "Mass Communication, Popular Taste, and Organized Social Action."

86 Jackie Orr has also done serious work on the emergence of psychology as a social instrument in the work of Talcott Parsons. See Orr, *Panic Diaries*.

87 Terranova, "Another Life," 235.

88 In this reading, I am not invoking the subjectively grounded analysis of affect in cybernetics, particularly in the work of Pitts, by individuals like Elizabeth Wilson following Silvan Tompkins. Affect here is not about subjectivity or about the repression of drives (although as the earlier discussion about psychoanalysis demonstrates, older models haunt this epistemology). Wilson, *Affect and Artificial Intelligence*. For other authors on affect in media and the performativity of subjects see: Cartwright, *Moral Spectatorship*, Bruno, *Atlas of Emotion*, Sedgwick, *Touching Feeling*, Sedgwick and Frank, eds., *Shame and Its Sisters*.

89 Clough, "New Empiricism," Holland, "Deterritorializing "Deterritorialization," Terranova, "Another Life."

90 Pias, *Cybernetics*, 31.

91 Pias, *Cybernetics*, 35.

92 Von Foerster, *Cybernetics*, 144–45.

93 Deutsch, *Nerves of Government*, 72.

94 Deutsch, *Nerves of Government*, 206–7.

95 Deutsch, *Nerves of Government*, 72.

96 Deutsch, Papers of Karl Wolfgang Deutsch, "Review: A New Landscape Revisited [,] by Gyorgy Kepes," April 1960, unpublished—delivered October 18, 1952, HUGFP, 141.50, Manuscripts and Research Materials ca. 1940–1990, box 1, Harvard University Archives.

97 Deutsch, *Nerves of Government*, 201–6.

98 Deutsch, Papers of Karl Wolfgang Deutsch, "Review: A New Landscape Revisited [,] by Gyorgy Kepes," April 1960, unpublished—delivered October 18, 1952, HUGFP, 141.50, Manuscripts and Research Materials ca. 1940–1990, box 1, Harvard University Archives.

99 Deutsch's archives contain numerous corporate consultancy lectures and many talks and courses given at such sites as the National War College and the State Department. Papers of Karl Wolfgang Deutsch, Papers Relating to Conferences and Symposia, 1944–1990, HUGFP, 141.77, box 7, and "Review: A New Landscape Revisited [,] by Gyorgy Kepes," April 1960, unpublished—delivered October 18, 1952, HUGFP, 141.50, Manuscripts and Research Materials ca. 1940–1990, box 1, Harvard University Archives.

100 Rebecca Lemov writes extensively on Merton's fantasized methodological machine. Other historians of urban planning, for example, such as Andrew Shanken, have demonstrated how post-war visions of the city increasingly became statistical rather than emotional or anthropomorphic, part of this move toward calculation as an imagined mode of administration over space. Shanken, "Uncharted Kahn," Lemov, "Hypothetical Machines."

101 Chun, *Control and Freedom*, 8.

102 The historical record shows that Deutsch, along with many of his colleagues, was a major proponent of the civil rights movement, a central analyst and adviser in addressing urban race riots in the 1960s, and a regular advocate of détente and a voice against extremism and Red bashing, and a compassionate advocate against hard liners. His reports express empathy for postcolonial concerns and voice a regular critique of reactionary attitudes toward change in institutions like the State Department. Whether his work is the face of a new humanitarian empire, or a genuine alternative to the violence of the Cold War is up for judgment.

103 Bateson, *Steps to an Ecology of Mind*, 277.

104 For further information on the relationship between schizophrenia, creativity, difference, and genius see: Gottesman, *Schizophrenia Genesis*, Felman, *Writing and Madness*. Gilman, *Difference and Pathology*.

105 Bateson, *Steps to an Ecology of Mind*, 278.

106 Bateson, *Steps to an Ecology of Mind*, 278.

107 It might also be noted that in separating these pathologies, Deleuze and Guattari were merely following medical dictates of the time. Until the most recent *Diagnostic and Statistical Manual* for psychiatric disorders was issued, there were thought to be three forms of schizophrenia—disorganized, catatonic, and paranoid. The other two forms are not considered violent and often even were

associated with creativity throughout the early twentieth century. Only paranoid schizophrenia was considered violent. It is a marker of medical history that
throughout the 1960s it is the paranoid form that increasingly takes precedence
in defining the disease, and increasingly the disease classification is also expo-
nentially applied to people of color, at least in the United States. Metzl, *Protest
Psychosis*.

108 Deleuze and Guattari, *Thousand Plateaus*, 21–22.
109 Holland, "Deterritorializing 'Deterritorialization,'" 64.
110 This observation is owed to the work of Eugene Holland, "Deterritorializing
'Deterritorialization'" and Deleuze and Guattari, *Thousand Plateaus*.

CHAPTER 4. **Governing**

1 Miller, "Magical Number Seven, Plus or Minus Two."
2 Nelson, *Problems of Design*, 63.
3 Sandeen, *Picturing an Exhibition*, 125–26, Colomina, "Enclosed by Images."
4 Cited in Sandeen, *Picturing an Exhibition*, 134.
5 Cited in Sandeen, *Picturing an Exhibition*, 134.
6 Rancière, *Politics of Aesthetics*, 10, 13.
7 Foucault, *Birth of Biopolitics*.
8 Colomina, "Enclosed by Images."
9 For a sampling of writers addressing this particular exhibition, and the place of
design or media in the Cold War, see Colomina, *Domesticity at War*; Masey and
Morgan, *Cold War Confrontations*; Hixson, *Parting the Curtain*; Wulf, *Moscow
'59*; Kushner, "Exhibiting Art at the American National Exhibition in Moscow";
Castillo, *Cold War on the Home Front*; Bogart and Bogart, *Cool Words, Cold War*.
10 Lettvin et al., "What the Frog's Eye Tells the Frog's Brain"; Galison, "Ontology of
the Enemy"; Heims, *Cybernetics Group*; Pickering, *Cybernetic Brain*.
11 Lettvin et al., "What the Frog's Eye Tells the Frog's Brain," 237, 254.
12 Lettvin et al., "What the Frog's Eye Tells the Frog's Brain," 251.
13 Arbib, *Brains, Machines, and Mathematics*, 32–33.
14 Crary, *Techniques of the Observer*; Galison and Daston, *Objectivity*.
15 At the time there were numerous efforts to produce such models of cognition
and perception. As Fritjof Capra writes, "in the 1950s scientists began to actually build models of such binary networks [like those of McCulloch and Pitts],
including some with little lamps flickering on and off at the nodes. To their
great amazement they discovered that after a short time of random flickering,
some ordered patterns would emerge in most networks. They would see waves
of flickering pass through the network, or they would observe repeated cycles.
Even though the initial state of the network was chosen at random, after a while
those ordered patterns would emerge spontaneously, and it was that spontaneous emergence of order that became known as 'self-organization.'" Capra,
Web of Life, 83–84, Johnston, *Allure of Machinic Life*, 301–8.
16 A lot of work was done at the time on scanning, flickering, and other significant
visual phenomena. For example, the work of Frank Rosenblatt under the influ-

ence of McCulloch on machine vision, cognition, and scanning had an impact on the conception and design of perception and cognition in electronics and machine intelligence and in developing scanning technologies. Rosenblatt, "Perceptron," of 1958 anticipates Lettvin et al., "What the Frog's Eye Tells the Frog's Brain," of 1959, and both articles anticipated future work in contemporary machine learning and vision models where symbolic processing and neural nets are now returned to use, even though at the time ideas of perceptrons were later debunked in 1969 by Marvin Minsky and Seymour Papert in a book titled *Perceptrons*, in an ongoing set of debates over the nature and approach to machine intelligence.

17 McCulloch, *Embodiments of Mind*, 34.

18 McCulloch elaborated on these concepts in many of his lectures, including those preceding the actual conduct of this experiment; "Physiology of Thinking and Perception."

19 Miller et al., *Plans and the Structure of Behavior*. For more on the relationship between cybernetics and cognitive science, and on the rise of cognitive science see Harnish, *Minds, Brains, Computers*; Gardner, *Mind's New Science*; Dupuy, *Mechanization of the Mind*; Boden, *Mind as Machine*; Edwards, *Closed World Computers*.

20 Miller, "Magical Number Seven, Plus or Minus Two," 81.

21 Miller, "Magical Number Seven, Plus or Minus Two," 81–83.

22 Miller, "Magical Number Seven, Plus or Minus Two," 81.

23 Miller, "Magical Number Seven, Plus or Minus Two," 82–83.

24 Miller, "Magical Number Seven, Plus or Minus Two," 92–94.

25 Miller, "Magical Number Secen, Plus or Minus Two," 91–94.

26 Miller et al., *Plans and the Structure of Behavior*, 3.

27 Eames, "Language of Vision," 13–14.

28 Numbers are unclear, but official figure given by the United States Government was 2.7 million visitors. Keylor, "Waging the War of Words," 92.

29 Cited in Kirkham, *Charles and Ray Eames*, 323. For further information on the experience of this installation also see Castillo, *Cold War on the Home Front*, v–vii. For information on the psychological warfare aspects see Belmonte, *Selling the American Way*, 50.

30 Cited in Kushner, "Exhibiting Art at the American National Exhibition in Moscow," 6.

31 See also Nilsen, *Projecting America, 1958*, Mathews, "Art and Politics in Cold War America."

32 Documented in Kirkham, *Charles and Ray Eames: Designers of the Twentieth Century*, 323.

33 I saw *Glimpses* reenacted at the Eames Office, Santa Monica, CA in August 2009.

34 Sandeen, *Picturing an Exhibition*, 139, IBM, "Ramac Answers Questions in Moscow." Press release. August 5, 1959. Courtesy of IBM Corporate Archives.

35 Sekula, "Traffic in Photographs"; Sontag, *On Photography*, 32; Barthes, "Great Family of Man."

36 Other critics have argued otherwise. Commenting that the show "looked back"

on its viewers in feedback loops, the critic Bruce Stimson argues that *Family of Man* undid the ego, and forced an effort to affiliate with the Other with progressive potential. Stimson, *Pivot of the World*.

37 Eames, "Language of Vision." See also the film scripts and records for *Glimpses*, Script Outlines, WCRE, box 202, folder 4, Manuscripts Division, Library of Congress.

38 Youngblood, *Expanded Cinema*, 208–13.

39 Whitney, *Digital Harmony*, 43.

40 Whitney, "Animation Mechanisms," 26. Correspondences with John Whitney, Sr., WCRE, folder 1, box 117, Manuscripts Division, Library of Congress.

41 The first major cinematic release use of his machines was, perhaps not incidentally, Saul Bass's introduction to Alfred Hitchcock's *Vertigo* in 1957. Whitney, *Digital Harmony*, 83–97, 129–45.

42 John Whitney to Charles Eames, August 24, 1972, WCRE, folder 1, box 117, Manuscripts Division, Library of Congress.

43 Eames, "Language of Vision." See also film scripts and production notes for *Glimpses*, Script Outlines, WCRE, box 202, folder 4, Manuscripts Division, Library of Congress.

44 Doane makes this argument in her article on the close-up, but an important related argument is also put forth by Lisa Gitelman and Jonathan Auerbach related to scale and governance in microfilm during the Cold War. Doane, "Close-Up"; Gitelman and Auerbach, "Microfilm, Containment, and the Cold War."

45 Neuhart, *Eames Design*, 241.

46 Eames and Gingerich, "Conversation with Charles Eames," 326, 333.

47 Charles Eames also expanded on these themes in his Norton Lectures at Harvard in 1970. See Eames Norton Lectures at Harvard, 1970, WCRE, box 217, folder 10, Manuscripts Division, Library of Congress.

48 Eames and Gingerich, "Conversation with Charles Eames," 331.

49 Nilsen, *Projecting America, 1958*. It should also be noted that after years of Stalin and socialist realism, it is not clear how familiar a Soviet public would be with constructivist work.

50 Eames and Gingerich, "Conversation with Charles Eames," 334.

51 Benjamin, *Illuminations*, Colomina, "Enclosed by Images." Foucault, *The History of Sexuality*, 133–60, 140.

52 Press Write up, Publicity File 1959, WCRE, box 231, folder 3, Manuscripts Division, Library of Congress.

53 Nilsen, *Projecting America, 1958*.

54 Sekula, "Body and the Archive"; Stoler, *Along the Archival Grain*; Foucault, *Birth of Biopolitics*.

55 Turner, "*The Family of Man* and the Politics of Attention in Cold War America."

56 Turner, "*The Family of Man* and the Politics of Attention in Cold War America," 83.

57 These observations are based on the documentation of the installation available in WCRE, lot 13234-1-3, lot 13393, Photography and Print Division, Library of Congress.

58 Bogart and Bogart, *Cool Words, Cold War*, 55.

59 Colomina, "Enclosed by Images," 9.

60 Eames and Gingerich, "Conversation with Charles Eames," 334.

61 Morris, "War Drive," 105.

62 Morris, "War Drive."

63 Marx, *Grundrisse*, 163.

64 The entire discussion of Marx here is indebted to Morris, "Review of James T. Siegel, *Fetish, Recognition, Revolution*," 165.

65 Mekas, "Expanded Arts," Felicity Scott, *Architecture or Techno-Utopia*, 188.

66 John Whitney, Sr., was also quite conservative, and massively homophobic, a reason why he stopped speaking to his brother, James Whitney, with whom he had collaborated in the 1940s. But he is still forefather to some of the more radical body, sex, and gendered art of the period in the world of the "expanded arts diagram." This history was relayed to me in private communication with the film and music historian James Tobias, May 13, 2013.

67 Youngblood, *Expanded Cinema*, 220–21; emphasis mine.

68 Youngblood, *Expanded Cinema*, 215.

69 Youngblood, *Expanded Cinema*, 18.

70 Whitney espoused an ideal of reconciling differences and producing "human" images even though at the same time he ceased to work with his own brother on account of his brother's homosexuality. Despite producing a machine cinema that ostensibly no longer worked though normative forms of viewership, and therefore theoretically no longer producing disciplinary forms of discrimination, Whitney himself did not appear to be able to accept this possibility that could emerge from his own image making. The story of the Whitney brothers, both whom pioneered new modes of animation demonstrates the multiplying possibilities, both positive and negative, for rethinking difference that is produced by such practices. Private communication with James Tobias, May 13, 2013.

71 Deleuze, *Cinema 2*, 278.

72 Deleuze, *Difference and Repetition*, 139. Zabet Patterson in her work on John Whitney's brother, James, alerted me to this reference, which she also likened to the cybernetic cinema of the Whitneys. Patterson, "From the Gun Controller to the Mandala," 50.

73 Deleuze, *Cinema 2*, 279.

74 Deleuze, *Cinema 2*, 264–65.

75 Miller, "Magical Number Seven, Plus or Minus Two," 96.

Conclusion

1 Colomina, "Enclosed by Images."

2 Joseph Dumit has written a very insightful analysis of a similar logic operating in the pharmaceutical industry: Dumit, *Drugs for Life*. Kaushik Sunder Rajan has also created a mirror discussion in his work on "bio-capitalism" that both contests and abets my account here: Rajan, *Biocapital*.

3 Cited in Anthony Vidler, "Toward a Theory of the Architectural Program," 59.

4 Jameson, "Future City," 75–76.

5 Halpern et al., "Test-Bed Urbanism."

6 Burdett and Rode, *Stupefying Smart City* and *Social Nexus*.

7 Sassen, "From Database Cities to Urban Stories Conference," Open Source Urbanism lecture video. Created September 15, 2011. Accessed December 26, 2012. http://postscapes.com/trackers/video/saskia-sassen-i-bring-open-source -urbanism-and-urbanizing-technology.

8 Burdett and Rode, *Social Nexus*.

9 Front page, available at the website of Senseable Cities Lab, accessed April 14, 2014, http://senseable.mit.edu/.

10 Front page, available at the website of Senseable Cities Lab, accessed April 14, 2014, http://senseable.mit.edu/.

11 Burdett and Rode, "Electric City," 2.

12 For an important critical discussion on open source in contemporary culture that considers both the benefits and ideologies of open source see Coleman, *Coding Freedom*.

13 Rodowick, *Gilles Deleuze's Time Machine*.

14 Vidler, "Toward a Theory of the Architectural Program," 66–67.

15 *Archigram*, 6–8.

16 *Archigram*, 8.

17 *Archigram*, 36.

18 *Archigram*, 41.

19 Vidler, "Toward a Theory of the Architectural Program," 66–67.

20 Andrew Pickering also briefly discusses the relationship between cybernetics and Archigram, with a similar interest in demonstrating the emergent and progressive potentials of cybernetic thought. Pickering, *Cybernetic Brain*, 365–68.

21 Sadler, *Archigram*, 197.

22 *Archigram*, 110. The excerpt of the poem it should be noted is not by David Greene—it is by Richard Brautigan written in 1967.

Epilogue

1 This biography is synthesized from Noguchi, *Isamu Noguchi*; Lyford, "Noguchi, Sculptural Abstraction, and the Politics of Japanese American Internment"; Altschuler, *Noguchi*.

2 Noguchi, "I Become a Nisei." File Source: Bio: Poston, AZ: Relocation, Isamu Noguchi Writings-Readers Digest Submission-1942, folder 8 of 14. Archives of the Isamu Noguchi Foundation and Garden Museum, New York.

3 Noguchi, "I Become a Nisei," 1.

4 Noguchi, "I Become a Nisei," 1.

5 Noguchi, "I Become a Nisei," 4.

6 Noguchi, "I Become a Nisei," 6.

7 Noguchi, "I Become a Nisei," 4–5.

8 Noguchi, *Isamu Noguchi*, 177–79.

9 Lyford, "Noguchi, Sculptural Abstraction, and the Politics of Japanese American Internment."

10 Noguchi, "Hiroshima Memorial to the Dead: A Project," Project Series: Hiroshima Memorial for the Dead, 1951–53, Archives of the Isamu Noguchi Foundation and Garden Museum.

11 Isamu Noguchi to John (?), February 4, 1953, 2, Project Series: Hiroshima Memorial for the Dead, 1951–53, Archives of the Isamu Noguchi Foundation and Garden Museum.

12 Thomas J. Watson, Jr., to Isamu Noguchi, May 24, 1988, Archives of the Isamu Noguchi Foundation and Garden Museum.

13 "The New IBM Research Center in Yorktown," project prospectus, 1959–60, 1, 6, IBM Corporate Archives, Somers, New York. Courtesy of IBM Corporate Archives.

14 Noguchi, *Isamu Noguchi*, 172.

15 Noguchi, *Isamu Noguchi*, 172. Article on Noguchi for *Think Magazine*, IBM Armonk, (1974?), Projects: Armonk, NY, "IBM Gardens," 1975–77, folder 3 of 4, Archives of the Isamu Noguchi Foundation and Garden Museum.

16 Noguchi, *Isamu Noguchi*, 172. Correspondence with Herman Goldstine (August 26, 1964), Projects: Armonk, NY, "IBM Gardens," 1975–77, folder 1 of 4, Archives of the Isamu Noguchi Foundation and Garden Museum.

17 Article on Noguchi for *Think Magazine*, IBM Armonk, (1974?), Projects: Armonk, NY, "IBM Gardens," 1975–77, 2, folder 3 of 4, Archives of the Isamu Noguchi Foundation and Garden Museum.

18 Apostolos-Cappadona and Altshuler, *Isamu Noguchi*, 66.

19 Apostolos-Cappadona and Altshuler, *Isamu Noguchi*, 71.

20 Noguchi, *Isamu Noguchi*, 172.

21 Altschuler, *Noguchi*, 77.

22 Apostolos-Cappadona and Altshuler, *Isamu Noguchi*, 68.

23 The Israel museum labels the garden a mix of "cultures" including the Near East, Far East, and Western Cultures, Israel Museum, "The Billy Rose Garden." The museum, however, emphasizes that it is built as an "organic" Mediterranean village in architecture and master plan. The architect of the building, Al Mansfield, was a Russian who trained in Germany, and was heavily influenced by the Bauhaus. Israel Museum Al Mansfield page, accessed April 14, 2014, www.english.imjnet.org.il/htmls/page_1514.aspx?co=14946&bsp=14296 and www.english.imjnet.org.il/page_1554.

24 Noguchi, *Isamu Noguchi*, 174. His archive is also full of continuing correspondences and visits to the museum until his death.

25 Apostolos-Cappadona and Altshuler, *Isamu Noguchi*, 71.

26 Weizman, *Hollow Land*, 25–56.

27 Siegel, *Fetish, Recognition, Revolution*.

28 Apostolos-Cappadona and Altshuler, 119.

Bibliography

Archives

The Archigram Archival Project, University of Westminster, London, UK.

Department of Circulating Exhibitions Records, Museum of Modern Art Archives, Museum of Modern Art, NY.

Papers of Karl Deutsch, HUGFP 141 and HUGBD 322, Collections of the Harvard University Archives, Harvard University, Cambridge, MA.

The Eames Office, Santa Monica, CA.

Charles and Ray Eames Papers, Manuscript Division, Library of Congress, Washington, DC.

The Work of Charles and Ray Eames, Prints and Photographs Reading Room, Library of Congress, Washington, DC.

Film Study Center, Museum of Modern Art Archives, Museum of Modern Art, New York.

IBM Corporate Archives, Somers, NY.

Gyorgy Kepes Papers. Archives of American Art, Smithsonian Institution, New York, NY.

Jerome Y. Lettvin Papers, MC525, Massachusetts Institute of Technology Institute Archives and Special Collections, Cambridge, MA.

Kevin Lynch Papers, MC 208, Massachusetts Institute of Technology Institute Archives and Special Collections, Cambridge, MA.

George Maciunas/Fluxus Foundation, Inc. New York, NY.

Warren S. McCulloch Papers, Mss.B.M139, American Philosophical Society, Philadelphia, PA.

Papers of George Armitage Miller (1949–1957), HUG 4571.5, Collections of the Harvard University Archives, Harvard University, Cambridge, MA.

The Archives of the New York City Department of Parks and Recreation, New York, NY.

New York World's Fair 1964–65 Corporation Records, MssCol 2234, Manuscripts and Archives Division, The New York Public Library, New York, NY.

The Noguchi Museum Archives, Queens, New York, NY.

Norbert Wiener Papers, MC 22, Massachusetts Institute of Technology Institute, Archives and Special Collections, Cambridge, MA.

Sources

Abraham, Tara. "(Physio)logical Circuits: The Intellectual Origins of the McCulloch-Pitts Neural Networks." *Journal of the History of the Behavioral Sciences* 38 (Winter 2002): 3–25.

Adorno, Theodor W. "Culture Industry Reconsidered." In *Media Studies: A Reader*, edited by Paul Marris and Sue Thornham, 31–37. New York: New York University Press, 2000.

Akera, Atsushi. *Calculating a Natural World: Scientists, Engineers, and Computers during the Rise of U.S. Cold War Research*. Cambridge, MA: MIT Press, 2007.

Altschuler, Bruce. *Noguchi, Modern Masters*. New York: Abbeville Press, 1994.

Amadae, S. M. *Rationalizing Capitalist Democracy: The Cold War Origins of Rational Choice Liberalism*. Chicago: University of Chicago Press, 2003.

Amin, Ash, and Patrick Cohendet. *Architectures of Knowledge: Firms, Capabilities, and Communities*. Oxford: Oxford University Press, 2004.

Anable, Aubrey. "The Architecture Machine Group's *Aspen Movie Map*: Mediating the Urban Crisis in the 1970s." *Television and New Media* 13, no. 6 (2012): 498–519.

Apostolos-Cappadona, Diane, and Bruce Altschuler. *Isamu Noguchi: Essays and Conversations*. New York: Abrams, 1993.

Appadurai, Arjun. "The Ghost in the Financial Machine." *Public Culture* 23, no. 3 (2011.): 517–40.

Arbib, Michael A. *Brains, Machines, and Mathematics*. New York: McGraw Hill, 1964.

Arbib, Michael A. "Warren McCulloch's Search for the Logic of the Nervous System." *Perspectives in Biology and Medicine* 43, no. 2 (Winter 2000): 23.

Arendt, Hannah. *The Human Condition*. 2nd ed. Chicago: University of Chicago Press, 1958.

Arnheim, Rudolf. *Art and Visual Perception: A Psychology of the Creative Eye*. Berkeley: University of California Press, 1974.

Arrow, Kenneth Joseph. *Social Choice and Individual Values*. Cowles Foundation for Research in Economics, Yale University, monograph 12. New York: Wiley, 1963.

Arthur, Charles. "The Thinking City." BBC *Science Technology Future Focus* 237 (January 2012): 55–58.

Aspray, William. *John von Neumann and the Origins of Modern Computing*. Cambridge, MA: MIT Press, 1990.

Bacon, Francis. *New Atlantis and the City of the Sun: Two Classic Utopias*. London, 1627. Project Gutenberg edition posted October 23, 2008. Accessed April 14, 2014. www.gutenberg.org/ebooks/2434.

Bacsó, Zsuzsa. *Gyorgy Kepes and the Light Art*. The International Kepes Society. Posted 2011. Accessed April 14, 2014. http://kepes.society.bme.hu/art-science/Zsuzsa_Bacso_-_Gyorgy_Kepes_Light_Art.pdf.

Banham, Reyner. *Theory and Design in the First Machine Age*. Cambridge, MA: MIT Press, 1980.

Barthes, Roland. "The Great Family of Man." In *Mythologies*, edited by Roland Barthes, 100–102. New York: Hill and Wang, 1957.

Barthes, Roland. *The Rustle of Language*. New York: Farrar, Strauss, and Giroux, 1986.

Barton, Allen H. "Paul Lazarsfeld as Institutional Inventor." *International Journal of Public Opinion Research* 13 (2001): 3.

Bateson, Gregory. *Steps to an Ecology of Mind*. Chicago: University of Chicago Press, 2000.

Bateson, Gregory, D. D. Jackson, J. Haley, and J. Weakland. "Towards a Theory of Schizophrenia." *Behavioral Science* 1 (1956): 251–64.

Beck, Ulrich. *Risk Society: Towards a New Modernity*. Translated by Mark Ritter. 1st ed. London: Sage, 1992.

Behrens, Roy R. "Art, Design, and Gestalt Theory." *Leonardo* 31, no. 4 (1998): 299–303.

Beller, Jonathan. *The Cinematic Mode of Production: Attention Economy and the Society of the Spectacle*. Hanover, NH: University Press of New England, 2006.

Belmonte, Laura A. *Selling the American Way: U.S. Propaganda and the Cold War*. Philadelphia: University of Pennsylvania Press, 2010.

Beniger, James R. *The Control Revolution Technological and Economic Origins of the Information Society, Information Society*: Cambridge, MA: Harvard University Press, 1986.

Benjamin, Walter. *Illuminations*. Translated by Harry Zohn. Edited by Hannah Arendt. New York: Shocken Books, 1968.

Bennett, Jane. *Vibrant Matter: A Political Ecology of Things*: Durham, NC: Duke University Press, 2010.

Bentham, Jeremy. "Panopticon; or the Inspection House: Containing the Idea of a New Principle of Construction Applicable to Any Sort of Establishment, in Which Persons of Any Description Are to Be Kept under Inspection." In *The Panopticon Writings*, edited by Miran Bozovic, 29–95. London: Verso, 1995.

Bergson, Henri. *Creative Evolution*. Westport, CT: Greenwood Press, 1975.

Bergson, Henri. *Matter and Memory*. New York: Zone Books, 1988.

Bletter, Rosemarie Haag, Queens Museum, et al. *Remembering the Future: The New York World's Fair from 1939–1964*. With an introduction by Robert Rosenblum. New York: Rizzoli, 1989.

Bleuler, Eugen. *Dementia Praecox; or, The Group of Schizophrenias*. New York: International Universities Press, 1950.

Boden, Margaret A. *Mind as Machine: A History of Cognitive Science*. New York: Oxford University Press, 2006.

Bogart, Leo, and Agnes Bogart. *Cool Words, Cold War: A New Look at USIA's Premises for Propaganda*. American University Press Journalism History Series. Lanham, MD: University Press of America, 1995.

Bowker, Geoffrey C. "How to Be Universal: Some Cybernetic Strategies: 1943–70." *Social Studies of Science* 23, no. 1 (1993): 107–27.

Bowker, Geoffrey C. *Sorting Things Out: Classification and Its Consequences*. Cambridge, MA: MIT Press, 1999.

Braidotti, Rosi. *Nomadic Subjects: Embodiment and Sexual Difference in Contemporary Feminist Theory*: New York: Columbia University Press, 2011.

Brand, Stewart. *How Buildings Learn: What Happens after They're Built*. New York: Penguin Books, 1995.

Braun, Marta. *Picturing Time: The Work of Etienne-Jules Marey (1830–1904)*. Chicago: University of Chicago Press, 1992.

Bruno, Giuliana. *Atlas of Emotion: Journeys in Art, Architecture, and Film*. New York: Verso, 2002.

Bryson, Norman. *Vision and Painting: The Logic of the Gaze*. New Haven, CT: Yale University Press, 1986.

Burdett, Ricky, and Philipp Rode. "The Electric City." In *Urban Age Electric City Conference*, edited by Ricky Burdett and Philipp Rode, 2–4. London: London School of Economics, 2012.

Burks, Arthur W., Herman H. Goldstine, and John von Neumann. "Preliminary Discussion of the Logical Design of an Electronic Computing Instrument." In *Collected Works of John von Neumann*, edited by A. H. Taub, 5:34–79. New York: Macmillan, 1963.

Caillois, Roger. "Mimicry and Legendary Psychasthenia." *Minatoure* 7 (1935).

Canales, Jimena. *Tenth of a Second*. Chicago: University of Chicago Press, 2009.

Capra, Fritjof. *Web of Life*. New York: Anchor Books, 1996.

Card, James. "Silent Film Speed." *Image* (October 1955): 55–56.

Carpenter, Wava. "The Eames Lounge: The Difference between a Design Icon and Mere Furniture." In *Design Studies: A Reader*, edited by David Brody and Hazel Clark, 479–84. New York: Berg, 2009.

Carson, John. *The Measure of Merit: Talents, Intelligence, and Inequality in the French and American Republics, 1750–1940*. Princeton, NJ: Princeton University Press, 2007.

Cartwright, Lisa. *Moral Spectatorship: Technologies of Voice and Affect in Postwar Representations of the Child*. Durham, NC: Duke University Press, 2008.

Cartwright, Lisa. *Screening the Body: Tracing Medicine's Visual Culture*. Minneapolis: University of Minnesota Press, 1995.

Castells, Manuel. *The Rise of the Network Society*. 2nd ed. Vol. 1. of *The Information Age: Economy, Society and Culture*. New York: Wiley-Blackwell, 2000.

Castillo, Greg. *Cold War on the Home Front: The Soft Power of Midcentury Design*. Minneapolis: University of Minnesota, 2010.

Chun, Wendy Hui Kyong. *Control and Freedom: Power and Paranoia in the Age of Fiber Optics*. Cambridge, MA: MIT Press, 2006.

Chun, Wendy Hui Kyong. "The Enduring Ephemeral, or the Future Is a Memory." *Critical Inquiry* 35 (Autumn 2008): 23.

Chun, Wendy Hui Kyong. *Programmed Visions: Software and Memory*. Cambridge, MA: MIT Press, 2011.

Clarke, Bruce. "Steps to an Ecology of Systems: *Whole Earth* and Systemic Holism." In *Addressing Modernity: Social Systems Theory and U.S. Cultures*, edited by Hannes Bergthaller and Shinko Carston, 259–88. Amsterdam: Rodopi, 2011.

Clough, Patricia Ticineto. *Autoaffection: Unconscious Thought in the Age of Teletechnology*. Minneapolis: Minneapolis: University of Minnesota Press, 2000.

Clough, Patricia. "The New Empiricism: Affect and Sociological Method." *European Journal of Social Theory* 12, no. 2 (2009): 43–61.

Clough, Patricia, and Jean Halley. *The Affective Turn: Theorizing the Social*. Edited by

Patricia Ticineto Clough and Jean O'Malley Halley. Durham, NC: Duke University Press, 2007.

Coleman, Gabriella. *Coding Freedom: The Ethics and Aesthetics of Hacking.* Princeton, NJ: Princeton University Press, 2012.

Colomina, Beatriz. *Domesticity at War.* Cambridge, MA: MIT Press, 2007.

Colomina, Beatriz. "Enclosed by Images: The Eameses' Multimedia Architecture." *Grey Room* 2 (Winter 2001): 5–29.

Colomina, Beatriz. *Privacy and Publicity: Modern Architecture as Mass Media.* Cambridge, MA: MIT Press, 1994.

Cook, Peter, ed. *Archigram.* New York: Princeton Architectural Press, 1999.

Corbusier, Le. *The City of To-morrow and Its Planning.* 8th ed. New York: Dover, 2000. First published in 1929.

Crary, Jonathan. *Suspensions of Perception Attention, Spectacle, and Modern Culture.* Cambridge, MA: MIT Press, 1999.

Crary, Jonathan. *Techniques of the Observer: On Vision and Modernity in the Nineteenth Century.* Cambridge, MA: MIT Press, 1990.

Crary, Jonathan, and Stanford Kwinter. *Incorporations.* New York: Urzone, 1992.

Crowther-Heyck, Hunter. *Herbert A. Simon: The Bounds of Reason in Modern America:* Baltimore: Johns Hopkins University Press, 2005.

Dagognet, François. *Etienne-Jules Marey: A Passion for the Trace.* New York: Zone Books, 1992.

Daston, Lorraine. *Classical Probability in the Enlightenment.* Princeton, NJ: Princeton University Press, 1988.

Daston, Lorraine. *The Rule of Rules, or How Reason Became Rationality.* Berkeley: University of California Press, 2011.

Daston, Lorraine. *Wonders and the Order of Nature, 1150–1750.* New York: Zone Books, 1998.

Daston, Lorraine, and Elizabeth Lunbeck. *Histories of Scientific Observation.* Chicago: University of Chicago Press, 2011.

Deleuze, Gilles. *Anti-Oedipus: Capitalism and Schizophrenia.* Edited by Félix Guattari. New York: Viking Press, 1977.

Deleuze, Gilles. *Bergsonism.* Translated by Hugh Tomlinson and Barbara Habberjam. New York: Zone Books, 1988.

Deleuze, Gilles. *Cinema 1: The Movement-Image.* Translated by Barbara Habberjam and Hugh Tomlinson. Minneapolis: University of Minnesota Press, 1983.

Deleuze, Gilles. *Cinema 2: The Time-Image.* Translated by Hugh Tomlinson and Robert Galeta. Minneapolis: University of Minnesota Press, 1989.

Deleuze, Gilles. *Difference and Repetition.* Translated by Paul Patton. New York: Columbia University Press, 1994.

Deleuze, Gille. *Foucault.* Translated by Sean Hand. Minneapolis: University of Minnesota Press, 1988a.

Deleuze, Gilles. "Postscript on the Societies of Control." *October* 59 (Winter 1992): 3–7.

Deleuze, Gilles, and Félix Guattari. *A Thousand Plateaus: Capitalism and Schizophrenia.* Translated by Brian Massumi. St. Paul: University of Minnesota Press, 1987.

Derrida, Jacques. *Archive Fever: A Freudian Impression*. Translated by Eric Prenowitz. Chicago: University of Chicago Press, 1996.

Deutsch, Karl. *The Nerves of Government: Models of Political Communication and Control*. London: Free Press of Glencoe, 1963.

Dickerman, Leah. *Inventing Abstraction, 1910–1925*. New York: Museum of Modern Art, 2013.

Doane, Mary Ann. "The Close-Up: Scale and Detail in the Cinema." *Differences* 14, no. 3 (2003): 89–110.

Doane, Mary Ann. *The Emergence of Cinematic Time: Modernity, Contingency, the Archive*. Cambridge, MA: Harvard University Press, 2002.

Doane, Mary Ann. "Freud, Marey, and the Cinema." *Critical Inquiry* 22 (1996).

Dumit, Joseph. "Circuits in the Mind." Personal communication/lecture, 2007.

Dumit, Joseph. *Drugs for Life: How Pharmaceutical Companies Define Our Health*. Durham, NC: Duke University Press, 2012.

Dupuy, Jean Pierre. *The Mechanization of the Mind: On the Origins of Cognitive Science*. Princeton, NJ: Princeton University Press, 2000.

Eames, Charles. "Language of Vision: The Nuts and Bolts." *Bulletin: The American Academy of Arts and Sciences* 28 (October 1974): 13–45.

Eames, Charles, and Owen Gingerich. "A Conversation with Charles Eames." *American Scholar* 46, no. 3 (1977): 313–26.

Eames, Charles, and Ray Eames. "Communication Primer." Filmed 1953. The Eames Office film available at the Prelinger Archive, 21:29. https://archive.org/details/communications_primer.

Easterling, Keller. *Enduring Innocence*. Cambridge, MA: MIT Press, 2005.

Edwards, Paul N. *The Closed World Computers and the Politics of Discourse in Cold War America: Inside Technology*. Edited by Wiebe E. Bijker, Bernard Carlson, and Trevor Pinch. Cambridge, MA: MIT Press, 1997.

Felman, Shoshana. *Writing and Madness*. Palo Alto, CA: Stanford University Press, 2003.

Festinger, Leon. *Conflict, Decision, and Dissonance*. Stanford, CA: Stanford University Press, 1964.

Festinger, Leon. "Informal Social Communication." *Psychological Review* 57 (1950a): 271–82.

Festinger, Leon. *Social Pressures in Informal Groups; A Study of human Factors in Housing*. New York: Harper, 1950b.

Festinger, Leon. *A Theory of Cognitive Dissonance*. Stanford, CA: Stanford University Press, 1957.

Festinger, Leon. "A Theory of Social Comparison Processes." *Human Relations* 7 (1954): 117–40.

Festinger, Leon. *When Prophecy Fails*. Minneapolis: University of Minnesota Press, 1956.

Fisch, Michael. "Tokyo's Commuter Suicides and the Society of Emergence." *Cultural Anthropology* 28, no. 2 (2013): 320–43.

Fishman, Robert. *Urban Utopias in the Twentieth Century: Ebenezer Howard, Frank Lloyd Wright, and Le Corbusier*. Cambridge, MA: MIT Press, 1982.

Foerster, Heinz Von. *Cybernetics: Circular Causal, and Feedback Mechanisms in Bio-logical and Social Systems, Transactions of the Sixth Conference, March 15–16, 1951, New York, NY*. New York: Josiah Macy Jr. Foundation, 1950.

Foley, Duncan. "The Strange History of the Economic Agent." *The New School Economics Review* 1, no. 1 (2004): 82–94.

Forgács, Éva. *The Bauhaus Idea and Bauhaus Politics*. Translated by John Batki. English ed. Budapest: Central European University Press, 1995.

Forrester, Jay. *Industrial Dynamics*. New York: MIT Press and John Wiley and Sons, 1961.

Foucault, Michel. *The Birth of Biopolitics: Lectures at the College De France, 1978–1979*. Lectures at the Collège de France. New York: Macmillan Palgrave, 2010.

Foucault, Michel. *Discipline and Punish: Birth of the Prison*. Translated by Alan Sheridan. New York: Vintage Books, 1995.

Foucault, Michel. *The History of Sexuality*. New York: Vintage Books, 1988.

Foucault, Michel. "Nietzsche, Genealogy, History." In *Michel Foucault: Aesthetics, Method, and Epistemology*, vol. 2, edited by James D. Faubion, 369–89. New York: New Press, 1994.

Freud, Sigmund. "A Note upon the 'Mystic Writing-Pad.'" In *The Standard Edition of the Complete Psychological Works of Sigmund Freud*, edited by James Strachey, 225–32. London: Hogarth Press, 1961.

Friedman, Mildred S. *Nelson, Eames, Girard, Propst: The Design Process at Herman Miller*. Edited by Herman Miller Inc. and Walker Art Center. Minneapolis: Walker Art Center, 1975.

Fukuyama, Francis. *The End of History and the Last Man*. New York: Free Press, 1992.

Fuller, Buckminster. *Synergetics: Explorations in the Geometry of Thinking*. New York: Macmillian, 1975. Also available online from The Projects of R. W. Gray. Accessed April 14, 2014. www.rwgrayprojects.com/synergetics/synergetics.html.

Galanter, Eugene. "Writing Plans." In *The Making of Cognitive Science: Essays in Honor of George Miller*, edited by William Hirst. Cambridge: Cambridge University Press, 1988.

Galison, Peter. "The Ontology of the Enemy: Norbert Wiener and the Cybernetic Vision." *Critical Inquiry* 21 (1994): 228–66.

Galison, Peter, and Caroline A. Jones. *Picturing Science, Producing Art*. New York: Routledge, 1998.

Galison, Peter, and Lorraine Daston. *Objectivity*. New York: Zone Books, 2007.

Galloway, Alex. *Protocol*. Cambridge, MA: MIT Press, 2004.

Galloway, Alex, and Eugene Thacker. *The Exploit: A Theory of Networks*. Minneapolis: University of Minnesota Press, 2007.

Gans, Herbert J. *The Urban Villagers; Group and Class in the Life of Italian-Americans*. New York: Free Press, 1982.

Gardner, Howard. *The Mind's New Science: A History of the Cognitive Revolution*. New York: Basic Books, 1985.

Geoghegan, Bernard Dionysius. "From Information Theory to French Theory: Jakobson, Lévi-Strauss, and the Cybernetic Apparatus." *Critical Inquiry* 38 (Autumn 2011): 96–126.

Gibson, J. J., and E. Gibson. "Perceptual Learning: Differentiation or Enrichment?" *Psychology Review* 62, no. 1 (January 1955): 32–41.

Gibson, J. J. *The Senses Considered as Perceptual Systems.* Boston: Houghton Mifflin, 1966.

Gigerenzer, Gerd. *The Empire of Chance: How Probability Changed Science and Everyday Life.* Cambridge: Cambridge University Press, 1989.

Gilman, Sander. *Difference and Pathology: Stereotypes of Sexuality, Race, and Madness.* Ithaca, NY: Cornell University Press, 1985.

Gitelman, Lisa. *"Raw Data" Is an Oxymoron (Infrastructures).* Cambridge, MA: MIT Press, 2013.

Gitelman, Lisa, and Jonathan Auerbach. "Microfilm, Containment, and the Cold War." *American Literary History* 19, no. 3 (2007): 745–68.

Gödel, Kurt. *On Formally Undecidable Propositions of Principia Mathematica and Related Systems.* New York: Basic Books, 1962.

Goldstein, Rebecca. *Incompleteness: The Proof and Paradox of Kurt Gödel.* New York: Norton, 2005.

Goldstine, Herman H., and John von Neumann. *Planning and Coding of Problems for an Electronic Computing Instrument: Report on the Mathematical and Logical Aspects of an Electronic Computing Instrument.* Pt. 2. Vol. 1. Princeton, NJ: Institute for Advanced Study, 1948.

Golec, Michael. "A Natural History of a Disembodied Eye: The Structure of Gyorgy Kepes's Language of Vision." *Design Issues* 18, no. 2 (2002): 3–16.

Golumbia, David. *The Cultural Logic of Computation.* Cambridge, MA: Harvard University Press, 2009.

Goodyear, Anne Collins. "Gyorgy Kepes, Billy Klüver, and American Art of the 1960s: Defining Attitudes toward Science and Technology." *Science in Context* 17, no. 4 (2004): 611–35.

Gottesman, Irving. *Schizophrenia Genesis: The Origins of Madness.* New York: Freeman, 1991.

Gould, James L. "The Dance Language Controversy." *Quarterly Review of Biology* 51, no. 2 (1976): 357–65.

Gould, Stephen Jay. *The Mismeasure of Man.* New York: Norton, 1996.

Graham, Stephen, and Simon Marvin. *Splintering Urbanism Networked Infrastructures, Technological Mobilities and the Urban Condition.* New York: Routledge, 2001.

Greene, Jeremy A. *Prescribing by Numbers: Drugs and the Definition of Disease.* Baltimore: Johns Hopkins University Press, 2007.

Grosz, E. A. *The Nick of Time: Politics, Evolution, and the Untimely.* Durham, NC: Duke University Press, 2004.

Hacking, Ian. *The Emergence of Probability: A Philosophical Study of Early Ideas about Probability, Induction and Statistical Inference.* New York: Cambridge University Press, 1975.

Hacking, Ian. *The Taming of Chance.* Cambridge: Cambridge University Press, 1990.

Hald, Anders. *A History of Parametric Statistical Inference from Bernoulli to Fisher, 1713–1935.* New York: Springer, 2007.

Halpern, Orit. "Dreams for Our Perceptual Present: Temporality, Storage, and Inter-activity in Cybernetics." *Configurations* 13, no. 2 (2005): 36.

Halpern, Orit, Jesse LeCavalier, and Nerea Calvillo. "Test-Bed Urbanism." *Public Culture* 26 (March 2013): 272–306.

Hamilton, Mina. "Films at the Fair." *Industrial Design* (May 1964): 5.

Hansen, Mark B. N. *New Philosophy for New Media.* Cambridge, MA: MIT Press, 2004.

Haraway, Donna Jeanne. *Crystals, Fabrics, and Fields: Metaphors of Organicism in Twentieth-Century Developmental Biology.* New Haven, CT: Yale University Press, 1976.

Haraway, Donna. "A Cyborg Manifesto: Science, Technology, and Socialist-Feminism in the Late Twentieth Century." In *Simians, Cyborgs and Women: The Reinvention of Nature*, edited by Donna Haraway, 149–81. New York: Routledge, 1991.

Haraway, Donna Jeanne. *Primate Visions: Gender, Race, and Nature in the World of Modern Science.* New York: Routledge, 1989.

Harnish, Robert M. *Minds, Brains, Computers: An Historical Introduction to the Foundations of Cognitive Science.* Malden, Mass.: Blackwell, 2002.

Harvey, David. *The Condition of Postmodernity: An Enquiry into the Origins of Cultural Change.* Cambridge, MA: Blackwell, 1989.

Harwood, John. "The White Room: Eliot Noyes and the Logic of the Information Age Interior." *Grey Room* 12 (Summer 2003): 5–31.

Hayles, N. Katherine. *How We Became Posthuman: Virtual Bodies in Cybernetics, Literature, and Informatics.* Chicago: University of Chicago Press, 1999.

Healy, David. *The Creation of Psychopharmacology.* Cambridge, MA: Harvard University Press. 2002.

Heims, Steve J. *The Cybernetics Group.* Cambridge, MA: MIT Press, 1991.

Heims, Steve J. *John von Neumann and Norbert Wiener: From Mathematics to the Technologies of Life and Death.* Cambridge, MA: MIT Press, 1980.

Hixson, Walter L. *Parting the Curtain: Propaganda, Culture, and the Cold War, 1945–1961.* New York: St. Martin's Griffin, 1998.

Holland, Eugene. "Deterritorializing 'Deterritorialization': From the 'Anti-Oedipus' to 'A Thousand Plateaus.'" *SubStance* 20, no. 3, special issue 66: "Deleuze and Guattari" (1991): 55–65.

Hughes, Thomas. *Rescuing Prometheus.* New York: Pantheon Books, 1998.

Ingrao, Bruna. *The Invisible Hand: Economic Equilibrium in the History of Science.* Cambridge, MA: MIT Press, 1990.

Jackson, Don D. *Therapy, Communication, and Change.* Palo Alto, CA: Science and Behavior Books, 1968.

Jacobs, Jane. *The Death and Life of Great American Cities.* New York: Random House, 2002.

Jameson, Fredric. "Future City." *New Left Review* 21 (May/June 2003): 65–79.

Jameson, Fredric. *Postmodernism, or, The Cultural Logic of Late Capitalism.* Durham, NC: Duke University Press, 1991.

Jay, Martin. *Downcast Eyes: The Denigration of Vision in Twentieth-Century French Thought.* Berkeley: University of California Press, 1993.

Jazairy, El Hadi. "Toward a Plastic Conception of Scale." Scales of the Earth. *New Geographies* 4 (2011): 1.

Johnston, Hugh B. "From Old IBM to New IBM." *Industrial Design* 4, no. 3 (1957): 48–53.

Johnston, John. *The Allure of Machinic Life: Cybernetics, Artificial Life, and the New AI*. Cambridge, MA: MIT Press, 2008.

Jones, Caroline A. *Machine in the Studio: Constructing the Postwar American Artist*. Chicago: University of Chicago Press, 1996.

Julesz, Béla. "Binocular Depth Perception of Computer-Generated Patterns." *Bell Labs System Technical Journal* 39, no. 5 (1960): 1125–62.

Julesz, Béla. *Foundations of Cyclopean Perception*. Chicago: University of Chicago Press, 1971.

Kargon, Robert, and Arthur P. Molella. "The City as Communications Net: Norbert Wiener, the Atomic Bomb, and Urban Dispersal." *Technology and Culture* 45, no. 4 (2004.): 764–77.

Kay, Lily E. "Cybernetics, Information, Life: The Emergence of Scriptural Representations of Heredity." *Configurations* 5 (1997): 23–91.

Kay, Lily E. "From Logical Neurons to Poetic Embodiments of Mind: Warren McCulloch's Project in Neuroscience." *Science in Context* 14, no. 4 (2001): 591–614.

Kay, Lily E. *The Molecular Vision of Life: Caltech, the Rockefeller Foundation, and the Rise of the New Biology*. New York: Oxford University Press, 1993.

Kay, Lily E. *Who Wrote the Book of Life? A History of the Genetic Code*. Stanford, CA: Stanford University Press, 2000.

Keller, Evelyn Fox. *Refiguring Life: Metaphors of Twentieth-Century Biology*. New York: Columbia University Press, 1995.

Kentgens-Craig, Margret. *The Bauhaus and America: First Contacts, 1919–1936*. Cambridge, MA: MIT Press, 2001.

Kepes, Gyorgy, ed. *Education of Vision*. New York: Braziller, 1965a.

Kepes, Gyorgy. "Gyorgy Kepes." MIT video, 1:45. Undated. http://mit150.mit.edu/multimedia/gyorgy-kepes.

Kepes, Gyorgy. "Gyorgy Kepes Interview." MIT video, 1:00:24. December 1988. http://mit150.mit.edu/multimedia/gyorgy-kepes-interview-1988.

Kepes, Gyorgy. *Language of Vision*. 1st ed. Chicago: P. Theobald, 1944.

Kepes, Gyorgy. "Light and Design." *Design Quarterly* 68 (1967): 1–2, 4–32.

Kepes, Gyorgy. "The Lost Pageantry of Nature." *artscanada* 25 (1968): 5.

Kepes, Gyorgy, ed. *The New Landscape in Art and Science*. Chicago: Paul Theobald, 1956.

Kepes, Gyorgy, ed. *Sign, Image, Symbol*. New York: Braziller, 1966.

Kepes, Gyorgy, ed. *Structure in Art and in Science*. New York: Braziller, 1965b.

Kepes, Gyorgy. "Toward Civic Art." *Leonardo* 4, no. 1 (1971): 69–73.

Kepes, Gyorgy, ed. *The Visual Arts Today*. Middletown, CT: Wesleyan University Press, 1960.

Kepes, Gyorgy, and Marjorie Supovitz. *The MIT Years, 1945–1977: Paintings, Photographic Work, Environmental Pieces, Projects, at the Center for Advanced Visual Studies, Hayden Gallery, Hayden Corridor Gallery, Margaret Hutchinson Compton*

Gallery, Massachusetts Institute of Technology, Cambridge, Massachusetts, April 28–June 9, 1978. Cambridge, MA: Center for Advanced Visual Studies and MIT Committee on the Visual Arts, Massachusetts Institute of Technology, 1978.

Keylor, William R. "Waging the War Words: The Promotion of American Interests and Ideals Abroad During the Cold War." In *Power and Responsibility in World Affairs: Reformation Versus Transformation*, edited by Cathal J. Nolan, 79–102. Westport, CT: Praeger Publishers, 2004.

Kirkham, Pat. *Charles and Ray Eames: Designers of the Twentieth Century.* Cambridge, MA: MIT Press, 1995.

Kirkham, Pat. "Humanizing Modernism: The Crafts, 'Functioning Decoration,' and the Eameses." *Journal of Design History* 11, no. 1 (1998): 15–29.

Kittler, Friedrich. *Discourse Networks 1800/1900.* Translated by Geoffrey Winthrop-Young and Michael Wutz. Stanford, CA: Stanford University Press, 1990.

Kittler, Friedrich A. *Gramophone, Film, Typewriter.* Stanford, CA.: Stanford University Press, 1999.

Knight, Friedrich A. 1921. *Risk, Uncertainty, Profit.* Boston: Houghton Mifflin, 1921. Available at the Library of Economics and Liberty website. Accessed April 20, 2014. www.econlib.org/library/Knight/knRUP.html.

Koselleck, Reinhart. *Critique and Crisis: Enlightenment and the Pathogenesis of Modern Society.* Cambridge, MA: MIT Press, 1988.

Krauss, Rosalind. *The Originality of the Avant-Garde and Other Modernist Myths.* Cambridge, MA: MIT Press, 1985.

Kubler, George. *The Shape of Time; Remarks on the History of Things.* New Haven, CT: Yale University Press, 1962.

Kuhn, Thomas S. *The Structure of Scientific Revolutions.* Chicago: University of Chicago Press, 1996.

Kurzweil, Ray. *The Age of Spiritual Machines: When Computers Exceed Human Intelligence.* New York: Viking, 1999.

Kushner, Marilyn S. "Exhibiting Art at the American National Exhibition in Moscow, 1959 Domestic Politics and Cultural Diplomacy." *Journal of Cold War Studies* 4 no. 1 (2002): 6–26.

Lacan, Jacques. *The Seminars of Jacques Lacan.* Bk. II. *The Ego in Freud's Theory and the Technique of Psychoanalysis 1954–55.* Translated by Sylavana Tomaselli with notes by John Forrester. Edited by Jacques-Alain Miller. New York: Norton, 1991.

Lampland, Martha, and Susan Leigh Star. *Standards and Their Stories: How Quantifying, Classifying, and Formalizing Practices Shape Everyday Life.* Ithaca, NY: Cornell University Press, 2009.

Larkin, Brian. *Signal and Noise: Media, Infrastructure, and Urban Culture in Nigeria.* Durham, NC: Duke University Press, 2008.

Latour, Bruno. "Visualization and Cognition: Drawing Things Together." *Knowledge and Society: Studies in Sociology of Culture Past and Present* 6 (1986): 1–40.

Lazarsfeld, Paul F., and Robert K. Merton. "Mass Communication, Popular Taste, and Organized Social Action." In *The Communication of Ideas*, edited by L. Bryson, 95–118. New York: Harper, 1948.

Lee, Pamela. *Chronophobia: On Time in the Art of the 1960's*. Cambridge, MA: MIT Press, 2004.

Lemov, Rebecca. "Hypothetical Machines: The Science Fiction Dreams of Cold War Social Science." *Isis* 101, no. 2 (2010): 401–11.

Lemov, Rebecca. "Towards a Data Base of Dreams: Assembling an Archive of Elusive Materials, c. 1947–61." *History Workshop Journal* 67, no. 1 (2009): 44–68.

Lemov, Rebecca M. *World as Laboratory: Experiments with Mice, Mazes, and Men*. New York: Hill and Wang, 2005.

Lenoir, Timothy. "All but War Is Simulation: The Military-Entertainment Complex." *Configurations* 8 (2000): 289–335.

Leslie, Stuart W. *The Cold War and American Science: The Military-Industrial-Academic Complex at MIT and Stanford*. New York: Columbia University Press, 1993.

Lettvin, J. Y., H. R. Maturana, and W. S. McCulloch. "What the Frog's Eye Tells the Frog's Brain." In *Embodiments of Mind*, edited by Warren McCulloch, 230–55. Cambridge, MA: MIT Press, 1959.

Lewin, Kurt. "Action Research and Minority Problems." *Social Issues* 2, no. 4 (1946): 34–46.

Licklider, J. C. R. "Man-Computer Symbiosis." *IRE Transactions on Human Factors in Electronics* 1 (March 1960): 4–11.

Light, Jennifer. *From Warfare to Welfare: Defense Intellectuals and Urban Problems in Cold War America*. Baltimore: Johns Hopkins University Press, 2003.

Lindsay, Greg. "The New New Urbanism: Cisco's $30 Billion Dollar Bet." *Fast Company* 142 (February 2010): 88–95.

Lipset, David. *Gregory Bateson: The Legacy of a Scientist*. Englewood Cliffs, NJ: Prentice-Hall, 1980.

Liu, Lydia. *The Freudian Robot: Digital Media and the Future of the Unconscious*. Chicago: University of Chicago Press, 2010.

Lopez, Russ. *Building American Public Health: Urban Planning, Architecture, and the Quest for Better Health in the United States*: New York: Palgrave Macmillan, 2012.

Lupton, Ellen, and J. Abbott Miller, eds. *The ABC's of [Triangle, Square, Circle]: The Bauhaus and Design Theory*. New York: Herb Lubalin Study Center of Design and Typography, Cooper Union for the Advancement of Science and Art, 1991.

Lupton, Ellen, and J. Abbott Miller. "Language of Vision." In *Design, Writing, Research: Writing on Graphic Design*, 64. London: Phaidon, 1994.

Lury, Celia, Luciana Parisi, and Tiziana Terranova. "Introduction: The Becoming Topological of Culture." *Theory, Culture & Society* 29, no. 4/5 (2012): 3–35.

Lyford, Amy. "Noguchi, Sculptural Abstraction, and the Politics of Japanese American Internment." *Art Bulletin* 85, no. 1 (2003): 15.

Lynch, Kevin. *City Sense and City Design: Writings and Projects of Kevin Lynch*. Edited by Michael Southworth and Tridib Banerjee. Cambridge, MA: MIT Press, 1990.

Lynch, Kevin. *The Image of the City*. Cambridge, MA: MIT Press, 1960.

Lynch, Kevin. *What Time Is This Place?* Cambridge, MA: MIT Press, 1972.

Lyotard, Jean-François. *The Postmodern Condition: A Report on Knowledge*. Minneapolis: University of Minnesota Press. 1984.

MacKenzie, Donald A. *An Engine, Not a Camera: How Financial Models Shape Markets*. Cambridge, MA: MIT Press, 2006.

Manovich, Lev. 2001. *The Language of New Media*. Cambridge, MA: MIT Press.

Markov, A. A. 1954. *Theory of Algorithms*. Translated by Jacques J. Schorr-Kon and staff of Israel Program for Scientific Translation. Moscow: Academy of Sciences of the U.S.S.R. Reprint, Israel Program for Scientific Translation for National Science Foundation and U.S. Department of Commerce.

Marks, John D. *The Search for the "Manchurian Candidate": The CIA and Mind Control*. New York: Times Books, 1979.

Martin, Reinhold. *The Organizational Complex: Architecture, Media, and Corporate Space*. Cambridge, MA: MIT Press, 2003.

Marx, Karl. *Grundrisse*. Translated by Martin Nicklaus. Harmondsworth, England: Penguin, 1973.

Masco, Joseph. "'Sensitive but Unclassified': Secrecy and the Counterterrorist State." *Public Culture* 22, no. 3 (2010): 30.

Massumi, Brian. "Potential Politics and the Primacy of Preemption." *Theory & Event* 10, no. 2 (2007). http://muse.jhu.edu/journals/theory_and_event/toc/tae10.2.html.

Massumi, Brian. *A User's Guide to Capitalism and Schizophrenia: Deviations from Deleuze and Guattari*. Cambridge, MA: MIT Press, 1992.

Mathews, Jane de Hart. "Art and Politics in Cold War America." *American Historical Review* 81, no. 3 (1976): 762–87.

Matthews, J. Rosser. *Quantification and the Quest for Medical Certainty*. Princeton, NJ: Princeton University Press, 1995.

May, John. "Sensing: Preliminary Notes on the Emergence of Statistical-Mechanical Geographic Vision." *Perspecta* 40 (May 2008): 42–53.

McCulloch, Warren S. *Embodiments of Mind*. Cambridge, MA: MIT Press, 1970.

McCulloch, Warren S. "Recollections of the Many Sources of Cybernetics." *ASC Forum* 6, no. 2 (1974): 5–16.

McCulloch, Warren S., and Walter Pitts. "A Logical Calculus of Ideas Immanent in Nervous Activity." In *Embodiments of Mind*, edited by Warren McCulloch. 19–39. Cambridge, MA: MIT Press, 1970. Originally published in 1943.

McLuhan, Marshall. *The Gutenberg Galaxy; The Making of Typographic Man*. Toronto: University of Toronto Press, 1962.

McLuhan, Marshall. *Understanding Media: The Extensions of Man*. New York: McGraw-Hill, 1964.

Medina, Eden. *Cybernetic Revolutionaries: Technology and Politics in Allende's Chile*. Cambridge, MA: MIT Press, 2011.

Mekas, Jonas. "Expanded Arts." *Film Culture* 43 (1966).

Melley, Timothy. "Brainwashed! Conspiracy Theory and Ideology in the Postwar United States." *New German Critique* 35, no. 1 (2008): 145–64.

Mental Research Institute. 2011. MRI: Interactional Therapy, Training and Research. Mental Research Institute website. Accessed December 16, 2011. www.mri.org /about_us.html.

Metzl, Jonathan. *The Protest Psychosis: How Schizophrenia Became a Black Disease*. Boston: Beacon Press, 2010.

Miller, George. "The Magical Number Seven, Plus or Minus Two: Some Limits on Our Capacity for Processing Information." *Psychological Review* 63 (1956): 81–97.

Miller, George. "Reviewed Work(s): Cybernetics: Circular Causal and Feedback Mechanisms in Biological and Social Systems, Transactions of the Eighth Conference by Heinz von Foerster; Margaret Mead; Hans Lukas Teuber." *American Journal of Psychology* 66, no. 4 (1953): 3.

Miller, George, Eugene Galanter, and Karl H. Pribram. *Plans and the Structure of Behavior*. New York: Holt, Rinehart, and Winston, 1960.

Mills, Mara. "Deaf Jam: From Inscription to Reproduction to Information." The Politics of Recorded Sound. *Social Text* 102 (Spring 2010): 35–58.

Mills, Mara. "On Disability and Cybernetics: Helen Keller, Norbert Wiener, and the Hearing Glove." *Differences* 22, no. 2/3 (2011): 37.

Mindell, David. *Between Human and Machine: Feedback, Control, and Computing before Cybernetics*. Johns Hopkins Studies in the History of Technology. Baltimore: Johns Hopkins University Press. 2004.

Mirowski, Philip. *Machine Dreams: Economics Becomes a Cyborg Science*. New York: Cambridge University Press, 2002.

Mitchell, William J. *Me++: The Cyborg Self and the Networked City*. Cambridge, MA: MIT Press, 2003.

Mitchell, William J. *The Reconfigured Eye: Visual Truth in the Post-photographic Era*. Cambridge, MA: MIT Press, 1992.

Moholy-Nagy, László. *The New Vision; Fundamentals of Design, Painting, Sculpture, Architecture*. New York: Norton, 1938.

Morgan, Jack Masey, and Conway Lloyd. *Cold War Confrontations: U.S. Exhibitions and Their Role in the Cultural Cold War*. Baden, Switzerland: Lars Müller, 2008.

Morris, Rosalind. "Review of James T. Siegel, *Fetish, Recognition, Revolution* (Princeton: Princeton University Press, 1997)." *Indonesia* 67 (1999): 163–76.

Morris, Rosalind. "The War Drive: Image Files Corrupted." *Social Text* 25, no. 2 (2007): 39.

Mumford, Lewis. *Technics and Civilization*. New York, Harcourt, Brace & World, 1963.

Nakamura, Lisa. *Cybertypes: Race, Ethnicity, and Identity on the Internet*. New York: Routledge, 2002.

Negroponte, Nicholas. *The Architecture Machine*. Cambridge, MA: MIT Press, 1970.

Negroponte, Nicholas. *Soft Architecture Machines*. Cambridge, MA: MIT Press, 1975.

Nelson, George. *Problems of Design*. New York: Whitney Library of Design, 1979.

Nelson, Theodor H. *Literary Machines: The Report on, and of, Project Xanadu Concerning Word Processing, Electronic Publishing, Hypertext, Thinkertoys, Tomorrow's Intellectual Revolution, and Certain Other Topics Including Knowledge, Education and Freedom*. Sausalito, CA: Mindful Press, 1992.

Neuhart, John. *Eames Design: The Work of the Office of Charles and Ray Eames*. Edited by Charles Eames, Ray Eames, and Marilyn Neuhart. New York: Abrams, 1989.

Nilsen, Sarah. "Projecting America, 1958: Film and Cultural Diplomacy at the Brussels World Fair." Jefferson, NC: McFarland, 2011.

Noguchi, Isamu. *Isamu Noguchi: A Sculptor's World*. New York: Harper & Row, 1968.

NSC Study Group, Paul Nitze, John P. Davis, Robert Tufts, Robet Hooker, Dean Acheson, Chip Bohlen, Major General Truman Landon, Samuel S. Butano, and Robert Lovett. *NSC 68: United States Objectives and Programs for National Security*. Washington, DC: U.S. National Security Council, 1950.

Oetterman, Stephen. *The Panorama: History of a Mass Medium*. Translated by Deborah Lucas Schneider. New York: Zone Books, 1997.

Orr, Jackie. *The Panic Diaries: A Genealogy of Panic Disorder*. Durham, NC: Duke University Press, 2006.

Otter, Chris. *The Victorian Eye: A Political History of Light and Vision in Britain, 1800–1910*. Chicago: University of Chicago Press, 2008.

Packer, Randall, and Ken Jordan. *Multimedia: From Wagner to Virtual Reality*. New York: Norton, 2001.

Parikka, Jussi. *Insect Media*. Edited by Cary Wolfe. Posthumanities. Minneapolis: University of Minnesota Press, 2010.

Parisi, Luciana. *Contagious Architecture: Computation, Aesthetics, Space*. Cambridge: MIT, 2013.

Parks, Lisa. *Cultures in Orbit: Satellites and the Televisual*. Durham, NC: Duke University Press, 2005.

Parks, Lisa, and James Schwoch. *Down to Earth Satellite Technologies, Industries, and Cultures*. New Brunswick, NJ: Rutgers University Press, 2012.

Patterson, Zabet. "From the Gun Controller to the Mandala: The Cybernetic Cinema of John and James Whitney." *Grey Room* 36 (Summer 2009): 36–57.

Pias, Claus. *Cybernetics—The Macy Conference*. 2 vols. Vol. 1. Berlin: diaphanes, 2003.

Pickering, Andrew. *The Cybernetic Brain*. Chicago: University of Chicago Press, 2010.

Poovey, Mary. *A History of the Modern Fact: Problems of Knowledge in the Sciences of Wealth and Society*. Chicago: University of Chicago Press, 1998.

Poundstone, William. *Prisoner's Dilemma*. New York: Doubleday, 1992.

Rabinbach, Anson. *The Human Motor: Energy, Fatigue, and the Origins of Modernity*. Berkeley: University of California Press, 1992.

Rabinow, Paul. *Ethics: The Essential Works of Michel Foucault 1954–1984*. Vol. 1. London: Allen Lane, 1997.

Ragain, Melissa. "From Organization to Network: MIT's Center for Advanced Visual Studies." *X-TRA: Contemporary Art Quarterly* 14, no. 4 (2012), available at http://x-traonline.org/.

Rajan, Kaushik Sunder. *Biocapital: The Constitution of Postgenomic Life*. Durham, NC.: Duke University Press, 2006.

Rajchman, John. "Foucault's Art of Seeing." *October* 44 (Spring 1988): 29.

Ramón y Cajal, Santiago. *Texture of the Nervous System of Man and the Vertebrates*. New York: Springer, 1999.

Rancière, Jacques. *The Politics of Aesthetics: The Distribution of the Sensible*. Translated by Gabriell Rockhill. New York: Continuum, 2006.

Ratti, Carlo, and Anthony Townsend. "The Social Nexus." Paper presented at Urban Age Electric City Conference, London, December 6–7, 2012.

Reck, Erich. *From Frege to Wittgenstein: Perspectives on Early Analytic Philosophy*. Oxford: Oxford University Press, 2002.

"Research in the Round." *Architectural Forum* 114 (June 1961): 81–84.

Roach, Leigh Ann. "A Postivie, Popular Art: Sources, Structure, and Impact of Gyorgy Kepes's Language of Vision." Ph.D. diss., Florida State University, 2010.

Rodowick, David N. *Gilles Deleuze's Time Machine*. Durham, NC: Duke University Press, 1997.

Rose, Jacqueline. *Sexuality in the Field of Vision*. London: Verso, 1986.

Rosenblatt, Frank. "The Perceptron: A Probabilistic Model for Information Storage and Organization in the Brain." *Psychological Review* 65 (November 1958): 386–408.

Rosenblueth, Arturo, and Norbert Wiener. "Purposeful and Non-purposeful Behavior." *Philosophy of Science* 17, no. 4 (1950): 318–26.

Rosenblueth, Arturo, Norbert Wiener, and Julian Bigelow. "Behavior, Purpose, Teleology." *Philosophy of Science* 10, no. 1 (1943): 18–24.

Ruesch, Jurgen. *Disturbed Communication*. New York: Norton, 1957.

Ruesch, Jurgen, and Gregory Bateson. *Communication; The Social Matrix of Psychiatry*. New York: Norton, 1951.

Russell, Bertrand. *The Principles of Mathematics*. London: Routledge, 2009.

Sadler, Simon. *Archigram: Architecture without Architecture*. Cambridge, MA: MIT Press, 2005.

Samuel, Lawrence R. *The End of the Innocence: The 1964–1965 New York World's Fair*. Syracuse, NY: Syracuse University Press, 2007.

Sandeen, Eric J. *Picturing an Exhibition: The Family of Man and 1950s America*. Albuquerque: University of New Mexico Press, 1995.

Sassen, Saskia. *The Global City: New York, London, Tokyo*. Princeton, NJ: Princeton University Press, 2001.

Schachter, Simon. "Leon Festinger." In *Biographical Memoirs*, edited by National Academy of Sciences Office of the Home Secretary, 99–105. Washington, DC: National Academies Press, 1994.

Schievelbusch, Wolfgang. *The Railway Journey: The Industrialization of Time and Space in the 19th Century*. Berkeley: University of California Press, 1986.

Schorske, Carl E. *Fin-de-siècle Vienna: Politics and Culture*. New York: Knopf, 1980.

Schull, Natasha Dow, and Caitlin Zaloom. "The Shortsighted Brain: Neuroeconomics and the Governance of Choice in Time." *Social Studies of Science* 41, no. 4 (2011): 515–38.

Scott, Felicity. *Architecture or Tecno-Utopia: Politics after Modernism*. Cambridge, MA: MIT Press, 2007.

Scott, John L., L. Moholy-Nagy, and Gyorgy Kepes. "Civilian Action Goes into Action." *Civilian Defense* (June 1942): 5.

Seagran, Toby, and Jeff Hammerbacher. 2009. *Beautiful Data: The Stories behind Elegant Data Solutions*. Sebastapol, CA: O'Reilly.

Sedgwick, Eve Kosofsky. 2003. *Touching Feeling: Affect, Pedagogy, Performativity*. Durham, NC: Duke University Press.

Sedgwick, Eve Kosofsky, and Adam Frank. 1995. *Shame and Its Sisters: A Silvan Tomkins Reader*. Durham, NC: Duke University Press.

Segel, Edward, and Jeffrey Heer. *Beautiful Data: How to Tell Stories with Data*. Stanford Visualization Group 2012. Accessed April 14, 2014. Available at http://beautifuldata.tumblr.com/

Sekula, Allan. "The Body and the Archive." In *The Contest of Meaning: Critical Histories of Photography*, edited by R. Bolton, 343–89. Cambridge, MA: MIT Press, 1992.

Sekula, Allan. "The Traffic in Photographs." *Art Journal* 41, no. 1 (1981): 10.

Sennett, Richard. "The Stupefying Smart City." Paper presented at Urban Age Electric City Conference, London, London School of Economics and Deutsche Bank. Alfred Herrhausen Society. December 6–7, 2012.

Shanken, Andrew M. *194x: Architecture, Planning, and Consumer Culture on the American Home Front*. Minneapolis: University of Minnesota Press, 2009.

Shanken, Andrew M. "Uncharted Kahn: The Visuality of Planning and Promotion." *Art Bulletin* 88 (June 2006): 310–27.

Shanken, Edward. "Cybernetics and Art: Cultural Convergence in the 1960s." In *From Energy to Information*, edited by Bruce Clarke and Linda Dalrymple Henderson, 155–77. Palo Alto, CA: Stanford University Press, 2002.

Shannon, Claude, and Warren Weaver. *The Mathematical Theory of Communication*. Urbana-Champagne: University of Illinois Press, 1963. Originally published 1949; reprinted 1998.

Shorter, Edward. *Before Prozac: The Troubled History of Mood Disorders in Psychiatry*. Oxford: Oxford University Press, 2008.

Siegel, James. *Fetish, Recognition, Revolution*. Princeton, NJ: Princeton University Press, 1997.

Siegel, Ralph M. "Choices: The Science of Bela Julesz." *PLoS Biology* 2, no. 6 (2004). Accessed April 14, 2014. www.ncbi.nlm.nih.gov/pmc/articles/PMC423145/.

Silverman, Kaja. *The Threshold of the Visible World*. New York: Routledge, 1996.

Simon, Herbert A. "A Behavioral Model of Rational Choice." *Quarterly Journal of Economics* 69, no. 1 (1955): 99–118.

Simon, Herbert A. *The Sciences of the Artificial*. Cambridge, MA: MIT Press, 1996.

Simon, Herbert A., Massimo Egidi, Riccardo Viale, and Robin Marris. *Economics, Bounded Rationality and the Cognitive Revolution*. Northhampton, MA: E. Elgar, 1992.

Smith, Shawn Michelle. *American Archives: Gender, Race, and Class in Visual Culture*. Princeton, NJ: Princeton University Press, 1999.

Sontag, Susan. *On Photography*. New York: Farrar, Straus and Giroux, 1977.

Star, Susan Leigh. "The Ethnography of Infrastructure." *American Behavioral Scientist* 43 (1999): 377–91.

Star, Susan Leigh, and Geoffrey C. Bowker. "How to Infrastructure." In *Handbook of New Media: Social Shaping and Social Consequences of ICTs*, edited by Leigh A. Lievrouw and Sonia M. Livingstone, 230–44. Thousand Oaks, CA: Sage, 2006.

Steiner, Hadas A. *Beyond Archigram: The Structure of Circulation*. New York: Routledge, 2009.

Sterne, Jonathan. *The Audible Past: Cultural Origins of Sound Reproduction*. Durham, NC: Duke University Press, 2003.

Stimson, Blake. *The Pivot of the World: Photography and Its Nation*. Cambridge, MA: MIT Press, 2006.

Stoler, Ann Laura. *Along the Archival Grain: Epistemic Anxieties and Colonial Common Sense*. Princeton, NJ: Princeton University Press, 2010.

Szanton, Peter L. *Working with a City Government: Rand's Experience in New York*. New York: Rand Institute, 1970.

Tagg, John. *The Burden of Representation: Essays on Photographies and Histories*. Amherst: University of Massachusetts Press, 1988.

Terranova, Tiziana. "Another Life: The Nature of Political Economy in Foucault's Genealogy of Biopolitics." *Theory, Culture, and Society* 26, no. 6 (2009): 234–62.

Terranova, Tiziana. *Network Culture*. London: Pluto Press, 2004.

Thurschwell, Pamela. "Ferenczi's Dangerous Proximities: Telepathy, Psychosis, and the Real Event." *Differences* 11, no. 1 (1999): 150–78.

Tone, Andrea. *The Age of Anxiety: A History of America's Turbulent Affair with Tranquilizers*. New York: Basic Books, 2009.

Treichler, Paula, Lisa Cartwright, and Constance Penley. *The Visible Woman: Imaging Technologies, Gender, and Science*. New York: New York University Press, 1998.

Trifonova, Temenuga. "Matter-Image or Image-Consciousness: Bergson contra Sartre." *Janus Head* 6, no. 1 (2003): 80–114.

Tufte, Edward. 2006. *Beautiful Evidence*. New Haven: Graphics Press.

Turing, Alan Mathison. *A. M. Turing's ACE report of 1946 and Other Papers*. Cambridge, MA: MIT Press, 1986a.

Turing, Alan Mathison. *The Essential Turing: Seminal Writings in Computing, Logic, Philosophy, Artificial Intelligence, and Artificial Life, Plus the Secrets of Enigma*. [*The Essential Turing: The Ideas That Gave Birth to the Computer Age*]. Edited by Jack Copeland. Oxford: Oxford University Press, 2004.

Turing, Alan Mathison. "On Computable Numbers, with an Application to the Entscheidungsproblem." *Journal of Math* 58 (1936): 345–63.

Turing, Alan Mathison. "Proposal for Development in the Mathematics Division of an Automatic Computing Engine (ACE) (1946)." In *A. M. Turing's ACE Report of 1946 and Other Papers*. Charles Babbage Institute Reprint Series for the History of Computing, edited by B. E. Carpenter and R. W. Doran, 10–105. Cambridge, MA: MIT Press, 1986b.

Turner, Fred. "*The Family of Man* and the Politics of Attention in Cold War America." *Public Culture* 24, no. 1 (2012): 55–84.

Turner, Fred. *From Counterculture to Cyberculture: Stewart Brand, the Whole Earth Network, and the Rise of Digital Utopianism*. Chicago: University of Chicago Press, 2006.

Vidler, Anthony. "Toward a Theory of the Architectural Program." *October* 106 (Fall 2003): 59–74.

Virilio, Paul. *The Vision Machine*. Bloomington: Indiana University Press, 1994.

von Foerster, Heinz, ed. *Cybernetics: Circular Causal, and Feedback Mechanisms in Biological and Social Systems, Transactions of the Sixth Conference*. New York: Josiah Macy Jr. Foundation, 1949.

von Foerster, Heinz, Margaret Mead, and H. L. Teuber, eds. *Cybernetics: Transactions of the Eighth Conference, March 15–16, 1951, New York, NY*. New York: Josiah Macy Jr. Foundation, 1952.

von Foerster, Heinz, Margaret Mead, and H. L. Teuber, eds. *Cybernetics: Transactions of the Seventh Conference*. New York: Josiah Macy Jr. Foundation, 1950.

von Neumann, John. *First Draft of a Report on the EDVAC*. Contract no. W-670-ORD-4926 between U.S. Army Ordnance Department and University of Pennsylvania. Philadelphia: Moore School of Electrical Engineering. June 30, 1945.

von Neumann, John. "The General and Logical Theory of Automata." In *Papers of John von Neumann on Computing and Computer Theory*, edited by William Aspray and Arthur Burks, 491–31. Pasadena, CA: Tomash, 1986. Originally published in 1948.

von Neumann, John, and Oskar Morgenstern. *Theory of Games and Economic Behavior*. Edited by Oskar Morgenstern. Princeton, NJ: Princeton University Press, 1980.

Vrachliotis, Georg. "Popper's Mosquito Swarm: Architecture, Cybernetics and the Operationalization of Complexity." *European Review* 17 (2009): 13.

Waldrop, M. Mitchell. *The Dream Machine: J. C. Licklider and the Revolution That Made Computing Personal*. New York: Viking, 2001.

Wark, McKenzie. *Gamer Theory*. Cambridge, MA: Harvard University Press, 2007.

Wark, McKenzie. *A Hacker Manifesto: McKenzie Wark*. Cambridge, MA: Harvard University Press, 2004.

Watson, Thomas, Jr. "Good Design Is Good Business." In *The Art of Design Management: Design in American Business*, Thomas Schutte, ed. Philadelphia: University of Pennsylvania Press. 57–79. Courtesy of the IBM Corporate Archives. 1975.

Weintraub, E. Roy. *Toward a History of Game Theory*. Durham, NC: Duke University Press, 1992.

Weizman, Eyal. *Hollow Land: Israel's Architecture of Occupation*. London: Verso, 2007.

Wenner, Adrian M., and Patrick H. Wells. *Anatomy of a Controversy: The Question of a "Language" among Bees*. New York: Columbia University Press, 1990.

Whitney, John. "Animation Mechanisms." *American Cinematographer* (January 1971): 26–32.

Whitney, John. *Digital Harmony: On the Complementarity of Music and Visual Art*. Peterborough, NH: Byte Books, 1980.

Wiener, Norbert. *Collected Works*. Cambridge, MA: MIT Press, 1976a.

Wiener, Norbert. *Cybernetics; Or, Control and Communication in the Animal and the Machine*. New York: MIT Press, 1961.

Wiener, Norbert. *Ex-Prodigy: My Childhood and Youth*. Cambridge, MA: MIT Press, 1953.

Wiener, Norbert. *The Human Use of Human Beings: Cybernetics and Society*. New York: Avon Books, 1988.

Wiener, Norbert. *I Am a Mathematician: the Later Life of a Prodigy; An Autobiographical Account of the Mature Years and Career of Norbert Wiener and a Continuation of the Account of His Childhood in "Ex-Prodigy."* Garden City, NY: Doubleday, 1956.

Wiener, Norbert. "Problems of Organization." In *Collected Works*, edited by Norbert Wiener, 396–97. Cambridge, MA: MIT Press, 1976b.

Wiener, Norbert and Society for Advancement of Management New York Chapter, eds. *Cybernetics and Society; Based on Proceedings of the Meeting of the New York*

Chapter of the Society for the Advancement of Management Held in New York on November 16, 1950. Principal Paper. New York: Executive Techniques, 1951.

Wilder-Mott, C., and John H. Weakland, eds. *Rigor and Imagination: Essays from the Legacy of Gregory Bateson*. New York: Praeger, 1981.

Wilson, E. O. *Sociobiology: The New Synthesis*. Cambridge, MA: Harvard University Press, 2000. Originally published 1975.

Wilson, Elizabeth A. *Affect and Artificial Intelligence*. Seattle: University of Washington Press, 2010.

Wulf, Andrew. *Moscow '59: The "Sokolniki Summit" Revisited*. Los Angeles: USC Center on Public Diplomacy at the Annenberg School, 2010.

Youngblood, Gene. *Expanded Cinema*. New York: Dutton, 1970.

Index

Note: Italics indicate a figure; n denotes a footnote

Bauhaus: curriculum, 92; "form follows function" adage, 140, 259; and Gestalt psychology, 285n25; influence of on Kepes, 85–86, 284n21

behavior: as product of communication, 84, 180–81, 184; cognition as basis for, 16; cognitive dissonance model of, 168; cybernetic calculation and programming of, 43–45, 67, 172, 174, 177, 185; feedback and, 36, 43–46, 186–87, 190; memory and, 61, 66–67, 68; modification, 191–93; and nonreasonable rationality, 29, 172–73, 176; as response of a system to stimulus, 55–56, 177, 187–88

behavioral sciences: introduction of data-driven methodologies in postwar, 148; reconceptualization of truth in postwar, 84; research on learning, 105; subjectivity as tool in, 157, 159

Bell Labs: involvement of in defense research during World War II, 42; research at following World War II, 46, 62, 96

Bergson, Henri: Bergsonian time, 36, 48–49; on cinema and temporality, 54; Deleuze on Bergsonian time, 55–57; disagreement of with Freud, 53–54; influence of on cybernetic thought, 53; on matter, 52; *Matter and Memory*, 52, 54, 55; on perception and reality, 53, 59–60; and process philosophy, 52

biopolitics: Foucault on, 5, 7–8, 25–26; and reconfiguration of people and activities into data, 26, 184, 202–4, 236–37, 240, 272n14

black boxes, 44–46, 62, 77, 155

Brewster, David, 62, 65

Burdett, Ricky, 243–44

Bush, Vannevar, 72–73, 190

CAVS (Center for Advanced Visual Culture), 86, 95–96, 97, 98, 140, 273n10

choice: Arrow on political versus economic, 296–97n66; and communication, 46–47, 106; and decision-making, 175, 176, 178, 187, 193, 226, 280n22; the Eameses on, 102–4; infor-

mation inundation as enhancement of, 102, 124, 129, 138, 221–22, 279n17; and market behavior, 174; spectatorship as, 212

cinema: as collectable object, 92; close-up in, 220–21; and conflict between reality and representation, 69; Deleuze on, 20, 55–57, 235–36, 281n48; the Eameses' use of, 105, 224–25; early speeds, 281n45; "expanded," 230–31; experimentation by Whitney using, 218–20, 233–34, 302n41, 303n70; influence of on cybernetic research, 51–52, 105, 108–9, 215–20; innovations at Moscow exhibition, 212–25; sound in, 233–34; technology of, 54; temporality in, 54, 235–36

circuits: brain as basis for constructing machine circuitry, 151–53, 157–59, 186, 294n33; brain as basis for social science models, 187–88, 189–90; brain's neural, 56, 166, 168–69, 170–71, 184, 294n36; for information sharing, 222, 224, 228; machine, 160–61

Cisco Systems, 2–3, 18, 31, 239, 242, 271–72n8

cities: aerial views of, 79, 81, 88, *179*, 213–14; Archigram's fantasy, 33, 244–47; Eames on management of, 134–36; imageability as hallmark of great, 115, 287–88n63; research on images of by Lynch, 81–82, 114–22, 141–42; Senseable Cities Lab (MIT), 242, 243; smart, 1–8, 9–10, 20, 26, 32–33, 239–44; as transferable infrastructures, 239–42, 271n6, 271–72n8, 273n15, 276–77n61; twentieth-century changes in, 113–14, 299n100. *See also* utopias

civil rights movement, 122–23, 181, 225, 229, 299n102

cognition: and apprehension, 83, 88, 108, 110, 241; consciousness as, 16–17, 20–21; cybernetic, 207–12, 300n15, 300–301n16; human perception and, 36, 62, 133, 156–57; McCulloch's work on, 72, 206–7; machine perception as, 29; optical, 200–201, 204–6; and rationality, 29–30; reconceptualization of

in cybernetics, 11–13, 17, 80, 203–4; in
response to an environment, 114–15,
120; subjective nature of, 207, 211, 238.
See also perception
cognitive psychology, 50, 53, 55–56, 57,
69, 284–85n22
cognitive science: and cybernetics, 162,
201n19, 206; Festinger's research on
cognitive dissonance, 168–69, 296n54;
multimedia inundation and perception, 204, 211, 222, 224; realignment of
concepts of perception and cognition
in, 201, 204, 207–8, 211
Cold War: C3I ideal, 44; influence of
propaganda on later biopolitical practices, 201, 203–4, 236–37, 298n85;
influence of on aesthetics, 83, 202,
203–4, 285n25, 285n29; influence of on
concepts of truth, 83–84, 203, 206; influence of on education and research,
101, 278n4; influence of on thought
and ethics, 76–77, 78, 273–74n16;
nuclear specter underlying, 226;
propagation of atomistic approach to
research, 47–48; redefinition of rationality during, 149–50, 169; reforms in
ethical attitudes toward technology,
99; shift from documentation to processes in knowledge production,
84–85; US Information Agency Exhibit (Moscow, 1959), 201–2, 212–15,
215–27, 227–29
Colomina, Beatriz, 124, 203, 224, 239
communication: as concept in psychoanalysis, 68; as concern in social sciences, 15, 19, 47–48, 84, 119–20, 143;
as influence on aesthetics and space,
257–59, 261–63; as basis for cybernetics, 68, 74, 75; as basis for prediction,
45–48; behavior as, 43–46, 279n11;
and choice, 102–5, 138, 296–97n66;
and cognition, 171, 201, 206–7, 216;
Communication Primer (1953) and
film, *104*, 106, *107*, 134; communication technology, 15–16; communicative objectivity, 1, 15, 28, 84, 95, 139,
140; as control, 41–42, 76–78; Derrida
on, 75–76, *76*; and digital media tech-

nologies, 59; Eames' and Nelson's employment of, 102, 105, 110, 133–35, 216,
224, 286n48; failure, 67, 70, 163, 166,
168; honeybee example of, 49–50; in
human or machine networks, 43–44,
155–56, 158–59, 298n85; Lynch's use
of as tool in urban planning, 115–17,
120, 142; mathematical theory of,
46–47, 102–6, 133, 209, 279–80n20,
283n85, 286n48; perfect as deterrent to
thought, 75–77; psychosis as research
tool for study of, 148, 154, 163, 166–67,
193, 197, 211–12; reformulation of in
cybernetics, 19, 25, 51–57, 62, 66–67,
140; research by Bateson, 162–63,
167, 191, 193–94, 295n44; research by
Deutsch, 178–84, 186–87; research by
Festinger, 168–69, 296n54; research
by Kepes, 80, 81, 83, 86, 139–40; research by Miller, 209–12; research by
Wiener, 12–13, 17, 45, 53, 60–61, 67–68,
279n17; temporality and, 48–51, 193–
94; variables in information, 209–10;
visual as pedagogical tool, 100, 102,
105–6, 110. *See also* feedback
communication sciences, 1, 17, 36, 83, 117,
273n16
communication theory: research by
Eames and Nelson in, 102–3, 134, 216;
research by Miller on repetition, 209,
210–11; research by Wiener in, 12, 46;
significance of temporality in, 193
Communism, 19, 78, 110, 213, 225, 274n29
computer: computer-aided design, 12;
EDVAC, 160; IBM's 360 series, 126, 128;
stored program, 160–61
Congrès Internationaux d'architecture
Moderne (CIAM), 32, 33
consciousness: Bergson's work on reality
and, 53, 54–55, 56, 60; as cognition, 17,
169; cybernetic reformulations of, 60,
146–47, 159, 170, 172–74, 185; filtering power of, 70; in psychoanalysis,
52, 186–87; and reason, 150–51; versus
memory, 70
control: in Turing machine, 160–62;
Bateson on rationality and, 193–94,
196; Charles Eames's concept of, 134–

control (*continued*)
35, 142; Chun on paranoia of, 190–91; and communication, 13, 17, 25, 41–43, 155, 197, 203; concept of in cybernetics, 160–62, 279n11; McCulloch's concept of, 151; of space, 267–68

Crary, Jonathan, 18, 20

cybernetics: and advantages of excess data, 80, 99; as tool for analyzing behavior, 44–46; as tool for controlling and predicting, 25, 29, 36, 47; as tool for managing communication, 41–42, 44, 75–78; as tool for managing differing scales, 35; as tool for managing differing temporalities, 36, 41, 48–51, 55; as tool for managing information, 47, 51–52, 60–65, 71; coined by Norbert Wiener, 25, 39–40, 68; and computer-aided design, 12, 14; concerns of for human possibility, 78; and creation of smart cities, 10; etymology of term, 25, 41; influence of on epistemological values, 142–43, 150–51, 159–60; influence of on how individuals experience world, 65, 148; its transformative influences of, 27–31, 39, 41, 47, 51; military uses of during World War II, 43–44; perception and memory as concerns in, 52–55, 57, 60–61, 66–71, 74; and prediction, 26, 49, 57–59; representation as concern in, 67, 71–77; sensory prosthetics, 74; and visual perception, 62–65

Dalí, Salvador, 64

Daston, Lorraine, 18, 27, 150

data: abundance of as enhancement of choice, 124–25, 129, 132, 180–81, 201; as constraint on visions of future, 1, 5–6, 13, 34, 37–38, 196, 244; as product of observer-environment interaction, 83, 85, 142; as tool for managing uncertainty, 26, 29–30, 71, 204; biography as, 39–40; capture and management, 69, 204–6, 207–8, 209–10, 222–24; as communicative objectivity, 1; control of in Turing machine, 160–62; as credibility and power, 147; data-driven research, 148; Deutsch's employment of, 180–84, 187–90, 191; Eames's pedagogy for data management, 101–2, 105, 108, 110, 129–33, 204; error as product of improper storage and recall of, 94; Fuller on informational overload, 14; inundation, 80, 93–94, 129, 136, 184, 212, 226–27; Kepes's pedagogy for data management, 87–94, 105, 138; Lynch's data gathering methods, 116, 120, 121–22, 142; management of large amounts of, 5, 7, 11, 61, 83, 103, 133; Memex machine, 72–73; and memory, 158, 186, 187; Miller's research on data cognition, 207–12; mining, 30, 36, 194, 196–97; patterns in, 11, 14, 16, 43, 62–64, 173, 226; and politics, 25–26, 30–31; presentation at IBM "Thinking Machine" pavilion, 123–25, 129–33, 136–37; recombination of, 87, 94, 101, 121–22, 133, 178; reconfiguration of people and humanity as, 26, 236–38, 242, 272n12; and scale, 35; situational information, 47, 96, 276n61, 280n22; spatial versus temporal qualities of, 22; storage, 83, 85, 87, 108–9, 193–94; as used in smart cities, 3–7, 31, 34, 36; useful data as beautiful, 5, 185, 190–91; and visualization of information, 14–15, 17–18, 21–23, 151, 173

decision-making theory, 15–16, 175, 176–77, 201, 210–11, 296–97n66

Deleuze, Gilles: on Bergson's ideas concerning cinema, 55–59, 281n48; collaboration with Guattari, 159, 162, 194, 195, 196; and concept of assemblage, 276n56; concept of "dividual," 272n12; and idea of temporality, 57–58, 236; on perception and recollection, 55, 235–36; research collaboration with Guattari on paranoia and schizophrenia, 194–95, 299–300n107; and time-image, 58, 230, 235, 236, 244, 249; on visualness and visibilities, 24

democracy, 110, 178, 191, 214, 244, 254, 296–97n66

Derrida, Jacques, 75–76, 78, 230, 241, 245

desire: and behavior, 184; as opportunistic behavior in cybernetics, 170–71, 174, 189; in psychoanalysis, 194, 197; and studies of the unconscious, 170, 172

Deutsch, Karl: advocacy of data inundation by, 180, 184, 190; on American race relations, 181–82, 299n102; disagreement with Arendt, 184, 297n75; on freedom, 186–87; on game theory, 186, 187, 196; on governance and communication, 15, 178–85, 186–90, 191; on reason, 187; on scanning, 15–16, 62, 189–90; on self-reference versus self-reflexivity, 180–84, 186–87, 197

digital media: as tool for managing humans, 5, 32, 34, 57; biopolitics and, 272n14; communication and, 46, 59, 102, 103; potential of to dehumanize humanity, 1, 74–78, 147–48, 190–91, 241–42; Wiener's contributions to development of, 39, 59, 66, 74

Doane, Mary Ann, 52–53, 69, 220–21, 281n37, 302n44

documentation: disadvantages of, 12–13; need for, 1, 25, 71; photography as, 70; reconceptualization of in cybernetics, 17, 27, 41, 49, 72–73, 83

Eames, Charles: "A Sample Lesson," 81, 100–101, 110; *Communication Primer* (1953), *104*, 106, *107*; "Communication Primer" film, 134; emphasis by on design to amplify information, 102; emphasis on importance of vision by, 101; *Glimpses of the United States* exhibit, 212–13, 215–20, 220–24, 227; objectification of process by, 109–10, *111–12*; pedagogy of, 81, 100–108, 110, 212; research interests of in patterns, 102, 105–6, 108–9, 132, 136; research interests of in scale, 106; on spectatorship, 212; "The People's Wall" exhibit, 123–25, 129–33, 136–37; use of images to communicate ideas by, 105–9, 220–21, 287n52; use of game theory by, 134–35, 142; use of information inundation by, 101, 102, 136; use of multimedia by, 230

Eames Office: educational projects, 101,

102, 105, 108–9, 222, 286n48; emphasis on communication through senses by, 216, 224; on human factor in design, 109; on importance of process, 109–10; pedagogical principles of, 103–6, 136; on spectatorship as choice, 212, 215, 221; use of diagrams by, 132–33, 222; work for IBM, 123–25, 129–32, 289–90n85

Eames, Ray: emphasis by on importance of choice, 103–4; involvement of in exhibitions and education, 101; objectification of process by, 109–10, *111–12*; use of images by to communicate ideas, 108–9, 215–16, 220–21

E.A.T. (Experiment in Art and Technology), 96, 140, 286n32

enemy: and black box, 44; ontology of, 44; prey and, 200, 203–6; and studies of response time, 43

Enlightenment, 29, 150, 176, 187, 191

environment: as influence on senses and memory, 82, 85, 141–42, 143; as interface, 13, 28; conceptualization of IBM fair pavilion as a complete, 125, 129–33, 137; interactive, 133; Kepes's research on perception and, 95–99, 140; Lynch's research on perception of, 114–22, 141–42; smart city, 4–7, 10–11, 26

epistemology: as concern in urban planning, 83, 96, 113, 121, 136, 143; as force of history, 142–43; considerations of in cybernetics, 28, 36, 51, 58–59, 65, 76; and human sensorium, 25, 28; and legitimation of subjective experience, 173, 222; postwar reformulation of, 147–50, 159–60, 175, 206, 208, 211; vision-based, 21, 24

ethics: Derrida's concerns regarding technology and, 76; Kepes's concerns regarding technology and, 95–96, 99, 140; rationality and social science, 150–51; Wiener's concerns regarding technology and, 76–77. *See also* moral values

evidence, 17, 23–24, 27, 155, 159, 167–68, 228

123–25, 129–33, 289–90n85; reconfiguration of by Thomas J. Watson,
125–29; research on visualization, 22;
Saarinen's design for Yorktown laboratory, 126–29, 290n86
images: alienness quality in some, 229–
30, 235–37; on Bergson's concept of
the body as collection of, 52, 54–57;
cathartic effect of some, 237; close-
up, 220–21; conditioning effect of
some, 23, 44–45, 228–29, 237, 281n37;
Deleuze on the mind's capture of,
55–58, 142–43, 235–36; Deutsch's use
of diagrams to depict relationships
and systems, 178, 181–84, 187–91;
Deutsch's use of scanning, 15–16,
188–89; Eames's and Nelson's use of
to teach pattern recognition, 105–8,
287n52; flow chart examples, *16*, *130*,
135, *217*; flow charts, 13, 15, 160, 162,
216; imageability, 115, 118, 240–41,
287–88n63; as interactive communication, 206, 212; Julesz's experiments on
vision, 63–64; Kepes's experimentation using light-based, 97–98; Kepes's
use of to teach principles of design,
86–95, 143; Lynch's concept of, 81–82,
114–18, 121–22, 141, 143; modern circulation of as an end in itself, 228–29;
nonimages, 239–41; Picasso's *Guernica*
and evocative power of, 139; recreated
as dynamic entities in cybernetics, 80;
Steichen's use of for *Family of Man* exhibit in Moscow, 214–15, 220, 221, 224,
226; as storage mechanisms, 64, 67, 71;
thought-related versus virtual, 235–37,
239–41, 244; time-, 58, 230, 233, 235–
36; translation of as ethical action,
236–38; use of for *Glimpses* exhibit in
Moscow, 212–13, 213–14, 218–22, 224,
226; use of for "Think" screen at IBM
pavilion, 123–25, 129–33, 137; utopian,
241, 244–48; Whitney's development
of computer-generated, 218–20, 231–
35, 303n70
Incheon Free Economic Zone, 1–2, *4*,
11, *240*
indexicality: as disadvantage in informa-

tion transmission, 40–41; and information accumulation, 84, 103, 133, 159;
of photographs and film, 70, 71, 73;
and visual information, 40–41, 62, 63,
206, 218
information: as source of human freedom, 190, 191, 296–97n66; channel
capacities to consume, 209–12; cognitive dissonance, 168; cognitive processing of, 83, 94–96, 106–10, 136–37,
201; and control, 143; Deutsch on
channels, 179–80, 187; dynamic,
probabilistic nature of, 102–3, 133–34;
environmental and sensorial, 118;
flows, 62, 80, 101, 121–22, 179–80,
222; impact of transmission on, 103;
improper processing of, 163, 166; influence on observer of assumed infinite, 84, 123–25, 138; inundation used
as teaching tool by Eames and Nelson,
81, 101–5, 216–24; mental suppression
of, 170, 172; modern circulation of,
228–30, 239; overload, 14, 87–88, 101–
2, 136–37, 216; pattern extraction as
tool for managing, 82, 204, 207, 224;
perception as processing of, 206; recoding and compression of, 201, 210,
211–12, 224; research by Bateson, 163,
166, 172; research by McCulloch and
Pitts, 170–71, 186; research by Miller
on, 201, 208–12; research by Simon,
176–77; research by Wiener on, 40–41,
279n17, 279–80n20; research on vision
and by Kepes, 80, 83, 139; situational,
47; storage of, 66, 148; and subjectivity, 173; suppression of, 170–71;
theory, 47, 51–52, 105, 273–74n16; variance, 209; visualization of, 14–15, 95
infrastructures: digital, 32, 35, 194; of
Songdo, 1–3, 5, 32–33; smart cities as
transferable, 239–42, 271–72n8, 276–
77n61
intelligence: as product of dumb interactions, 18, 20, 31–32, 176–77, 196, 211;
artificial, 10, 29, 53, 117, 175, 176; collective, 19, 147–48, 194; machine, 10,
11, 300–301n16; Negroponte on human
and machine, 11, 273n7; and reason,

intelligence (*continued*)

26; reformulation of concept of, 8, 148, 150, 162, 176; and smartness, 3, 4

interactivity: as framework for urban planning, 83, 84; as pedagogical tool for teaching good design, 93, 133; as reformulation of attention and distraction, 17, 78, 212, 225; as tool for social intervention, 29, 119, 148, 214, 243; feedback as, 60, 211, 273n7, 273n10; merger of observer with process of, 133, 136–38, 224, 226–27, 234–35; networked, 194, 196–97, 206, 224, 228; and perception, 28, 82, 117, 120, 122, 140–41; predictive, 173–75, 192; psychoanalytic, 163, 166–68, 295n44; real-time, 22, 75, 283n87; screen as point of, 73; with static objects and data as goal, 80, 194, 196–97; translation zone as native to all forms of, 68–69, 84; Wiener's conceptualization of, 66, 68, 73–74

interfaces: computer, 162, 226; as conduits of information, 13, 68, 73, 173, 226, 235; physical spaces as, 82, 85, 98, 117, 262; proliferation of in Songdo, 4–7; tension between archives and, 27, 40, 74, 78, 80, 143–44

Jameson, Fredric, 240–41, 252

Julesz, Béla, 62–64, 204

Kepes, Gyorgy: background of, 85–86; and Center for Advanced Visual Culture (CAVS) at MIT, 86, 95–96, 98, 140, 273n10; and concept of experiential vision, 83, 87–94; and concept of visual flow, 79–80, 81; on data, 15, 190; emphasis by on method in pedagogy, 85, 86–87, 93–94; on "felt order," 13–14, 137–38; his collection of images, 87–92; on importance of using art to enhance science, 95–96, 139–40, 190, 284n16, 285–86n30, 286n32, 290–91nn102–3; interests of in gestalt psychology, 92–93, 284n21, 284–85n22, 285nn25–26; interests of in pattern and perception, 82, 88, 92, 94–95, 143, 204; interests of in scale, 88, 97; on

memory and objectivity, 94–95; mentorship of Kevin Lynch by, 81, 110, 114–16, 119–21, 143; and Moholy-Nagy, 86, 87, 97; on Picasso's *Guernica*, 139; research into materiality of perception, 96–99; research with lighting, 96–98; *The Language of Vision*, 139, 284–85n22, 285n26

Kittler, Friedrich, 68–69, 77

knowledge: assumption of infinite information as influence on, 84; Bergson's concept of, 52–54, 60; Deutsch on scale and, 16–17; and governmentality, 25–26; Kepes and redefinition of, 99; psychoanalytic, 68; redefinition of, 5–6, 12–13, 17, 19, 27, 129; subjectivity of, 84, 99; and vision, 23–25, 81

Lacan, Jacques, 74, 167, 228, 283n92

language: Eames and Nolan on experience and emotion as, 81, 100–101; mathematical, 18, 43; noncognitive, 49–51; replacement of words with symbols, 18–19; of vision, 21–24, 80, 83, 88, 99

Language of Vision, 139, 284–85n22, 285n26

Le Corbusier, 9, 31, 113, 277n65

Lettvin, Jerome, 30, 42, 204, 293n24, 300–301n16

Licklider, J. C. R., 115, 117, 287n61, 287–88n63

light: Kepes's sculptures, 96–97; in messaging, 106; scientific research in perception, 43, 79

logic: derived through information overload, 105, 125, 136, 204; gates, 149, 158, 196, 294n33; machine, 132, 147, 159; and memory, 163; neurons and computational, 154, 159, 162; nonreasonable, 29, 153, 154, 218; psychotic rational, 146, 151, 172–73, 175; redefined by cybernetics; of scale, 35; and subjectivity, 177, 187; visual, 105, 191, 203

Lynch, Kevin: academic background of, 110–13; and concept of "imageability," 115, 120–21; conceptualization of environmental psychology by,

Miller, George (*continued*)
butions to cognitive science by, 207–8;
on number seven, 199, 208–9, 211–12,
237–38; on recoding and compression
of information, 201, 210, 211–12
mind: Deleuze on functioning of, 55;
Deleuze on relationship between body
and, 58–59; human thought as challenge to cybernetics, 147–48; reconceptualization of, 17, 29; research by
McCulloch and Pitts on, 56
MIT: Architecture Machine Group, 10,
11–12; Center for Advanced Visual
Culture, 86, 95–96, 97, 98, 140, 273n10;
involvement of in defense research
during World War II, 42; Media Lab,
10, 119, 242, 273n10, 273n15; Radiation Laboratory (RAD Lab), 42–43;
Research Laboratory for Electronics,
200, 204; Senseable Cities Lab, 242,
243
modeling: feedback systems as basis
for, 46; of human thought, 60–61,
69; mathematical of human behavior,
43–44; and problem of scale, 35; taxonomic, 67; of vision, 62–65; Wiener's
use of mathematical, 43–45
Moholy-Nagy, László, 79, 86, 87, 92, 97
moral values: dilemma of rationality
and, 196–97; endowment of vision-
based truth with, 94, 148, 186; influence of altered human perceptions
on, 18–19; influence of technological
change on, 13, 19, 76, 109, 237–38, 242;
Noguchi's art as interrogation of, 257,
268–69; and question of governance
through media, 227–28. *See also* ethics
Morgenstern, Oskar, 173–75, 177
Moses, Robert, 122–23, 126, 288–89n74
multimedia, employed by the Eameses,
105, 135, 212, 216, 222, 230

Negroponte, Nicholas, 10–12, 119, 273n7,
273n10, 273n15
Nelson, George: "A Sample Lesson,"
81, 102, 105, 110, 190; employment
of images to communicate ideas by,
105–8; *Glimpses* exhibit, 202, 213;

pedagogy of, 81, 100–108, 110, 212; on
vision, 202–3
nervous networks: as model for computer programming, 160, 173–75; diagrams of by McCulloch and Pitts, *164*;
Eames on overstimulation of, 203,
204, 224; and thought processes, 29,
55, 152, 154–60, 197, 294n33
networks: ARPANET and Internet, 39,
191, 287n61; binary, 300n15; interter-
ritorial, 2, 136, 239; as isolating yet
connecting influences, 28, 78, 84, 138,
194, 196–97; Kittler on discourse, 69;
of smart spaces, 3, 10; overabundance
of data as problem for, 61; sensory, 20,
26, 31–32, 276–77n61; smart cities as,
26, 33, 240, 246, 273n15; as tools for
social action, 242, 249, 268; as virtual
experience mechanisms, 13
neural nets: Dumit's question of minds
versus machines, 162; impact of research on other disciplines, 176, 178,
184, 187, 189, 297n67; and language,
163; predictive temporality of, 196; and
vision, 206, 300–301n16; von Neumann's reliance on, 282n72; Wiener's
work on, 13, 48; work on by McCulloch and Pitts, 12, 16, 29, 55–56, 72,
154–59, 207
neuroscience, 56, 147–48, 201, 204, 206,
294n33
New Landscape of Art and Science, 80, 85,
86–87, 88, *89–93*, 190
New York World's Fair (1964–1965): as
example of urban redevelopment,
122–23; GM's "World of Tomorrow"
pavilion, 123; IBM's "Information Machine" pavilion, 28, 123–25, 129–33;
racial tensions surrounding, 82, 122–
23, 126; Robert Moses and, 122–23
Noguchi, Isamu: aesthetic logic of IBM
North Garden (Garden of the Future),
259, *260*, 261, 262–63; aesthetic logic
of IBM South Garden (Garden of the
Past), 259, 261, 262–63; aesthetic logic
of Jerusalem garden, 266–67; artistic interests of, 252, 254–56; communications of with Billy Rose, 263;

Pitts, Walter: collaboration of with McCulloch on neural nets research, 29, 154–59, 162, 207, 294n33; neural net diagrams, *16, 146, 161, 164*; research by on mind and brain, *16,* 55–56, 169, 293–94n24

politics. *See* biopolitics

population: biopolitical reconfiguration of, 25–26, 29, 181–82, 229, 236, 298n85; reconceptualization of by cybernetics, 16, 20, 120, 203–4, 242, 249; redefined as data rather than as populace, 4, 5–6, 224, 240; smart cities as machines to control, 2; urban, 113, 120

poststructuralism, 74, 76, 77, 283n92

postwar. *See* Cold War; World War II

prediction: communication as basis for, 46, 49–50, 234; in cybernetics, 25, 41, 44, 49, 60, 238, 248; dependence of control on, 135, 160; information as basis for, 26, 30, 36, 61, 66, 173; mathematical modeling of, 43–46, 48, 51; memory as basis for, 189, 204; in neuronal nets, 154, 156, 157, 158, 196

prey or enemy, 200, 203, 204, 205–6

probability: in Bergsonian "vitalist" time, 49; in calculations of interactions, 46, 48; of time and history, 34, 36, 52, 55, 64, 156; probabilistic thought, 40–41, 46, 49; versus determinism, 58, 59–60; and Wiener's concept of incomplete determinism, 40–41

processing: as a thing, 56; information, 137, 179, 209; memory and response, 40, 42, 56, 60–61, 68; perception and, 61, 66; perception versus recall in, 71–72; signal, 42, 154, 208, 293–94n24

psyche: and action, 53–54; in cybernetic thought, 68–69; Kittler on, 68–69

psychoanalysis: Bateson's merging of with cybernetics, 162–63, 167, 295n44; Freudian, 69–70; Kittler on uses of, 68–69; Lacan on cybernetics and, 74; McCulloch's distrust of, 147–48, 149, 152–54, 169–70; and psyche, 69; redefinition of, 68, 171–72, 193–94; transformation of uses of, 29, 36, 51, 53; Wiener's reapplications of to machine processing, 67–68

psychology: Bateson's influence on, 162, 163, 295n44; behavioral, 105; and concept of interface, 68, 98; and cybernetics, 162, 169, 173, 197, 206; Deutsch's contributions to, 184, 186–87; and environment, 28, 82, 86, 119–21, 141, 296n54; and feedback, 218; Gestalt, 92, 140, 149, 284–85n22, 285n25; influence of on McCulloch's research, 147, 149, 151–52, 155, 159; and memory, 147, 201; Miller's contributions to, 207–9; organizational, 119, 175–77; and perception, 222, 224, 227–28, 284–85n22; research in vision, 21, 64–65, 87, 199–200, 205–6, 207; Simon's contributions to, 175–77

psychosis: as basis for defining rationality, 30, 137, 146–47, 149, 172–73, 175; as basis for methodologies in social sciences, 151, 177, 184; as inspiration for Turing machine, 145–46, 148–49, 151, 157–59, 161–62, 167; McCulloch's study of, 145–47, 149, 152, 155, 157–58, 170, 291n2; neurosis, 72, 170, *188*; psychoanalytic, 147, 167, 170–71, 191, 194, 197

race relations: biopolitical approaches to, 28–29; Deutsch's reanalysis of vision and race, 180–81; and redefinition of urban environment, 136, 141; tensions during Cold War, 28, 122–23

Rajchman, John, 24

RAND Corporation, 99, 126, 135–36, 175–76, 288n73, 297n67

rationality: affective, 133–34; cognition and, 147; cybernetic, 29–30, 146–51, 162, 173, 196–97; and decision-making, 178, 187, 201, 296–97n66; and objectivity, 133; Simon's redefinition of, 175–77; Terranova on biopolitical, 184–85; unreasonable, 29, 145, 150, 159–60, 192; von Neumann's and Morgenstern's concept of, 173–75

Ratti, Carlo, 242–43

reality: Barthes's concept of "reality effects," 49–51; and Bergson's concept of process philosophy, 52–53, 60; Freud on perception and, 53; Freud's

Shannon, Claude, 46–47, 102, *104, 156*, 279–80n20, 286n48

signals: machine management of, 160, 161, 189; neural net, 154, 156–57, 158, 171; physiologically misread, 152, 157; psychological, 163, 166, 181, 192, 227; signal processing, 42, 154, 208

Simon, Herbert, 26, 29, 175–77

smart spaces: as challenge to democratic government, 26; bases for, 4–5, 11–12, 20, 22, 31–32, 271n7, 276–77n61; controversies surrounding, 241–42, 243–44; Songdo as "smart" city, 3, 6, 9, 276–77n61

social sciences: changing concepts of truth and knowledge in postwar, 168–69; embrace of data as tool by postwar, 29, 83, 116, 190; embrace of measurement as tool by postwar, 29, 84, 175; influence of cybernetics on, 29–30, 71, 142, 185; influence of information communication and feedback on postwar, 19, 47, 117, 119, 147, 181, 184; reconceptualization of truth and reason in postwar, 150, 169, 171, 172–73, 176, 191; systems analysis as tool of postwar, 119–20

Songdo: as data management tool, 4–6, 26, 276–77n61; as financial loss for its developers, 32; as interface for its residents, 2–4; architecture and infrastructure of, 6–7, 31–34; as spatial product, 1–3; as utopian experiment, 9, 18, 31, 239, 242; multiple temporalities of, 34–35, 36

spectator: Deleuze on role of, 56, 57; Eames on role of, 102, 124–25, 129, 132, 212, 222–24; and immersive multimedia experience, 218, 221–26; and pedagogy of Eames and Nelson, 105, 106, 108, 125, 132–33; versus user, 240. *See also* observation

Steichen, Edward, 214–15, 220–22, 224, 225–26, 230

stimuli: capacity of to overload a system, 61, 62; as education, 102, 105; as elements of perception and memory, 55, 60, 65; pattern as tool for managing, 82, 137–38; research by Bergson on

management of, 53; research by Freud on management of, 52–53; research by Julesz on visual, 63

storage: in cybernetic feedback systems, 61; image as a medium of, 70–71, 73–74; research by Freud on memory, 52–53; research by Kittler on memory, 69; research by Wiener on problems of memory, 62, 66–67, 70; and transmission of information, 62

Stroud, John, 185–86

subjectivity: Lynch's use of as research tool, 115, 118, 120, 121, 141, 142; and memory, 52, 54; of perception as influence on thought, 60

teleology: and action, 51–57; and purposeful feedback, 43, 44–46, 234; Wiener's research on, 45–46, 55–56, 67, 155, 293n23

temporality: as concern of response time, 43; as product of consciousness, 70; Barthes on, 49–50; Bergsonian perception of, 49, 52; and communication, 48–51; cybernetic assumptions concerning, 41, 148, 151, 155–56, 159; Deleuze and concept of, 57–58, 236; Freud on, 53; honeybee example of, 49–50; Kepes on, 139–40, 140–41; Lynch on as dimension of urban environment, 82, 118, 141–42; Newtonian concepts of, 48–49, 279n17; in process philosophy, 52; reformulations of concept of, 12, 17, 20; and representation of experience, 52–55, 69, 262; Wiener on, 48–49

territory: reconceptualization of in biopolitics, 16, 26, 27, 142–43, 251, 268; redefinition of since twentieth-century, 181–82, 184–85, 191, 203–4, 243–44

time: as dimension of space, 141–42; Bergsonian "vitalist," 36, 48–49, 53–54, 60; causality and, 45; consciousness and, 70; cybernetic concerns with, 27, 51–52, 60–61, 67; Deleuze's ideas regarding, 57–58; feedback as manipulation of, 36, 44–45; Freud's research on memory through, 52–53; historical, 27, 49; Newtonian "mechanical," 48; photographic and cinematic, 69, 71, 73; probabilistic nature of, 49, 52, 55; real, 22, 75–77; response, 43–44; scanning as reordering of, 15–16; time-image, 22, 57–58; Wiener's research on cybernetic, 49, 53, 55

truth: idea of norm as, 151, 153–54; McCulloch's conceptualizations of evidence and, 153–54, 155–56, 158–59; and memory, 158; perception and, 68, 129, 133, 148; and rationality, 148; realignment of objectivity versus, 83, 84, 94, 171–72; reformulations of bases defining, 1, 14–15, 17, 19, 23–25, 27–28; subjectivity as standard for, 118, 121, 173, 184, 190–91, 196, 207; versus knowledge, 168–69

Turing, Alan: research on human sense perception by, 149, 206; research on machine and brain functions by, 150, 160, 161, 177

Turing machine, 145, 149, 160

uncertainty, 25, 26, 32, 121, 275n52, 288n72

unconsciousness: Bergson on, 53; Freud's research on consciousness and, 52, 53, 69–70; Lacan on, 74; in psychoanalysis, 149, 170–72; and repression, 172, 180

urban planning: as form of communication, 83, 85, 86, 114, 115; Brasília, 3, 9; Chandigarh, 9; cities as commodities, 3; computerization of, 10, 11–12, 28; excess of information as tool in, 83, 88, 135; Lynch's ideas regarding, 81–82, 114, 117–18, 141–42; principles of before Lynch, 113; RAND Corporation on application of military research to, 136; revised priorities of, 5, 7, 10

utopias: Lynch's redefinition of concept of, 114, 120; as reflections of their times, 9, 113; Songdo and concept of, 9, 31. See also Archigram; cities

Vidler, Anthony, 240–41, 244
virtual, 40–41, 57, 235, 257, 278n1
visibilities, 24

vision: aspedagogical tool, 99, 101–2, 105–8; cognition and, 13, 81, 82–83, 110, 139, 143–44; as defined by Kepes and Fuller, 15; definitions of, 21, 23–24; depth perception, 62–65; Deutsch on human, 15–16; engineering of machine, 62, 64–65; experiments by Julesz, 64; Farocki's experimental tableau of, 23; Kepes's ideas concerning, 79, 83, 85–95, 97–99; machine, 62, 218, 222–24, 226, 300–301n16; movement and, 79–80; and perception, 62–64, 79–80; physiological, 23–24; and reality, 14; reconceptualizations of, 17, 27, 29, 36, 206; Silverman's observations on generational, 23; stereoscopic, 62; technologies of, 14, 136; three-dimensional, 63–64; versus visualization, 21; versus visualness, 24; Wiener's work on human, 7, 62

visuality, 6, 9, 23–24, 138, 202–3

visualization, 21–24; as mechanism for consciousness, 185; as technique for communicating information, 28–29, 95, 147–48, 184, 190–91, 248; definition of, 21–23; and drive to store data, 194; mapping of movements, 242–43; modern obsessions with, 1, 5, 7, 13, 14, 16–17, 201; Wiener's interest in data, 12

von Frisch, Karl, 49–50

von Neumann, John: and EDVAC stored program computer, 160, 282n72; on neurons and human behavior, 185; research on digital computing architecture, 47; research on logical automata by, 146, 175; research on "rationality," 173–75, 177

Weaver, Warren, 43, 46–47, 102–3, 104, 156, 279–80n20

Wheatstone, Charles, 62, 65

Whitney, John Sr., 218–20, 230–31, 233–36, 303n66, 303n70

Wiener, Norbert: aspirations of for a future world, 12–13, 40, 60; association with McCulloch, 293–94nn23–24; "Behavior, Purpose, and Teleology," 45, 67, 155; on cybernetics, 25, 51, 68; Ex-Prodigy, 12–13; feedback systems

Wiener, Norbert (*continued*)
    and learning, 45–46, 48–49, 60–61; on
    human-machine interface, 72–73; on
    incomplete determinism, 40; influ-
    ence of Bergson on, 51–55, 58, 59–60;
    influence of Freud on, 40–41; influ-
    ence of on cybernetics, 7, 12–13, 27,
    40–41, 74, 278–79n8; on management
    of information, 40–41, 279–80n20;
    mathematical modeling of communi-
    cation with Weaver by, 45–46, 73–74,
    279–80n20; mathematical modeling of
    human behavior by, 43–44; research
    concerning teleology, 45; research
    on memory, 51, 53, 66–71; research on
    prediction, 48–49, 59; research on
    temporality, 48–49, 50–51, 55, 58–60,
    185; research on relationship between
    thought and action, 13, 15, 36, 44,
    60–61, 279n17; use of diagrams as re-
    search tools by, 12, 13, 44, 76–77; on
    vision and knowledge, 12–13, 62
World War II: aerial warfare studies, 23,
    43–45; as redefining influence on life,
    85–86, 95, 249, 251, 285n29; impact of
    on scientific research, 42–44, 51, 61, 81,
    204, 278n4; military research contrac-
    tors, 41–43; and rise of cybernetics, 39,
    41, 51, 58–60, 78, 204; transformative
    effects of on philosophy and ethics, 12,
    20, 59, 76–77; transformative effects of
    on psychological research, 150, 152–53,
    162–63, 197, 294n36; usefulness of "the
    enemy" as research tool, 43–44, 48, 81,
    200–201, 204–6, 234–35, 295n44
Wright, Frank Lloyd, 113

Printed and bound by CPI Group (UK) Ltd, Croydon, CR0 4YY

16/04/2025